VORTEX METHODS AND VORTEX MOTION

VORTEX METHODS AND VORTEX MOTION

EDITED BY

KARL E. GUSTAFSON
UNIVERSITY OF COLORADO AT BOULDER

JAMES A. SETHIAN
UNIVERSITY OF CALIFORNIA AT BERKELEY

siam® Philadelphia, Pennsylvania

Society for Industrial and Applied Mathematics

All figures in Chapter 3 courtesy of Thinking Machines Corporation, Cambridge, Massachusetts.

Library of Congress Cataloging-in-Publication Data

Vortex methods and vortex motion / edited by Karl E. Gustafson, James A. Sethian.
p. cm.
Includes bibliographical references and index.
ISBN 0-89871-258-0
1. Vortex-motion. I. Gustafson, Karl E. II. Sethian, James Albert.
QA925.V67 1991
632'.0595--dc20

Contents

LIST OF CONTRIBUTORS

Thomas F. Buttke, Courant Institute of Mathematical Sciences, New York University, New York, NY 10012

Alexandre Joel Chorin, Department of Mathematics, University of California, Berkeley, CA 94720

Peter Freymuth, Department of Aerospace Engineering Sciences, University of Colorado, Boulder, CO 80309-0429

Karl E. Gustafson, Department of Mathematics, University of Colorado, Boulder, CO 80309-0426

Ole H. Hald, Department of Mathematics, University of California, Berkeley, CA 94720

Robert Leben, Colorado Center for Astrodynamics Research, University of Colorado, Boulder, CO 80309

James A. Sethian, Department of Mathematics, University of California, Berkeley, CA 94720

PREFACE

The development of vortex motions in fluid flow is responsible for some of the most fascinating aspects of fluid mechanics, such as mixing, shearing, transport, and instability. Such issues arise in a variety of flow regimes, ranging from fundamental mathematical questions in laminar, transitional, and turbulent flow to sophisticated engineering settings and devices. In recent years vortex methods have emerged as a new class of powerful numerical techniques to analyze and compute vortex motions. In this book, we bring together a set of papers under the common theme of vortex methods and vortex motion.

We begin (Chap. 1) with a review article which surveys the design and application of vortex methods. While by no means exhaustive, this article is meant to serve as a general introduction to some of the algorithmic and computational aspects of using vortex methods to compute complex flows. Chapter 2 addresses theoretical and numerical issues about the convergence properties of the method. Chapter 3 is a collection of photographs of numerical simulations using the vortex methods, showing such phenomena as flow in cavities, in closed regions, and around valves.

We then shift the emphasis to the development and motion of large vortex structures. We begin with actual laboratory visualizations of physical vortex motion (Chap. 4). These photographs reveal both the beauty and complexity of such flows. Chapter 5 presents a framework for a theory of vortex motions. Four stages and principles of vortex motion are discussed and illustrated by examples from cavity flows and airfoil flows. Chapter 6 describes some physical and numerical experiments for the vortex dynamics of the hovering mode believed to underlie the still-air flight of dragonflies and hummingbirds.

Next we present two papers describing new directions for vortex calculations. The first (Chap. 7) provides discussion of the flow of superfluid helium, which can serve as a simplified model for studying some aspects of classical turbulence. Chapter 8 relates vortex methods, vortex motions, polymer systems, and statistical mechanics. As well as addressing some of the most fundamental aspects of three-dimensional turbulence, this chapter lays the groundwork for a practical vortex algorithm for computing complex three-dimensional flow.

We would like to take this opportunity to thank all of the contributors to this book, as well as J. Corey Gray, Production Manager at SIAM. Finally, we wish to thank Tricia Manning, Managing Editor at SIAM, whose unfailing humor, careful organization, and welcome advice guided us throughout this process.

Karl E. Gustafson
University of Colorado at Boulder

James A. Sethian
University of California at Berkeley

1 A BRIEF OVERVIEW OF VORTEX METHODS

James A. Sethian

Introduction

In the past 10 to 15 years, vortex methods have been applied to a variety of fluid flow problems. The essential aspect of these methods is the focus on the *vorticity* of a flow, which is the local "twist" of the fluid. In an incompressible fluid, the time evolution of vorticity dictates the essential physics of the unfolding flow.

Velocity Versus Vorticity. Traditionally, flow is described by the Navier–Stokes equations, which are written in terms of the fluid velocity at any point and express Newton's law that force equals mass times acceleration. Since the vorticity ξ is defined to be the curl of the velocity ($\xi = \nabla \times \mathbf{u}$), the vorticity may be derived from knowledge of the fluid velocity. For our purposes, the reverse observation will be equally important: the velocity may be determined from the vorticity by means of a Biot–Savart integral.

Equations of motion in terms of velocity are especially convenient when boundary conditions are introduced. Imagine flow in a vessel or around an object. Suppose we divide the velocity $\mathbf{u}$ into a component u_n normal to the wall and a component u_τ tangential to the wall. Then the statement $u_n = 0$ on solid walls means fluid cannot pass through the obstacle, and $u_\tau = 0$ enforces the fact that fluid in direct contact with the wall must itself not slip. However, the boundary conditions for the vorticity are not as straightforward.

Given this, why have vortex methods, which approximate the solution to the equations of motion written in terms of vorticity evolution, proven so valuable? The answer is because fluid vorticity often lies in concentrated regions of the flow. The situation is analogous to the gravitational field induced by the planets. The mass of the solar system is concentrated in relatively few places, and it is easier to focus on the positions and strengths of these masses at any time rather than the evolving gravitational field at any point in space. Thus, a vorticity formulation reduces the mathematical description to its essential components. Furthermore, the physical meaning of the tangential boundary conditions becomes apparent; by requiring that solid walls "grab" onto the flow, a large shearing vorticity must be imparted to the flow. Thus, the boundary conditions both confine the flow ($u_n = 0$) and add vorticity ($u_\tau = 0$).

In two special cases, a vorticity formulation becomes particularly simple. If the flow is *inviscid*, that is, one in which vorticity does not diffuse due to viscosity or "friction" in the flow, patches of vorticity are self-advected and move without diffusing. In addition, in the absence of any vorticity ($\xi = 0$), the problem reduces to one of potential flow requiring the solution of Laplace's equation in the given domain.

Numerical Approximations. There are only a few explicit solutions to the equations of motion. Consequently, the goal of a numerical algorithm is to provide an efficient approximation to the solution of these equations, together with a systematic way of refining the approximation to increase the accuracy of the results. An approximation based on a vorticity formulation can focus computational resources by following the limited amount of vorticity in the system, rather than computing the velocity $\mathbf{u}$, which exists everywhere.

The first numerical calculation of a flow using an approximation to the vorticity equations is due to Rosenhead [191], who studied the evolution of a vortex sheet. Imagine one inviscid fluid on top of another, each initially moving with constant (but different) speed. Since $\xi = \nabla \times \mathbf{u}$, the vorticity is initially zero everywhere except along the infinitely thin boundary between the two. This interface is unstable to small perturbations and rolls up into large spiral structures as the flow evolves. As an example, the flow above and below an airfoil meet as they exit the trailing edge, with the top flow moving faster than the bottom one. The *vortex sheet* that comes off the wing rolls up into a complicated pattern.

Since the velocity may be recovered from the vorticity, and the flow is assumed inviscid, all of the dynamics are contained in the evolution of this infinitely thin sheet of vorticity. (In a real physical flow, the sheet has a finite thickness.) Rosenhead [191] approximated the motion of the vortex sheet by discretizing the infinite line of vorticity into a finite number of discrete point vortices. Each point vortex induces a velocity field of the form $1/r$, where r is distance to the vortex. By superposition, the evolution of a finite collection of point vortices initially placed along the line where $\xi \neq 0$ could be calculated. Thirty years before the advent of large-scale computing, Rosenhead computed the motion of this finite collection of points by hand, and produced a reasonable portrait of the rollup of a vortex sheet.

As clever as this work is, it contains two subtle complications. First, point vortices induce a singular velocity field; as $r \to 0$, their velocities become infinite. Second, vortex sheets develop a singularity as they evolve, even starting from smooth initial data. Thus, singular objects are being used to follow an object that develops a singularity. Attempts to improve on this calculation by using more discrete vortices have revealed numerous instabilities and breakdowns (see, for example, the paper by Birkhoff [39]). The source of these instabilities was not evident. For the most part, vortex techniques, i.e., the systematic resolution of vorticity into a finite number of objects which would be more accurate with more elements, were considered unworkable.

Modern Vortex Methods. The introduction of modern vortex methods as a practical technique began with the calculations of flow past a cylinder by Chorin [57]. Several factors contributed to the success of these calculations. First, discrete vortex elements, called vortex "blobs," were used which induced a bounded velocity field with finite kinetic energy. Second, the diffusion of vorticity was simulated by means of a random walk imposed on the vortex trajectories. Third, the no-slip ($u_\tau = 0$) tangential velocity boundary condition was enforced by the creation of new vorticity; while numerous techniques had previously created single discrete point vortex elements at boundary separation points in response to a Kutta condition (see Clements and Maul [67]), this represented the first approximation to the application of the no-slip condition around the entire body. Finally, the method was applied to viscous flow past a cylinder, where integrated quantities such as the drag could be calculated and compared to physical experiments.

Since that time, vortex methods have become a practical tool for computing a wide variety of engineering flows. In recent years, they have been applied to external flows past cylinders, wedges and blunt objects, flow inside pistons, flame propagation and turbulent combustion, ship design, thermal bubbles, blood flow in the heart, flow past buildings, and flow in the mantle. They provide tools for studying what happens in more pathological problems, such as vortex sheets in two dimensions and vortex rings and filaments in three dimensions. In addition, they link together a variety of other techniques for approximating vorticity, such as point vortex methods, vortex-in-cell algorithms, and discrete Kutta-condition models. Finally, mathematical interest in proving convergence of these algorithms has provided the setting for significant theoretical insight into the underlying equations.

Outline. In this paper, we will review some past work in the evolution of vortex methods. No review can be exhaustive or all-inclusive, and undoubtably we have missed important work. However, we hope to survey the major themes. With few exceptions, we have focused on reviewed articles. We apologize in advance for omissions and hope that such work can be brought to our attention.

A collection of excellent review articles on a variety of topics related to vortex methods and vortex motion have appeared in recent years, and have served as guides for this review. An early review by Clements and Maul [67] discusses vortex shedding by discrete vortices for Kutta conditions. Leonard [142], [143] has written two comprehensive reviews of work in vortex methods. Saffman and Baker [195] review some more theoretical aspects of vortex motion. Vortex shedding from bluff bodies in a variety of settings has been covered by Bearman [34] and Bearman and Graham [36], and vortex flows in aerodynamics by Smith [214]. Some additional theoretical work in vortex filaments has been provided by Widnall [234], and vortex breakdown by Hall [112] and Leibovich [141]. A full discussion of the motion of few point vortices and vortex patches is given by Aref [9] and Aref and Kambe [13], and more general perspectives by Aref [10] and Zabusky [237]. A review of the field of computational fluid mechanics is provided by Rizzi and Engquist [190], and the field of particle simulations in electromagnetism, which is related to general particle methods, by Buneman et al. [48]. An excellent overview of the importance of vorticity is provided by Saffman [194], and a clear exposition on the mathematical theory behind vorticity formulations and vortex motion is given by Majda [150]. Finally, a recent review by Majda [149] discusses the interplay between numerical and theoretical components of vortex methods and computational techniques.

The outline of this review article is as follows.

1. Equations of Motion

In this section, we discuss the formulation and approximations underlying vortex methods. This explanation benefits from considerable hindsight and follows the discussion by Anderson and Greengard [6] and Majda [149].

The Navier–Stokes and the Vorticity Transport Equations. The **Navier–Stokes equations** for incompressible, viscous flow in a region D with boundary ∂D are

$$\frac{D\mathbf{u}}{Dt} = -\nabla P + \frac{1}{R}\nabla^2\mathbf{u} \quad \text{in } D, \tag{1.1}$$

$$\begin{aligned} \nabla\cdot\mathbf{u} &= 0 \quad \text{in } D, \\ \mathbf{u} &= 0 \quad \text{on } \partial D. \end{aligned} \tag{1.2}$$

Here, $P = P(\mathbf{x}, t)$ is the pressure, R is the Reynolds number, $\mathbf{u} = \mathbf{u}(\mathbf{x}, t)$ is the three-dimensional velocity vector at $\mathbf{x} \in R^3$, and (D/Dt) is the material derivative

$$\frac{D}{Dt} = \frac{\partial}{\partial t} + \mathbf{u}\cdot\nabla.$$

Equation (1.1) is an expression of Newton's law $\mathbf{F} = m\,\mathbf{a}$, while (1.2) states that the flow is incompressible. The boundary conditions require that both the normal and tangential components of $\mathbf{u}$ vanish on solid walls.

We define the *vorticity* vector ξ as $\xi = \nabla \times \mathbf{u}$. The curl of (1.1), together with a few vector identities, result in the **Vorticity Transport Equation**

$$\frac{D\xi}{Dt} = (\xi\cdot\nabla)\mathbf{u} + \frac{1}{R}\nabla^2\xi. \tag{1.3}$$

Equation (1.3) states that the material derivative $D\xi/Dt$ of vorticity depends on two terms: *vortex stretching* $(\xi\cdot\nabla)\mathbf{u}$ and *vorticity diffusion* $(1/R)\nabla^2\xi$. In the limit of infinite Reynolds number, the diffusion term disappears, and only the normal boundary condition can be imposed (see Chorin and Marsden [63]). In the special case of two-dimensional flow, the vorticity vector ξ is normal to the velocity $\mathbf{u}$ and the term $(\xi\cdot\nabla)\mathbf{u}$ vanishes. These comments are summarized below.

	3D	2D
Viscous	$\frac{D\xi}{Dt}=(\xi\cdot\nabla)\mathbf{u}+\frac{1}{R}\nabla^2\xi$	$\frac{D\xi}{Dt}=\frac{1}{R}\nabla^2\xi$
	$\mathbf{u}=0 \quad \text{on } \partial D$	$\mathbf{u}=0 \quad \text{on } \partial D$
Inviscid	$\frac{D\xi}{Dt}=(\xi\cdot\nabla)\mathbf{u}$	$\frac{D\xi}{Dt}=0$
	$u_N=0 \quad \text{on } \partial D$	$u_N=0 \quad \text{on } \partial D$

In order to "close" (1.3), we must recover $\mathbf{u}$ from the vorticity. Given that

$$\nabla\cdot\mathbf{u}=0, \qquad \xi=\nabla\times\mathbf{u}, \tag{1.4}$$

there exists a vector function $\psi(x)$ such that $\mathbf{u}=\nabla\times\psi$ and

$$\nabla^2\psi=-\xi. \tag{1.5}$$

In three dimensions, ψ is known as the velocity potential; in two dimensions, ψ is called the stream function, and both ξ and ψ are normal to the plane of motion. We see that ψ satisfies a Poisson equation with the vorticity distribution as the right-hand side. This means that we may write ψ, and consequently, the velocity $\mathbf{u}$, in terms of ξ by making use of the fundamental solution to the Laplace operator ∇^2. Recall that the solution to (1.5) is given by

$$\psi(\mathbf{x},t)=\int L(\mathbf{x}-\mathbf{z})\xi(\mathbf{z})\,d\mathbf{z}, \tag{1.6}$$

where

$$L(\mathbf{x})=\left\{\begin{array}{ll}\frac{-1}{2\pi}\log|\mathbf{x}|, & \mathbf{x}\in R^2\\ \frac{1}{4\pi}\frac{1}{|\mathbf{x}|}, & \mathbf{x}\in R^3\end{array}\right\}.$$

Since $\mathbf{u}=\nabla\times\psi$, we have that

$$\mathbf{u}(\mathbf{x},t)=\int K(\mathbf{x}-\mathbf{z})\xi(\mathbf{z})\,d\mathbf{z}, \tag{1.7}$$

where the kernel K is defined by

$$K(\mathbf{x})=\left\{\begin{array}{ll}\frac{1}{2\pi}\frac{(-x_2,x_1)}{|\mathbf{x}|^2}, & \mathbf{x}\in R^2\\ \frac{1}{4\pi|\mathbf{x}|^3}\left\{\begin{array}{ccc}0 & x_3 & -x_2\\ -x_3 & 0 & x_1\\ x_2 & -x_1 & 0\end{array}\right\}, & \mathbf{x}\in R^3\end{array}\right\}, \tag{1.8}$$

where $\mathbf{x} = (x_1, \cdots, x_N)$ for $N = 2, 3$. Thus we have the following closed system for the evolution of ξ:

$$\frac{D\xi}{Dt} = (\xi \cdot \nabla)\mathbf{u} + \frac{1}{R}\nabla^2 \xi, \tag{1.9}$$

$$\mathbf{u}(\mathbf{x}, t) = \int K(\mathbf{x} - \mathbf{z})\xi(\mathbf{z})\, d\mathbf{z}, \tag{1.10}$$

where K is given in (1.8). We note that the kernel K is singular in both two and three dimensions.

Lagrangian Perspective: Particle Trajectories. Equations (1.9), (1.10) are an *Eulerian* formulation of the vorticity equations. That is, if we are given $\xi(\mathbf{x}, 0)$ at each point $\mathbf{x}$ in the domain at $t = 0$, they describe the change of ξ at any point $\mathbf{x}$.

Suppose, on the other hand, we view the initial condition $\xi(\mathbf{x}, 0)$ as describing the vorticity of the particle initially located at $\mathbf{x}$, and follow the motion of particles located at all possible starting points. This leads to a *Lagrangian* formulation for the particle trajectories, which we now formulate. Throughout this section, we shall assume that the flow is inviscid.

Let $\mathbf{X}(\alpha, t)$ give the position $\mathbf{X}$ at time t of that particle initially located at α. Then the particle trajectory is described by the initial value problem

$$\begin{aligned} \frac{d\mathbf{X}(\alpha,t)}{dt} &= \mathbf{u}(\mathbf{X}(\alpha,t),t) \\ &= \int K(\mathbf{X}(\alpha,t) - \mathbf{z})\xi(\mathbf{z})\, d\mathbf{z} \\ &= \int K(\mathbf{X}(\alpha,t) - \mathbf{X}(\beta,t))\xi(\mathbf{X}(\beta,t),t)\, d\beta, \end{aligned} \tag{1.11}$$

$$\mathbf{X}(\alpha, 0) = \alpha, \tag{1.12}$$

where we have replaced the dummy variable $\mathbf{z}$ into the trajectory form by the change of variable $\mathbf{z} = \mathbf{X}(\beta, t)$ and have used the incompressibility of the flow to imply that particle trajectories conserve volume and thus the Jacobian of the transformation is one.

Equations (1.11), (1.12) describe the evolution of a particle. The integrand depends on the positions of all other particles, as well as their velocity. Thus, we must provide a recipe for computing the vorticity $\xi(\mathbf{X}(\beta, t), t)$. For three-dimensional flow, recall that

$$\frac{D\xi}{Dt} = (\xi \cdot \nabla)\mathbf{u}.$$

Thus, one straightforward technique, suggested in Anderson and Greengard [6], is to add an additional equation for the rate of change of vorticity for the particle following the trajectory $\mathbf{X}(\alpha, t)$, namely

$$\begin{aligned} \frac{d\xi(\mathbf{X}(\alpha,t),t)}{dt} &= \xi(\mathbf{x}(\alpha,t),t) \cdot \nabla\mathbf{u} \\ &= (\xi(\mathbf{X}(\alpha,t),t) \cdot \nabla) \int K(\mathbf{X}(\alpha,t) - \mathbf{X}(\beta,t))\xi(\mathbf{X}(\beta,t),t)\, d\beta. \end{aligned} \tag{1.13}$$

Alternatively, as suggested by Beale and Majda [32], one can use that

$$\xi(\mathbf{X}(\alpha,t),t)=[\nabla_\alpha \mathbf{X}(\alpha,t)]\cdot\xi(\mathbf{X}(\alpha,0),0) \tag{1.14}$$

(see Chorin and Marsden [63]). The difference is that (1.13) is an expression for the time derivative of the vorticity in term of the velocity, whereas (1.14) directly describes the vorticity in terms of the spatial derivative of the particle trajectory.

In two-dimensional flow, the additional equation necessary to calculate the vorticity of the moving particles is trivial. Since $D\xi/Dt = 0$, we may directly write

$$\xi(\mathbf{X}(\alpha,t),t)=\xi(\mathbf{X}(\alpha,0),0). \tag{1.15}$$

Modern Vortex Methods: Approximating the Lagrangian Particle Trajectories. The above equations, while mathematically correct, are difficult to approximate numerically. The difficulty occurs because the kernel is singular in both two and three space dimensions. Thus, if particles come close together, they can exert extremely large velocities on one another. The idea behind modern vortex methods is to smooth the kernel to limit this singular interaction. The first vortex blob calculations set the magnitude of the velocity inside a given cutoff size δ to be constant, thus eliminating the singularity. A large number of smoothed kernels have been constructed since then, providing vortex methods of various orders of accuracy. A fuller discussion of the techniques for choosing the smoothing function is discussed in Chapter 2. For the rest of this section, we shall simply assume that the kernel K has been replaced by an appropriate smoothed kernel K_δ.

The basic idea behind vortex methods is to discretize in both space and time the initial value problem described as follows:

(1.16)

(Particle Update): $$\frac{d\mathbf{X}(\alpha,t)}{dt}=\int K_\delta(\mathbf{X}(\alpha,t)-\mathbf{X}(\beta,t))\xi(\mathbf{X}(\beta,t),t)\,d\beta;$$

(Vorticity Update): $$\left\{\begin{array}{ll} \dfrac{d\xi(\mathbf{X}(\alpha,t),t)}{dt}=(\xi(\mathbf{X}(\alpha,t),t)\cdot\nabla)\displaystyle\int K_\delta(\mathbf{X}(\alpha,t) & \\ \qquad -\mathbf{X}(\beta,t))\xi(\mathbf{X}(\beta,t),t)\,d\beta & \text{in 3D} \\ \qquad\text{or} & \\ \xi(\mathbf{X}(\alpha,t),t)=[\nabla_a \mathbf{X}(\alpha,t)]\cdot\xi(\mathbf{X}(\alpha,0)) & \\ \xi(\mathbf{X}(\alpha,t),t)=\xi(\mathbf{X}(\alpha,0),0) & \text{in 2D} \end{array}\right\};$$

(Initial Conditions): $$\left\{\begin{array}{ll} \mathbf{X}(\alpha,t)=\alpha & \text{initial positions of particles} \\ \xi(\mathbf{X}(\alpha,0),0) & \text{initial vorticity of particles} \end{array}\right\}.$$

Given an initial vorticity distribution $\xi(\mathbf{x},0)$, we begin by constructing a lattice Λ^h in R^N, $N = 2, 3$, with meshsize h. Let $\mathbf{j}h$ be the meshpoints of Λ^h, where $\mathbf{j}$ is any N-tuple with integer coefficients. The motion of the set of particles originally located on the nodes of this lattice may be approximated by a discrete approximation to the following:

(1.17)

(Particle Update):
$$\frac{d\mathbf{X}(\mathbf{j}h,t)}{dt} = \sum_{\mathbf{i}h\in\Lambda^k} K_\delta(\mathbf{X}(\mathbf{j}h,t)-\mathbf{X}(\mathbf{i}h,t))\xi(\mathbf{X}(\mathbf{i}h,t),t)h^N;$$

(Vorticity Update):
$$\left\{\begin{array}{ll} \dfrac{d\xi(\mathbf{X}(\mathbf{j}h,t),t)}{dt} = (\xi(\mathbf{x}(\mathbf{j}h,t),t)\cdot\nabla) \displaystyle\sum_{\mathbf{i}h\in\Lambda^k} K_\delta(\mathbf{X}(\mathbf{j}h,t) & \\ \qquad -\mathbf{X}(\mathbf{i}h,t))\xi(\mathbf{X}(\mathbf{i}h,t),t)h^N & \text{in 3D} \\ \text{or} & \\ \xi(\mathbf{X}(\mathbf{j}h,t),t) = [\nabla_{\mathbf{j}h}^{h}\mathbf{X}(\mathbf{j}h,t)]\cdot\xi(\mathbf{X}(\mathbf{j}h,0),0) & \\ \xi(\mathbf{X}(\mathbf{j}h,t),t) - \xi(\mathbf{X}(\mathbf{j}h,0),0) & \text{in 2D} \end{array}\right\};$$

(Initial Conditions):
$$\left\{\begin{array}{ll} \mathbf{X}(\mathbf{j}h,t)=\mathbf{j}h & \text{initial positions of particles on grid} \\ \xi(\mathbf{X}(\mathbf{j}h,0),0) & \text{initial vorticity of particles on grid} \end{array}\right\}.$$

Here, the operator $\nabla_{\mathbf{j}h}^{h}$ refers to a finite-difference operator on the mesh.

Equations (1.17) form a finite number of coupled differential equations for initial vorticity with compact support. Finally, the time derivative is approximated by a suitable finite-difference operator to provide a complete algorithm for updating the positions of the particles in time. This completely specifies the vortex method for inviscid flow.

2. Developments in Numerical Approximations to the Equations of Motion

In this section, we discuss some of the advances made in numerical approximations based on (1.17).

Free Space Inviscid Flows: Smoothing and Accuracy. We begin by discussing inviscid flow in free space. In the previous section, the equations of motion for viscous, incompressible flow in free space were approximated by a finite set of ordinary differential equations for the particle trajectories. The kernel K given in (1.8) is singular, and thus particles that come close together exert large velocities on each other. The idea behind modern vortex methods is to smooth the kernel to limit this singular interaction. How does one construct this smoothed kernel K_δ? One possibility is as follows. Consider two-dimensional flow, and let $f(\mathbf{x})$ be a radially symmetric function such that $\int_{R^2} f(x)\,dx = 1$. Define

(2.1)
$$f_\delta(\mathbf{x}) = \frac{1}{\delta^2} f\left(\frac{\mathbf{x}}{\delta}\right).$$

As $\delta \to 0$, $f_\delta(\mathbf{x})$ approaches the Dirac delta function. Thus, we define the kernel by the convolution

$$K_\delta(\mathbf{x}) = K * f_\delta(\mathbf{x}) = \int_{R^2} K(\mathbf{x}-\mathbf{x}') f_\delta(\mathbf{x}')\, d\mathbf{x}'. \tag{2.2}$$

As $\delta \to 0$, $K_\delta \to K$.

The accuracy of the vortex method depends on several factors: (1) h, the size of the mesh size used in the discretization of the initial vorticity distribution; (2) δ, the amount of smoothing in (2.1); (3) the choice of smoothing function f; and (4) the timestep Δt used to solve the ordinary differential equations for the particle trajectories. Since the introduction of smoothed vortex methods, considerable theory has been developed studying accuracy as a function of these parameters and convergence.

Hald and del Prete [111] constructed the first proof of the convergence of vortex methods in two dimensions. They considered several types of smoothed kernels, including the one used in practice by Chorin, and established short-time convergence. This was followed by the seminal paper in vortex theory, Hald's 1979 proof of long-time convergence of the vortex approximation (see Hald [109]). In that paper, Hald analyzed the relation between h, δ and the rate of convergence. His main result is as follows (Hald [109]):

> (Hald 1979): If $h = \delta^2$, then for a certain class of smoothing functions, the difference between the computed particle trajectories and the exact particle trajectories is of $O(h^2)$.

Thus, as one uses more vortex elements to resolve an initial vorticity approximation (that is, as $h \to 0$), linking the smoothing size to the number of vortices yields convergence of the position of the vortices to their exact trajectories.

On the basis of this work, Beale and Majda [32], [33], showed that carefully designed vortex methods for smooth inviscid flow could provide results of any desired degree of accuracy in both two and three dimensions. Their three-dimensional vortex method (eq. 1.14) updates three-dimensional vortex stretching by computing derivatives along the particle paths. These papers were responsible for starting the major interest in the theory of vortex methods. The proofs were technical, and much of the work that followed has been aimed at simplifying the arguments. Cottet [72], using work on particle methods for the Vlasov–Poisson equation by Cottet and Raviart [73], gave a stronger and simpler consistency argument, followed by an even simpler consistency argument by Anderson and Greengard [6]. Anderson and Greengard also suggested a different three-dimensional stretching algorithm, namely, the one given in (1.13), in which the kernel is explicitly differentiated. In addition, they provide the compact form of the Beale–Majda three-dimensional scheme given in (1.14), as well as a convergence proof which takes into account the timestep error along with the integration along particle trajectories. Hald [110] then showed convergence for an extremely wide class of two-dimensional methods, requiring only Hölder continuity in the vorticity field, as well as fourth-order convergence for the classical Runge–Kutta techniques for the integration of ordinary differential equations. Proofs of the three-dimensional method with explicit differentiation were also provided by Beale [29] and Greengard [102].

In tandem with these theoretical investigations were numerical convergence studies of the actual accuracy obtained in practice (see, for example, those performed by Sethian and Ghoniem [208]). Beale and Majda [30] showed how to construct explicit velocity kernels for high order accurate methods, and verified the increased accuracy on a test problem of the evolution of a radially symmetric patch of vorticity. Perlman [174] made a careful, detailed study of the optimal choice of δ compared to h in practice for short- and long-time flow.

Viscous, Incompressible Flow. The extension of vortex methods to viscous flow requires treatment of the viscous diffusion term $(1/R)\nabla^2\xi$ in (1.9). The Lagrangian formulation summarized in (1.17) assumes an inviscid formulation and yields a set of ordinary differential equations for the particle trajectories. The addition of viscosity generates a set of *stochastic* differential equations, in which the diffusion term joins the right-hand side of (1.16), and represents a probabilistic perturbation away from the deterministic particle paths described by the inviscid case. The motion of a continuum of particles along trajectories with both a deterministic and probabilistic component constitutes an *exact* solution to the viscous equations of motion. No approximation is involved until the infinite system is reduced to a finite number of particles. A good introduction to the stochastic differential equation view of vortex methods was given by Chang [51].

Numerical approximation to the infinite set of stochastic ordinary differential equations for the vorticity evolution usually begins with "operator splitting." For simplicity, consider the vorticity transport equation (1.9) for two-dimensional flow, namely

$$\partial_t \xi = -(\mathbf{u}\cdot\nabla)\xi + \left(\frac{1}{R}\right)\nabla^2\xi, \tag{2.3}$$

where we have decomposed the total derivative into its two components. Suppose we discretize the problem in time but not in space, i.e., we consider the evolution of a system of particles with discrete timestep Δt. Let ξ^n be the approximate vorticity distribution at time $n\Delta t$. Suppose we decompose the vorticity update $\partial_t\xi$ into two terms as follows. Let ξ^{n+1}_{adv} be the solution at timestep $(n+1)\Delta t$ of the pure advection equation

$$\partial_t \xi = -(\mathbf{u}\cdot\nabla)\xi \tag{2.4}$$

with $\xi^n_{\text{adv}} = \xi^n$. The solution to this equation is approximated by integration along particle trajectories using (1.16). Similarly, let ξ^{n+1}_{vis} be the approximate solution to the diffusion equation

$$\partial_t \xi = \frac{1}{R}\nabla^2\xi \tag{2.5}$$

with $\xi^n_{\text{vis}} = \xi^n$. If we let $\xi^{n+1} = \xi^{n+1}_{\text{adv}} + \xi^{n+1}_{\text{vis}}$ at the end of the timestep, we should get a solution to (2.3) as $\Delta t \to 0$. This technique is known as "operator splitting." Some analysis of the effect of operator splitting was provided by Beale and Majda [31].

How do we solve the vorticity diffusion equation (2.5)? Since the advection term has been formulated in a Lagrangian setting, it is desirable to provide a Lagrangian approach to the heat equation. We recall that an infinite collection of particles undergoing Brownian motion yields a solution to the heat equation (see Feller [86]). Thus, to accomplish both advection and diffusion, Chorin [57] updated the positions of the infinite system of particles by (1) advancing them by their induced velocity field, and (2) adding an appropriately chosen random step.

It might seem, at first glance, that such a method is highly inaccurate, since a relatively large number of particles are needed to adequately approximate the diffusion equation. However, note that the random motion is being applied to approximate the evolution of *vorticity*, not velocity. The Biot–Savart integration acts to smooth the solution, and accurate solutions can therefore be obtained with a reasonable number of particles. A variety of numerical experiments have been performed to examine the error associated with a random walk solution to the diffusion equation in the context of vortex methods (see, for example, Ghoniem and Sherman [100], Milinazzo and Saffman [161], Sethian and Ghoniem [208]). Detailed calculations by Roberts [187] verify convergence of Chorin's original random walk algorithm applied to a model problem.

One can think of other ways to approximate the diffusion equation (2.4). One possibility is a finite difference approximation. However, the introduction of a grid and the process of interpolating vorticity to the grid and back can result in considerable smoothing, known as "artificial viscosity." Thus, the major feature of a Lagrangian representation may be compromised, and the size of the Reynolds number can be limited by the grid resolution. Another proposed approach is core spreading, in which the core of each vortex element spreads as an exact solution to the diffusion equation. This technique was used by Kuwahara and Takami [136] in their early vortex work. However, Greengard [103] has shown that while the vorticity is correctly diffused in such techniques, it is advected incorrectly, and the limit of the computed solution will not be a solution of the Navier–Stokes equations.

Convergence of the random vortex method is measured in terms of expected values and variances. Due to the stochastic nature of the equations, probabilistic techniques come into play. As a test problem, Hald [108] studied an approximation based on a random walk to a particular reaction-diffusion equation, and showed that the expected value of the computed solution tended to the correct solution, and that the variance tended to zero as the number of particles was increased. Marchiro and Pulvirenti [153] gave a proof of weak convergence of the two-dimensional random vortex method. This was followed by Goodman [101] who provided a stronger proof of convergence. Some recent analysis of the random vortex method has been recently developed by Long [147], who has studied realistic convergence rates for the random vortex method in both two and three space dimensions.

Boundary Conditions

Normal Boundary Conditions. For both inviscid and viscous flow, the addition of the normal boundary condition on solid walls is not difficult. Given a region D, suppose we require that $\mathbf{u} \cdot \mathbf{n} = 0$ on the boundary ∂D, where $\mathbf{n}$ is the inward normal vector. Let $\mathbf{u}_{\text{vor}}(\mathbf{x}, t)$ be the velocity field obtained from the distribution of vorticity. Suppose we find a potential flow $\mathbf{u}_{\text{pot}} = \nabla\phi$ such that $(\mathbf{u}_{\text{pot}} + \mathbf{u}_{\text{vor}}) \cdot \mathbf{n} = 0$ on ∂D. Then superposition of the vorticity flow $\mathbf{u}_{\text{vor}}$ with the potential flow $\mathbf{u}_{\text{pot}}$ yields a flow which satisfies the normal boundary condition by construction and has the same vorticity (since $\nabla \times \mathbf{u}_{\text{pot}} = \nabla \times \nabla\phi = 0$).

For a simple geometry, finding the potential function ϕ is fairly straightforward. Although this potential function must be found at every time step, it is usually a minor part of the calculation, compared to the vortex-vortex interaction. Some of the earliest calculations used the method of images, such as calculations of flow past a cylinder (see Cheer [52], Chorin [57]). Other possibilities are conformal mapping to the upper halfplane followed by the method of images (see Ghoniem, Chorin, and Oppenheim [94]) and fast Poisson solvers (Sethian [206]). In the case of complex geometries, considerable work may be involved. Finally, appropriate inflow and outflow conditions may be difficult to pose (see the discussion in § 3).

Tangential Boundary Conditions. The addition of the tangential no-slip condition ($\mathbf{u} \cdot \tau = 0$) ($\tau$ is the unit tangent to solid walls) on ∂D for viscous flow adds considerable complication. A thin transition zone, known as the *boundary layer*, must develop between the vanishing tangential velocity at the wall and the rapidly moving flow some distance away from the wall (see Batchelor [27] and Schlicting [203]). By way of explanation, let $\mathbf{u} = (u, v)$ and suppose that the x axis is tangential to the wall. Then $\partial u/\partial y$, which is the derivative of the tangential velocity in the direction normal to the wall, must be large. On the other hand, $\partial v/\partial x$, which is the derivative of the normal velocity along the wall, is small, since, by the addition of the potential flow, v is zero along the x-axis. Since vorticity is defined by $\xi = \partial v/\partial x - \partial u/\partial y$, the tangential boundary condition is a major source of vorticity in the flow. At large Reynolds numbers, particularly in the case of bluff bodies, streamlines break away near the sides of the body and enclose the fluid behind the body in an unsteady manner. This phenomenon is known as "boundary layer separation" and leads to the downstream shedding of large physical vortices.

One way to approximate this effect is to release physical vortices into the flow from a predetermined separation point. A "Kutta condition" can be used to impose an a priori estimate of the strength of the shed vortices. One such technique views the separation point as a source of an array of line vortices. For example, this model was used to study the periodic wake past a cylinder (see Abernathy and Kronauer [1], Gerrard [91], Sarpkaya [201]). An early detailed set of calculations which created physical vortices for flow past bluff bodies is described in Clements [66], and a review of such techniques may be found in Clements and Maul [67]. Further discussion of boundary layer theory can be found in Batchelor [27] and Schlicting [203].

In Chorin [57], vorticity creation was extended to the entire boundary through the following technique. The length of boundary was broken into intervals, and within each interval a smoothed vortex "blob" was created at each time step. The strength of the blob was chosen to nullify the tangential velocity, and hence, by construction, satisfy the no-slip condition. There are two advantages to this technique. First, no a priori assumption about the location of the separation point is made, since the entire body acts as a line source of vorticity. Second, since vortex elements with smoothed velocity fields are used, more vortices may be used to resolve the vortex sheet along the boundary while avoiding the instabilities that plague point vortex approximations.

The technique of adding vortex blobs at the boundary in response to the no-slip condition is not very accurate. Part of this inaccuracy occurs because a small motion displacement in newly created vortex elements can cause significant variation in the tangential velocity at the next time step. Thus, large counterrotating vortices must be created at the next time step to enforce the no-slip condition. This can result in a large number of vortices near the boundary with one sign essentially balanced by a large number of vortices of opposite sign. Since the vortex-vortex interaction is expensive, this soon proves unworkable.

There are still further sources of inaccuracy if one tries to use vortex blobs to satisfy the no-slip boundary condition, as can be seen from the following argument (see Chorin [60]). The no-slip condition requires that we create a vortex sheet along the x-axis of strength U. Then we might imagine that the velocity at (x, y) right above the wall should be $+U$, and at $(x, -y)$ right below the wall should be $-U$. By the definition of the vorticity in the Navier–Stokes equation, we have that $\xi(x, y) = \partial v/\partial x - \partial u/\partial y$. Note that $\xi(x, y) \neq \xi(x, -y)$, thus substantial inaccuracy results from using vortex blobs to approximate this boundary condition. In order to remedy this situation, an alternative was introduced in Chorin [61]. Close to smooth solid walls, the Navier–Stokes equations can be replaced by the Prandtl boundary layer equations, which are derived under the assumption that $\partial v/\partial x \ll \partial u/\partial y$ and that diffusion of vorticity occurs mostly in a direction normal to the wall. The vorticity in the boundary layer is then discretized by finite length vortex sheets. The jump in the tangential velocity across the sheet determines its strength. Just as for the vortex blobs, the velocity of each sheet can be constructed by summing the influence of all sheets located in a narrow neighborhood nearby (see Chorin [61] and Sethian [206] for the equations of motion). In Chorin [61], this algorithm was used to approximate flow past a semi-infinite flat plate. Cheer [52] coupled vortex sheets in the boundary to vortex blobs in the interior to study flow past a cylinder. Sheets became blobs when they moved more than a certain distance away from the wall, and vice versa; in both cases, conservation of circulation was maintained. Other applications using this hybrid algorithm of vortex sheets and vortex blobs include flow past a backward facing step by Ghoniem, Chorin, and Oppenheim [94] and flow in a closed cavity by Sethian [206]. A hybrid algorithm which used vortex sheets in the boundary coupled to finite differences in the interior was studied by Shestakov [211]. An extension of the solid wall vortex sheet creation algorithm to three-dimensional flow is described in Chorin [60].

There is some controversy over the use of the Prandtl boundary layer equations near the wall coupled to the vortex blob Navier–Stokes solver in the interior. Two issues are often raised. First, the Prandtl boundary layer equations do not apply when separation occurs, thus, at least formally, they cannot be used in such situations. Second, the edge of the boundary layer is not well defined, and the use of a fixed transition point from sheets to blobs may not match the size of the physical boundary layer. We believe that an appropriate viewpoint is to interpret the use of the Prandtl boundary layer equations and vortex sheet discretization as merely a numerical device to get vorticity off the boundary and into the main part of the flow. The size of this zone does not correspond to the size of the physical boundary layer. Along these lines, interesting studies have been performed by Teng [224], [225], in which vortices whose ellipticity varies with distance from the wall were used to approximate vorticity throughout the flow, thus providing a smooth transition from the wall to the interior.

Proving convergence of a random vortex method with creation of vorticity at the boundary is a difficult problem. Some theoretical work on this topic is provided by Benfatto and Pulvirenti [37]. Hald [107] studied a particular vorticity creation algorithm for the equations of free convection near a boundary discussed by Ghoniem and Sherman [100] in their survey paper. Hald showed that both the pointwise and least squares error of the computed solution tended to zero as the timestep went to zero and the number of particles increased. For the random sheet method discussed above, numerical convergence tests have been performed by Puckett [180]. The convergence properties of the complete random vortex element method applied to a test problem of flow over a backward facing step have been studied in Sethian and Ghoniem [208].

Fast Summation Methods. The approximation of the vorticity into N discrete vortex elements leads to an N-body problem, namely, that each vortex interacts with each of the others. This is an $O(N^2)$ algorithm, which is expensive for large N. Several authors have investigated fast summation algorithms to lower the operation count.

A straightforward approach is to solve directly for the stream function by introducing a finite-difference grid. In this technique, the discrete vorticity field is interpolated to neighboring gridpoints, and a fast Poisson solver is used to find the stream function at gridpoints. Finite-difference approximations then yield the velocity field at the gridpoints, which is then interpolated back to the vortex locations. This "vortex-in-cell" technique has an operation count of $O(N \log N)$, and was used in some of the earliest vortex calculations (see Christiansen [64], Christiansen and Zabusky [65]). It has a long history, and is a major component of many N-body calculations in a variety of areas (see Eastwood and Hockney [85]). It is related to the particle-in-cell techniques invented by Harlow [113], [114]. More recent applications of vortex-in-cell techniques to problems in fluid mechanics are described in Leonard [143], Baker [18], and Tryggvason [226].

The process of interpolating the vorticity onto a mesh can introduce significant inaccuracy into the computation of vortex interaction. In the close interaction between neighboring vortices, the smearing of vorticity onto an intermediate mesh can introduce artificial viscosity which dominates the real physical viscosity in the system. As a compromise between the direct N-body method and the grid-based fast Poisson solution, various techniques have been developed to divide the computation into near- and far-field effects. One set of techniques uses fast solvers for the contributions of far away effects, and direct summation for the nearby elements. For a summary of such P^3M (particle-particle, particle-mesh) techniques, see Hockney and Eastwood [119]. One such algorithm in the context of vortex methods was given in Anderson [4] using, in part, ideas in Mayo [155]. Other work in this type of fast summation method may be found in Appel [8], Barnes and Hut [24], and Boris [41]. Recently, a fast summation technique using a multipole expansion was developed by Greengard and Rokhlin for the far particles (see Carrier, Greengard, and Rokhlin [50], Greengard and Rokhlin [104]). Further advances in rapid summation techniques have come from the use of computers with parallel architectures, significantly reducing the computational time (see Baden [17]).

3. Physical Problems

In this section, we provide references for the application of vortex methods to particular physical problems. Each of these applications has a vast literature and is a topic of ongoing interest as well as, in some cases, controversy. It is unreasonable, and probably nearly impossible, to hope that all the issues can be straightened out and presented in a concise, linear manner. Our goal is to summarize the topics of interest in each area, sketch some of the variations on vortex methods that have been used to attack the problems, and provide a few references.

Vortex Sheets. Many flows are characterized by sharp changes of velocity in narrow zones. For example, in parallel shear flow, the vorticity is concentrated in a narrow transition region between the two moving fluids. In the limit, the thickness of this shear layer goes to zero, and suggests the study of "vortex sheets." In this two-dimensional flow problem, the vorticity is zero everywhere except along an infinitely thin line or curve. As an example, flow around the trailing edge of a wing forms a vortex sheet whose strength depends on the given wing design. The ensuing motion and rollup of the vortex sheet affects both the drag on the wing and the flight of following aircraft.

The first numerical study of vortex sheets using vortex methods was due to Rosenhead [191], who discretized a vortex sheet into a small number ($N \approx 10$) of point vortices placed along an initial curve with periodic boundary conditions. The results obtained using this technique were physically reasonable. This was followed by a calculation by Westwater [232] of a finite vortex sheet carrying elliptically varying circulation (the "elliptically loaded wing"). In both of these calculations, the positions of the point vortices were updated in time using Euler's method and the vortex sheet was reconstructed by interpolating through the point vortices. Later attempts to achieve better results using more points and more accurate time integration schemes failed. Contrary to expectation, the calculations produced irregular point vortex motion (see, for example, the calculations by Birkhoff [39]). As discussed earlier, this cast doubt on the validity of vortex methods. The difficulties stemmed from both the use of singular point vortices and the subtle nature of the vortex sheet evolution equation. In particular, a workable numerical algorithm must be able to handle potential singularities in an evolving vortex sheet.

Since these initial calculations, a variety of techniques have been tried to obtain a more accurate picture of the rollup phenomenon and to understand the nature of the singularity that develops. The equations of motion for the evolving sheet are usually written in terms of a complex curve in two-dimensional flow (see Birkhoff [39]). The curve's motion is given by a nonlinear integrodifferential equation involving a Cauchy principal value integral. The various approaches may be broadly divided into two groups; those that attempt a higher order accurate discretization of the singular integral, and those that instead desingularize the integral before discretization.

A good starting point is the review article by Saffman and Baker [195]. Some other papers concerning numerical, asymptotic, and theoretical aspects of vortex sheets are Anderson [5], Baker [18], [19], Baker, Meiron, and Orszag [20], Chorin and Bernard [62], DiPerna and Majda [79], [80], Fink and Soh [88], Guiraud and Zeytounian [106], Higdon and Pozrikidis [117], Kaden [125], Krasny [133]–[135], Kuwahara and Takami [136], Mangler and Weber [152], Moore [165]–[167], Pullin [182], [183], Pullin and Phillips [185], Schwartz [204], and van der Vooren [231].

Vortex Pairing/Mixing/Shear Layers/Two Fluids. The goal of many vortex calculations is to understand fundamental questions about mixing, shearing, and transport. As a practical example, consider two chemical solutions which must be adequately mixed before use. What is the most efficient means to accomplish this mixing? Some possibilities are agitators placed in the containers or aimed injection streams. The large vortex structures created by such mechanisms determine the mixing, and some configurations are more efficient than others. This leads to fundamental questions of vortex interactions. A wide array of theoretical, numerical, and experimental investigations have been undertaken, for example, Brown and Roshko [46] have provided detailed experimental analysis of large coherent structures in turbulent shear layers.

One set of model problems analyze the stability of complicated shapes and distributions of vorticity, including the interaction of patches of constant vorticity. Of particular interest is the effect of mutually induced velocity on neighboring patches and the intertwining of boundary contours. Another model problem is the vortex street, which idealizes the wake shed behind a stationary cylinder placed in a uniform flow. Here, the stability and later coalescence of vorticity initially organized in two infinite parallel rows of opposite-signed vortices is studied to analyze the rollup of vortex structures and oscillating wake patterns.

To begin, some experimental studies of vortex pairing have been performed by Winant and Browand [236], and of vortex stability in a rotating fluid by Griffiths and Linden [105]. The interaction of a pair of vortices has been studied in several different settings. Saffman [193] studied the approach of such a pair to a plane surface, and Saffman and Szeto [199] studied steady equilibrium shapes for two equal uniform vortices rotating about each other. Pierrehumbert [175], using a relaxation method, found a family of steady pairs which can move through a fluid without change of shape; Saffman and Tanveer [200] enlarged the class of possible solutions. Dritschel [83] provided equilibrium shapes for two to eight corotating uniform vortices.

An early numerical point vortex study of vortex motion is due to Christiansen [64], who studied the merger of two corotating patches of vorticity. This was followed by a calculation by Christiansen and Zabusky [65] of the motion and merger of an asymmetric four vortex system. In order to study in more detail the motion of patches of constant vorticity in two dimensions, Zabusky, Hughes, and Roberts [238] introduced the method of "contour dynamics." The idea behind this method is as follows. Rather than follow the evolution of vorticity, which is the underlying notion in the vortex methods described in § 2, suppose that (1) vorticity is distributed on a compact set, and (2) vorticity may be subdivided into distinct regions of constant magnitude within each patch. Let γ_i be the contour that bounds patch i. By using the Biot–Savart law, an evolution equation for the moving contours γ_i can be derived (see Zabusky, Hughes, and Roberts [238]). The goal of contour dynamics algorithms, which reduces the dimension of the problem by one, is to accurately follow the motion of the contours, capturing the interaction of the patches of vorticity. The contours are followed by a marker particle technique in which the contour is represented by points distributed along the curve, and appropriate finite-difference approximations are used to compute the ensuing motion. Calculations using contour dynamics include the coherence of stretched vortices by Jacobs and Pullin [124] and Pullin and Jacobs [184], plasma cloud dynamics by Overman, Zabusky, and Ossakow [172], interacting patches by Dritschel [82] and Overman and Zabusky [171], the Kelvin–Helmholtz instability by Pozrikidis and Higdon [179], and the dynamics of vorticity fronts by Stern and Pratt [220]. As a by-product, detailed technology has been developed on the frontier of marker particle technology for front tracking (see Dritschel [81], Zabusky and Overman [239]). However, in the application of this technology to problems in fluid mechanics, there are significant limitations. Bounding contours have meaning for inviscid, two-dimensional flow; extensions to three dimensions are not obvious. Furthermore, any attempt to introduce viscosity to the bounding contours is an ad hoc adjustment in the contours, rather than in the vorticity itself.

As mentioned earlier, vortex streets have been used as model problems to analyze vortex rollup and the wake development. Baker, Saffman, and Sheffield [22] studied the structure of a linear array of hollow vortices of finite cross section, in which the vorticity of each vortex is concentrated just on the boundary. This was followed by a series of papers by Saffman and Schatzman [196]–[198] which investigate questions of vortex street stability as a function of spacing parameters, vortex strengths, and core size. The stability of two infinite parallel rows of finite core vortices was also studied by Kida [128], whose work was revisited by Meiron, Saffman, and Schatzman

[157]. Calculations of rollup into vortex streets using a large number of point vortices was performed by Aref and Siggia [14]. The stability of vortex arrays in three dimensions has been studied by Robinson and Saffman [189]. Other papers in these areas are Bromilow and Clements [44] and Ferziger [87].

Shear flows are significantly more complicated mechanisms for mixing. Imagine two fluids, moving at different speeds and parallel to a splitter plate separating the two. When the edge of the plate is reached, shearing is introduced as the differentially moving fluids come into contact. It has been studied in detail in the laboratory (see Batt [28], Bradshaw [43], Browand [45], Brown and Roshko [46], Freymuth [89], Freymuth, Bank, and Palmer [90], Ho and Huang [118], Pui and Gartshorne [181], Winant and Browand [236]). In a series of papers, Corcos, Lin, and Sherman (see Corcos and Lin [69], Corcos and Sherman [70], [71], Lin and Corcos [144]) have studied various aspects of the mixing layer and unstable free shear layers using a variety of mathematical techniques. Analysis of the dynamics of stretched vortices is provided by Neu [169], [170]. Some theoretical investigations may be found in Moore and Saffman [168] and Pierrehumbert and Widnall [177]. Vortex methods applied to shear flows and the mixing layer have achieved considerable success; we refer the interested reader to numerical calculations in Acton [2], Aref and Siggia [14], Ghoniem and Ng [98], Heidarinejad and Ghoniem [115], Inoue [122], Inoue and Leonard [123], Meiburg and Lasheras [156], Meng [159], and Meng and Thomson [160].

A variety of problems involve following the complicated interface that develops between two regions of flow, such as between fluids of differing density and/or viscosity. One example is the Rayleigh–Taylor instability, which occurs when a light fluid pushes a heavier fluid. Rather than remain flat, the interface can develop long spikes which can reconnect or break away into bubbles. Some examples where this instability can occur are in the collapse of a massive star, the laser implosion of deuterium-tritium fusion targets, and the electromagnetic implosion of a metal liner. A simple example is the novelty store toy in which fluids of differing densities are trapped between two glass plates. By upending the apparatus, the lighter fluid rises to the top and long spikes form in the interface. Bubbles can break off from the interface and merge with other bubbles. The interface between the two fluids becomes highly complex, breaking into numerous different parts with wildly varying shapes.

In their most complicated form, the equations of motion are the equations of full viscous, compressible flow plus interface effects. Some important factors controlling the growth of instability are: (1) the density ratio, which governs the growth of small amplitude perturbations; (2) surface tension, which stabilizes wavelengths shorter than a critical wavelength; (3) the viscosity, which reduces growth rate and regularizes the flow; (4) compressibility, which reduces growth rate; and (5) heterogeneity, which can excite instabilities of various wavelengths. For an overview of some work on the Rayleigh–Taylor instability, see Sharp [210].

Vortex methods have been applied to many of these problems. One technique is to reformulate the problem as a boundary integral, and discretize the boundary integral by a collection of vortexlike objects along the interface. The integral is desingularized to smooth the propagating interface. An early study of the Rayleigh–Taylor instability and the Saffman–Taylor instability may be found in Meng and Thomson [160]. Variations on vortex techniques and coupling to other numerical methods applied to Kelvin–Helmholtz and Rayleigh–Taylor problems may be found in Baker, Meiron, and Orszag [20], [21], Kerr [127], Pullin [183], and Tryggvason [226]. Applications to Hele–Shaw flow may be found in Tryggvason and Aref [227], [228]. The motion of a cylinder of fluid was studied in Rottman, Simpson, and Stansby [192] using vortex sheets, vortex blobs and vortex-in-cell methods, and inviscid flow about a rising cylinder was studied in Telste [223].

The extension of vortex methods to complicated flows with variable density is not straightforward. Begin with the Navier–Stokes equations (1.1):

$$\frac{D\mathbf{u}}{Dt} = -\frac{\nabla P}{\rho}. \tag{3.1}$$

Here we have ignored viscosity and included the density term. Taking the curl of (3.1) produces the vorticity transport equation. If the density is constant, the curl of the right-hand side vanishes, since ρ may be factored out and we have the vector identity $\nabla \times \nabla P = 0$. In general, however, the *baroclinic* term $\nabla \times \nabla P = 0$ is not zero, and is somewhat difficult to handle in the context of vortex methods. Under the assumption of small density variation, Anderson [5] used the Boussinesq approximation to the baroclinic term on the right-hand side and developed a vortex method for the transport of density gradients in the flow, similar to the advection of velocity gradients (that is, vorticity) by traditional vortex methods. Using this technique, he performed detailed calculations of a rising thermal. His calculations show the spiral arms of the rising thermals. This refinement approach between h and δ was later used in the study of vortex sheets by Krasny [134], [135]. The extension of vortex methods to problems with large density variations (the "baroclinic" term) remains an open question. Such an extension would be particularly valuable in a variety of phenomena, such as density jumps across propagating flames. One approach has been suggested by Ghoniem and Givi [96].

External and Internal Flows: Cylinders / Bluff Bodies / Chambers. As discussed earlier, boundary conditions on solid walls produce a thin boundary layer where velocity gradients can be large. Traditional finite-difference schemes often require prohibitively small meshes in order to accurately capture the mechanisms at the boundary. The grid-free approach of vortex methods offers an alternative in a variety of engineering applications.

There are three categories of flows involving solid bodies. First are external flows around bluff bodies. Some examples are flows past cylinder, airfoils, flat plates, and wedges. The second category contains inlet/outlet flows, in which confining walls trap much of the flow, and inlet/outlet conditions are required. Examples are flow over a backward-facing step and flow through choked nozzles. In the third category are internal flows in completely confined regions, such as pistons and driven cavities. In each of these areas, the fundamental questions are about the development of large eddies, wakes, and mixing.

In all cases, a potential flow must be found which, through superposition with the induced velocity field, satisfies the normal boundary condition on solid walls. In the case of flow past a cylinder (or, by conformal mapping, an airfoil), this might be accomplished through the method of images. In the case of flow in a rectangular region a fast Poisson solver can be used. However, given more complicated geometries, the solution of this potential problem can require significant attention. In the case of inlet/outlet flows, correct boundary conditions may not be obvious. Care must be taken at outlets to pose correct boundary conditions which are neither inconsistent with the flow dynamics nor unnecessarily restrict or influence flow upstream of the exit.

There is another source of difficulty. In response to the no-slip condition, vortices are created at the boundary which then enter the flow regime. In inlet/outlet flows, vortices can leave and no longer influence the region of interest. In external flows, one might assume that vortices far enough away from the solid body do not influence the flow and may be discarded with little effect. However, in internal flows, such as driven cavities, the number of vortices must increase. Eventually, this number can grow so large that computing the mutual vortex interaction is prohibitively expensive. This is one of the reasons why fast summation methods of the vortex interaction are of great importance, even for two-dimensional flows. Ideally, one would like to have a technique for combining vortices at some point to reduce the number. However, significant error is usually introduced when this is done.

As an introduction, we refer the interested reader to vortex calculations studying flow past cylinders (see Cheer [52], Chorin [57], Kimura and Tsutahara [129], Laitone [137], Smith and Stansby [216], Stansby and Dixon [219], van der Vegt and de Boom [230]), flow past airfoils (see Basu and Hancock [25], Basuki and Graham [26], Cheer [52], Katz [126]), flow past blunt bodies (see Bearman and Fackrel [35], Boldman, Brinich, and Goldstein [40], Clements [66], Smith and Stansby [215], Stansby [218]), flow over a backward-facing step (see Ghoniem, Chorin, and Oppenheim [94], Ghoniem and Gagnon [95], Ghoniem and Sethian [99], Sethian and Ghoniem [208]), flow past a flat plate (see Chorin [61], [60], Inoue [121], Kiya and Arie [130], Kiya, Sasaki, and Arie [131], Sarpkaya [202]), flow past a bed (see Smith and Stansby [217]), flow past buildings (see Summers, Hanson, and Wilson [221]), flow past immersible elastic boundaries (see Mendez [158]), flow past in heart valves (see McCracken and Peskin [148]), and flow in closed chambers (see Sethian [206], Sethian and Salam [209], Shestakov [211], Sung, Laitone, and Patterson [222]).

Reactive Flows: Combustion / Flame Propagation. In reactive flows that occur in turbulent combustion, one important problem is the interaction between the hydrodynamics and the propagating flame. Much of the efficiency of combustion comes from the large-scale mixing effects of the flow, which bring more reactants in contact with the flame, promoting shorter combustion times. The equations of turbulent combustion are complex, since the full Navier–Stokes equations must be coupled to complete reaction chemistry models of combustion. Fortunately, in many situations, the problem can be simplified. Since the chemistry time scale is fast relative to the hydrodynamic scale, the assumption of premixed turbulent combustion is often made. Under this model, the flame is taken as infinitely thin, pressure fluctuations are neglected, the Mach number is assumed small, and combustion is characterized by a single-step irreversible chemical reaction taking place at a constant rate. In this simplified model, the central issues are the feedback mechanisms between the propagating flame and the developing eddy structures.

Even with these assumptions, the underlying physical mechanisms remain formidable. The total velocity field is affected by a number of components. From a purely hydrodynamic side, the flow dynamics are controlled by the confinement geometry (which may be moving, as in a piston), initial vorticity distribution, and inlet/outlet conditions. The flame burns at a speed which may depend on the local curvature and pressure changes within the vessel. As the flame burns, it acts as a source of specific volume as reactants are converted to products, and an elliptic exothermic velocity field results which affects the entire flow dynamics. At the same time, vorticity is shed along the flame front, which adds a rotational velocity field that globally influences the flow.

Vortex methods have been used to isolate the components of this process to understand the controlling mechanisms and interactions. They have been coupled to flame propagation algorithms to study the effects of mixing and fluid transport on flame dynamics in a variety of combustion settings. In particular, the effects of such factors as Reynolds number, confinement geometries, inflow and outflow conditions, exothermic heat release, vorticity shedding along flames, pressure changes in confined vessels, and curvature-dependent flame speeds have been addressed. We refer the reader to Chorin [55], Ghoniem, Chorin, and Oppenheim [94], Ghoniem, Chen, and Oppenheim [93], Ghoniem and Givi [96], Ghoniem and Knio [97], Majda and Sethian [151], and Sethian [206], [207].

Three Dimensions: Filaments / Rings / Shear Layers. In three dimensions, vortex stretching occurs. Consequently, discrete vortex elements must be adaptively added to the calculation in order to maintain accuracy as filaments stretch and contort. This can create a large number of vortex elements, hence fast summation techniques to compute the vortex-vortex interactions are valuable.

Three popular model problems for study in three dimensions are vortex filaments, vortex rings, and shear layers. Vortex methods have been used to study the stability of an evolving vortex filament, the breakdown of the solution, and the concentration of vorticity into smaller sets. Some papers are the numerical calculations in Chorin [53], [54], [60], Siggia [212], and the theoretical discussion in Tsai and Widnall [229], Widnall [233], and Neu [169]. Further theoretical investigations into three-dimensional stability and calculations using vortex methods may be found in Moin, Leonard, and Kim [162], Liu, Tavantzis, and Ting [145], Pierrehumbert [176], Aref and Flinchem [12], and Pumir and Siggia [186].

For vortex rings, vortex methods have been used to study the instabilities that develop on a single ring, and the merger of two rings which exhibit "leapfrogging" as each ring induces a velocity on the other. Some work on the theoretical, numerical, and experimental issues can be found in Anderson and Greengard [7], Del Prete [75], Dhank and De Bernardinis [78], Knio and Ghoniem [132], Maxworthy [154], Leonard [142] and Moin, Leonard, and Kim [162].

Finally, the analysis of three-dimensional shear layers using vortex methods may be found in Ashurst and Meiburg [16] and Knio and Ghoniem [132], with some theoretical and experimental studies in Agui and Hesselink [3], Landman and Saffman [138], Lasheras and Choi [140], and Robinson and Saffman [188]. Some theoretical issues are discussed in Corcos and Lin [69], Corcos and Sherman [70], and Lin and Corcos [144], and analysis of the dynamics of stretched vortices may be found in Neu [170]. Additional three-dimensional vorticity calculations may be found in Chorin [60] and Couet, Buneman, and Leonard [74].

Turbulence Studies / Physics. A number of interesting experiments have been conducted to numerically analyze some of the more theoretical issues in turbulence, such as folding, intermittency, possible fractalization of vorticity, and the cascade of energy through various scales. Studies have included estimation of Hausdorff dimensions for flows using filament methods, energy statistics of large numbers of point vortices, subgrid lattice models to produce inertial range exponents and inertial range statistics, and chaotic motion of point vortices driven by unsteady agitators. As a starting point we refer the interested reader to Aref [11], Chorin [53], [54], [56], [58], [59], Pointin and Lundgren [178], and Siggia and Aref [213].

Lastly, we mention some related works of interest. First, there are similarities between some of the techniques in vortex methods and those in large plasma simulations. We refer the interested reader to Barnes et al. [23], Birdsall and Langdon [38], Brunel et al. [47], Cohen, Langdon, and Friedman [68], Denavit [76], Hewett and Langdon [116], and Langdon, Cohen, and Friedman [139]. A few papers involving particle/grid methods in other areas are Loewenstein and Matthews [146], Brackbill and Ruppel [42], Dukowicz [84], and Monaghan and Gingold [164]. Some general papers in the area of particle methods may be found in Monaghan [163], Eastwood and Hockney [85], Gingold and Monaghan [92], Hockney and Eastwood [119], and Williamson [235]. Finally, we mention the connection between vortex methods and superfluid turbulence explored by Buttke [49]. This relationship is analyzed in detail in Chapter 7.

4. Measuring the Accuracy of a Random Vortex Simulation

"Convergence" of a Random Method Applied to an Unstable Problem. The difficulty associated with measuring the accuracy of the random vortex method lies in both the method itself and the physical problem under study. For example, imagine flow past a cylinder. Suppose we compare a very local, sensitive quantity, such as the velocity at a particular point in the flow, with an integrated quantity, such as the drag. At low enough Reynolds numbers, measurements of both the velocity and the drag one body diameter downstream are repeatable from one laboratory experiment to the next. In this regime, the flow can be viewed as deterministic on the smallest scales. On the other hand, at a much higher Reynolds number, an oscillatory wake develops. Pointwise, instantaneous probe measurements vary wildly between physical experiments, whereas the drag is suitably averaged so that it remains unchanged from one experiment to the next. On the smallest scales, this flow is "random," on the largest, it is predictable.

What quantities, structures, and flow attributes can one find on the middle scales to measure this "repeatability" issue? Suppose one had a numerical method that supposedly "solved" the equations of motion describing the flow. What would one expect the method to do? Certainly it should predict the largest scales. More detailed than that, it is as much a question about the flow regime as it is about the numerical algorithm.

It was with this in mind that we have analyzed the accuracy of random vortex methods. In Sethian and Ghoniem [208], we studied two-dimensional viscous flow over a backward-facing step. The assumption of two-dimensional flow ignores such three-dimensional effects as vortex stretching. This is a significant limitation. However, the issue of repeatability of the smallest scales remains, and convergence of the two-dimensional vortex method is itself a complex question.

We began with a study of laminar flow, where the solution is smooth and steady and we can except convergence to pointwise, instantaneous values. Other methods are more economical in this regime, however the point of the study was to examine the convergence of the vortex method. We then increased the Reynolds number, through the transition regime where unsteady flow occurs into the turbulent regime. At each stage, we measured the effects of refining numerical parameters on the computed solution.

We chose to study flow over a backward-facing step because of the wealth of experimental data available. The results of numerous laboratory experiments (see Armaly,

Durst, and Pereira [15], Denham and Patrick [77], Honji [120], Periaux, Pironneau, and Thomasset [173]) are described in Sethian and Ghoniem [208] and summarized as follows. Within the laminar regime, the flow is stable, and there are experimentally measured values for the location of the reattachment point and the recirculation zone size and shape. Thus, "converged" numerical quantities and experimental data can be compared. In the transitional and turbulent regime, laboratory experiments yield integrated quantities such as the averaged recirculation zone length and averaged velocity profiles which can be compared with numerical calculations. In addition, there is a periodic and regular pattern to the generation and development of coherent eddy structures which should not change under refinement of numerical parameters.

Summary of Convergence Results. The results of extensive numerical experiments of this flow over a step are described in detail in Sethian and Ghoniem [208]. Details of the code itself may be found in Ghoniem, Chorin, and Oppenheim [94]. We may summarize the results our convergence study as follows.

In the laminar regime, where there is a steady, stable, recirculating profile, we found the following:

(1) The variance around computed means decays in inverse proportion to the number of vortex elements.
(2) The computed means converge pointwise to a time-independent profile.
(3) The calculations accurately predict the experimentally measured relationship between Reynolds number and the size and length of the recirculation zone.
(4) The accuracy primarily depends on the number of vortex elements. The time step, boundary layer resolution are of secondary importance.

In the transitional/turbulent regime, physical experiments indicate a periodic mechanism of eddy shedding downstream of the recirculation bubble, and coherent traveling eddy structures. Here, we found the following:

(1) The calculations accurately predicted the qualitative large-scale mechanisms of the flow dynamics, which remained unchanged as the numerical parameters were refined.
(2) The calculations produced numerically converged values for integrated, global quantities, such as eddy size, shape, shedding frequency, and average velocity profiles.
(3) As above, the accuracy primarily depends on the number of vortex elements. In addition, there was a critical empirical relationship between the number of elements and the time step.

We may interpret these results as follows. Within the laminar regime, the random vortex method accurately predicts pointwise, instantaneous measurements. Because it is a probabilistic method, one studies variances and means. Nonetheless, accurate calculation of the relevant quantities can be achieved. While vortex methods may not be the method of choice in this regime, they can be expected to converge to the correct solution under parameter refinement.

In the turbulent regime, our conclusions are somewhat more guarded. Results indicate that the larger scales (average velocity profiles, shedding frequencies, size and shape of eddies) are accurately modeled, and converge as numerical parameters are refined. In this sense, vortex methods can be used to predict periodic flow dynamics, as well as large-scale averaged, integrated quantities.

Note that one is getting more than might naturally come out of more traditional turbulence closure models approaches to the problem. Vortex methods allow one to compute the transitory, periodic dynamics of the flow, rather than just the average profile. Indeed, in many engineering situations, it is the large-scale transitory dynamics that are crucial for mixing and stress/pressure fluctuations. Turbulence-closure models can average out these effects, masking the most interesting features of the flow.

What does one get for refining numerical parameters? We suspect that the answer depends on the equations themselves, and summarize some the arguments discussed in Sethian [205]. Our experience indicates that one can obtain fairly good predictions of such integrated quantities as average velocity profiles and eddy shedding size and frequencies. On the other hand, a tremendous amount of computational effort might be necessary to get a significant improvement.

We suggest that this insensitivity to further numerical refinement lies in the equations of motions as much as the method. In turbulent flow, small-scale perturbations organize themselves and pass energy on to the larger scales. Most finite-difference approximations to turbulent flow oversmooth the calculation, inhibiting the natural growth of these perturbations. On the other hand, in a random vortex approximation, the number of vortex elements and particular string of random numbers chosen represent perturbations to the smaller scales. These perturbations organize themselves and pass energy to the larger scales. Thus, we argue that refining numerical parameters merely serves to inject a different set of small-scale perturbations to the flow. The most important requirement is to calculate the flow with a resolution greater (that is, on a smaller scale) than the "repeatability" scale of the flow. Put another way, if excitations below a certain size organize themselves into large structures somewhat independent of the particular excitation, then it might not be worth a tremendous amount of effort to excite the flow on a significantly finer scale.

5. Concluding Remarks

In this article, we have described some of the central themes in the design and application of vortex methods. In the next few years, a major focus of interest will be the practical application of vortex methods to realistic three-dimensional flows. Due to the shear complexity of such flows, the intricate folding and stretching of vorticity will require a large number of discrete vortex elements to adequately resolve the solution. Thus, research in fast summation methods for efficiently computing vortex interactions will be an important element of a practical vortex algorithm, as well as some sort of vortex amalgamation technique, in which vortex motion on the smallest scales is homogenized to reduce the total number of vortex elements.

Acknowledgments

The author thanks O. Hald and R. Krasny for helpful comments.

This work is supported in part by the Applied Mathematics Subprogram of the Office of Energy Research under contract DE-AC03-76SF00098. The author also acknowledges the support of the National Science Foundation and the Sloan Foundation.

References

[1] F. H. ABERNATHY AND R. E. KRONAUER, *The formation of vortex streets*, J. Fluid Mech., 13 (1962), pp. 1–20.

[2] E. ACTON, *The modeling of large eddies in a two-dimensional shear layer*, J. Fluid Mech., 76 (1976), pp. 561–592.

[3] J. C. AGUI AND L. HESSELINK, *Flow visualization and numerical analysis of a coflowing jet: a three-dimensional approach*, J. Fluid Mech., 191 (1988), pp. 19–45.

[4] C. ANDERSON, *A method of local corrections for computing the velocity field due to a distribution of vortex blobs*, J. Comput. Phys., 62 (1986), pp. 111–123.

[5] ———, *A vortex method for flows with slight density variations*, J. Comput. Phys., 61 (1985), pp. 417–444.

[6] C. ANDERSON AND C. GREENGARD, *On vortex methods*, SIAM J. Numer. Anal., 22 (1985), pp. 413–440.

[7] ———, *Vortex ring interaction by the vortex method*, in Vortex Methods, C. Anderson and C. Greengard, eds., Springer-Verlag, New York, 1360, 1988, pp. 23–36.

[8] A. W. APPEL, *An efficient program for many-body simulations*, SIAM J. Sci. Statist. Comput., 6 (1985), pp. 85–103.

[9] H. AREF, *Integrable, chaotic, and turbulent vortex motion in two-dimensional flows*, Ann. Rev. Fluid Mech., 15 (1983), pp. 345–389.

[10] ———, *The numerical experiment in fluid mechanics*, J. Fluid Mech., 173 (1986), pp. 15–41.

[11] ———, *Stirring by chaotic advection*, J. Fluid Mech., 143 (1984), pp. 1–21.

[12] H. AREF AND E. P. FLINCHEM, *Dynamics of a vortex filament in a shear flow*, J. Fluid Mech., 148 (1984), pp. 477–497.

[13] H. AREF AND T. KAMBE, *Report on the* IUTAM *symposium: fundamental aspects of vortex motion*, J. Fluid Mech., 190 (1988), pp. 571–595.

[14] H. AREF AND E. D. SIGGIA, *Evolution and breakdown of a vortex street in two dimensions*, J. Fluid Mech., 109 (1981), pp. 435–463.

[15] B. F. ARMALY, F. DURST, AND J. C. F. PEREIRA, *Experimental and theoretical investigations of flow over a backward-facing step*, J. Fluid Mech., 127 (1983), pp. 473–496.

[16] W. T. ASHURST AND E. MEIBURG, *Three-dimensional shear layers via vortex dynamics*, J. Fluid Mech., 180 (1988), pp. 85–116.

[17] S. B. BADEN, *Very large vortex calculations in two dimensions*, in Vortex Methods, C. Anderson and C. Greengard, eds., Vol. 1360, Springer-Verlag, New York, (1988), pp. 96–120.

[18] G. R. BAKER, *The "cloud-in-cell" technique applied to the roll up of vortex sheets*, J. Comput. Phys., 31 (1979), pp. 76–95.

[19] ———, *A test of the method of Fink and Soh for following vortex sheet motion*, J. Comput. Phys., 100 (1980), pp. 209–220.

[20] G. R. BAKER, D. I. MEIRON, AND S. A. ORSZAG, *Generalized vortex methods for free-surface flow problems*, J. Fluid Mech., 123 (1982), pp. 477–501.

[21] ———, *Vortex simulations of the Rayleigh–Taylor instability*, Phys. Fluids, 23 (1980), pp. 1485–1489.

[22] G. R. BAKER, P. G. SAFFMAN, AND J. S. SHEFFIELD, *Structure of a linear array of hollow vortices of finite cross-section*, J. Fluid Mech., 74 (1976), pp. 469–476.

[23] D. C. BARNES, T. KAMIMURA, J. N. LEBOEUF, AND T. TAJIMA, *Implicit particle simulation of magnetized plasmas*, J. Comput. Phys., 52 (1983), pp. 480–502.

[24] J. BARNES AND P. HUT, *A hierarchical* $O(N \log N)$ *force calculation*, Nature, 34 (1986), pp. 446–449.

[25] B. C. BASU AND G. J. HANCOCK, *The unsteady motion of a two-dimensional aerofoil in incompressible inviscid flow*, J. Fluid Mech., 87 (1978), pp. 159–178.

[26] J. BASUKI AND J. M. R. GRAHAM, *Discrete vortex computation of separated airfoil flow*, AIAA J., 25 (1987), pp. 1409–1410.

[27] G. K. BATCHELOR, *An Introduction to Fluid Mechanics*, Cambridge University Press, New York, 1967.

[28] R. G. BATT, *Some measurements on the effect of tripping the two-dimensional shear layer*, J. Fluid Mech., 26 (1975), pp. 281–307.

[29] J. T. BEALE, *A convergent* 3-D *vortex method with grid-free stretching*, Math. Comp., 46 (1986), pp. 401–424.

[30] J. T. BEALE AND A. MAJDA, *High order accurate vortex methods with explicit velocity kernels*, J. Comput. Phys., 58 (1985), pp. 188–208.

[31] J. T. BEALE AND A. MAJDA, *Rates of convergence for viscous splitting for the Navier–Stokes equations*, Math. Comp., 37 (1981), pp. 243–259.

[32] ———, *Vortex methods* I: *convergence in three dimensions*, Math. Comp., 39 (1982a), pp. 1–27.

[33] ———, *Vortex methods* II: *higher order accuracy in two and three dimensions*, Math. Comp., 39 (1982b), pp. 29–52.

[34] P. W. BEARMAN, *Vortex shedding from oscillating bluff bodies*, Ann. Rev. Fluid Mech., 16 (1984), pp. 195–222.

[35] P. W. BEARMAN AND J. E. FACKRELL, *Calculation of two-dimensional and axisymmetric bluff-body potential flow*, J. Fluid Mech., 72 (1975), pp. 299–241.

[36] P. W. BEARMAN AND J. M. R. GRAHAM, *Vortex shedding from bluff bodies in oscillatory flow, a report on Euromech* 119, J. Fluid Mech., 99 (1980), pp. 225–245.

[37] G. BENFATTO AND M. PULVIRENTI, *Generation of vorticity near the boundary in planar Navier–Stokes flows*, Comm. Math. Phys., 96 (1984), pp. 59–95.

[38] C. K. BIRDSALL AND A. B. LANGDON, *Plasma Physics via Computer Simulation*, McGraw-Hill, New York, 1985.

[39] G. BIRKHOFF, *Helmholtz and Taylor instability*, in Proc. Symposium on Applied Mathematics, XIII, American Mathematical Society, Vol. 55, Providence, RI, 1962.

[40] D. R. BOLDMAN, P. F. BRINICH, AND M. E. GOLDSTEIN, *Vortex shedding for a blunt trailing edge with equal and unequal external mean velocities*, J. Fluid Mech., 75 (1976), pp. 721–735.

[41] J. BORIS, *A vectorized "Near neighbors" algorithm of order N using a monotonic logical grid*, J. Comput. Phys., 66 (1986), pp. 1–20.

[42] J. U. BRACKBILL AND H. M. RUPPEL, FLIP: *a method for adaptively zoned, particle-in-cell calculations of fluid flows in two dimensions*, J. Comput. Phys., 65 (1986), pp. 314–343.

[43] P. BRADSHAW, *The effect of initial conditions on the development of a free shear layer*, J. Fluid Mech., 26 (1966), pp. 225–236.

[44] I. G. BROMILOW AND R. R. CLEMENTS, *A numerical study of vortex interaction*, J. Fluid Mech., 146 (1984), pp. 331–345.

[45] F. K. BROWAND, *An experimental investigation of an incompressible, separated shear layer*, AIAA J., 2 (1975), pp. 245–261.

[46] G. R. BROWN AND A. ROSHKO, *On density effects and large structures in turbulent mixing layers*, J. Fluid Mech., 64 (1974), pp. 775–816.

[47] F. BRUNEL, J. N. LEBOEUF, T. TAJIMA, AND J. M. DAWSON, *Magnetohydrodynamic particle code: Lax–Wendroff algorithm with finer grid interpolations*, J. Comput. Phys., 43 (1981), pp. 268–288.

[48] O. BUNEMAN, C. W. BARNES, J. C. GREEN, AND D. E. NIELSEN, *Principles and capabilities of* 3-D, *E-M particle simulations*, J. Comput. Phys., 38 (1980), pp. 1–44.

[49] T. F. BUTTKE, *A numerical study of superfluid turbulence in the self-induction approximation*, J. Comput. Phys., 76 (1988), pp. 301–326.

[50] J. CARRIER, L. GREENGARD, AND V. ROKHLIN, *A fast adaptive multiple algorithm for particle simulations*, SIAM J. Sci. Statist. Comput., 9 (1988), pp. 669–686.

[51] C. CHANG, *Random vortex methods for the Navier–Stokes equations*, J. Comput. Phys., 76 (1988), pp. 281–300.

[52] A. Y. CHEER, *Numerical study of incompressible slightly viscous flow past blunt bodies and airfoils*, SIAM J. Sci. Statist. Comput., 4 (1983), pp. 685–705.

[53] A. J. CHORIN, *Estimates of intermittency, spectra, and blow-up in developed turbulence*, Comm. Pure Appl. Math., 14 (1981), pp. 853–866.

[54] ———, *The evolution of a turbulent vortex*, Comm. Math. Phys., 83 (1982), pp. 517–535.

[55] ———, *Flame advection and propagation algorithms*, J. Comput. Phys., 35 (1980), pp. 1–11.

[56] ———, *Numerical estimates of Hausdorff dimension*, J. Comput. Phys., 46 (1982), pp. 390–396.

[57] ———, *Numerical study of slightly viscous flow*, J. Fluid Mech., 57 (1973), pp. 785–796.

[58] A. J. CHORIN, *Solutions of lattice models by successive linkage*, Comm. Math. Phys., 99 (1985), pp. 501–515.

[59] ———, *Turbulence and vortex stretching on a lattice*, Comm. Pure Appl. Math., 19 (1986), Special Volume, pp. S47–S65.

[60] ———, *Vortex models and boundary layer instability*, SIAM J. Sci. Statist. Comput., 1 (1980), pp. 1–21.

[61] ———, *Vortex sheet approximation of boundary layers*, J. Comput. Phys., 27 (1978), pp. 428–442.

[62] A. J. CHORIN AND P. S. BERNARD, *Discretization of a vortex sheet with an example of roll-up*, J. Comput. Phys., 13 (1973), pp. 423–429.

[63] A. J. CHORIN AND J. E. MARSDEN, *A Mathematical Introduction to Fluid Mechanics*, Springer-Verlag, New York, 1979.

[64] J. P. CHRISTIANSEN, *Numerical simulation of hydrodynamics by the method of point vortices*, J. Comput. Phys., 13 (1973), pp. 363–379.

[65] J. P. CHRISTIANSEN AND N. J. ZABUSKY, *Instability, coalescence and fission of finite-area vortex structures*, J. Fluid Mech., 61 (1973), pp. 219–243.

[66] R. R. CLEMENTS, *An inviscid model of two-dimensional vortex shedding*, J. Fluid Mech., 57 (1973), pp. 321–336.

[67] R. R. CLEMENTS AND D. J. MAUL, *The representation of sheets of vorticity by discrete vorticity*, Progr. Aerospace Sci., 16 (1975), pp. 129–146.

[68] B. I. COHEN, A. B. LANGDON, AND A. FRIEDMAN, *Smoothing and spatial grid effects in implicit particle simulation*, J. Comput. Phys., 56 (1984), pp. 51–64.

[69] G. M. CORCOS AND S. J. LIN, *The mixing layer: deterministic models of a turbulent flow. Part 2. The origin of the three-dimensional motion*, J. Fluid Mech., 139 (1984), pp. 67–95.

[70] G. M. CORCOS AND F. S. SHERMAN, *The mixing layer: deterministic models of a turbulent flow. Part 1. Introduction and two-dimensional flow*, J. Fluid Mech., 139 (1984), pp. 29–65.

[71] ———, *Vorticity concentration and the dynamics of unstable free shear layers*, J. Fluid Mech., 73 (1976), pp. 241–264.

[72] G. H. COTTET, Ph.D. thesis, Universite Pierre et Marie Curie, Paris 6, France, 1982.

[73] G. H. COTTET AND P. A. RAVIART, *Particle methods for the one-dimensional Vlasov–Poisson equations*, SIAM J. Numer. Anal., 21 (1984), pp. 52–76.

[74] B. COUET, O. BUNEMAN AND A. LEONARD, *Simulation of three-dimensional incompressible flows with a vortex-in-cell method*, J. Comput. Phys., 39 (1981), pp. 305–328.

[75] V. DEL PRETE, *Numerical simulation of vortex breakdown*, Boll. Un. Mat. Stal. B (6), 2 (1983), pp. 241–253.

[76] J. DENAVIT, *Time-filtering particle simulations with $\omega_{pe}\Delta t \gg 1$*, J. Comput. Phys., 42 (1981), pp. 337–366.

[77] M. K. DENHAM AND M. A. PATRICK, Trans. Inst. Chem. Engrs., 52 (1974), pp. 361–367.

[78] M. R. DHANK AND B. DE BERNARDINIS, *The evolution of an elliptic vortex ring*, J. Fluid Mech., 109 (1981), pp. 189–216.

[79] R. J. DIPERNA AND A. MAJDA, *Concentrations in regularizations for 2-D incompressible flow*, Comm. Pure Appl. Math., 40 (1987), pp. 301–345.

[80] ———, *Oscillations and concentrations in weak solutions of the incompressible fluid equations*, Comm. Math. Phys., 108 (1987), pp. 667–689.

[81] D. G. DRITSCHEL, *Contour surgery: a topological reconnection scheme for extended integrations using contour dynamics*, J. Comput. Phys., 77 (1988), pp. 240–266.

[82] ———, *The nonlinear evolution of rotating configurations of uniform vorticity*, J. Fluid Mech., 172 (1986), pp. 157–182.

[83] ———, *The stability and energetics of corotating uniform vortices*, J. Fluid Mech., 157 (1985), pp. 95–134.

[84] J. K. DUKOWICZ, *A particle-fluid numerical model for liquid sprays*, J. Comput. Phys., 35 (1980), pp. 229–253.

[85] J. W. EASTWOOD AND R. W. HOCKNEY, *Shaping the force law in two-dimensional particle-mesh models*, J. Comput. Phys., 16 (1974), pp. 342–359.

[86] W. FELLER, *An Introduction to Probability Theory and Its Applications*, Vol. II, John Wiley, New York, 1971.

[87] J. FERZIGER, *Energetics of vortex rollup and pairing*, Phys. Fluids, 23 (1980), pp. 1–4.

[88] P. T. FINK AND W. K. SOH, *A new approach to roll-up calculations of vortex sheets*, Proc. Roy. Soc. London Ser. A, 362 (1978), pp. 195–209.

[89] P. FREYMUTH, *On transition in a separated laminar boundary layer*, J. Fluid Mech., 25 (1966), pp. 683–704.

[90] P. FREYMUTH, W. BANK, AND M. PALMER, *Further experimental evidence of vortex splitting*, J. Fluid Mech., 152 (1985), pp. 289–299.

[91] J. H. GERRARD, *Numerical computation of the magnitude and frequency of the lift on a circular cylinder*, Philos. Trans. Roy. Soc. London Ser. A, 261 (1967), pp. 137–162.

[92] R. A. GINGOLD AND J. J. MONAGHAN, *Kernel estimates as a basis for general particle methods in hydrodynamics*, J. Comput. Phys., 46 (1982), pp. 429–453.

[93] A. F. GHONIEM, D. Y. CHEN, AND A. K. OPPENHEIM, *Formation and inflammation of a turbulent jet*, AIAA J., 24 (1986), pp. 224–229.

[94] A. F. GHONIEM, A. J. CHORIN, AND A. K. OPPENHEIM, *Numerical modeling of turbulent flow in a combustion tunnel*, Philos. Trans. Roy. Soc. London Ser. A, 304 (1982), pp. 303–325.

[95] A. F. GHONIEM AND Y. GAGNON, *Vortex simulation of laminar recirculating flow*, J. Comput. Phys., 68 (1987), pp. 346–377.

[96] A. F. GHONIEM AND P. GIVI, *Lagrangian simulation of a reacting mixing layer at low heat release*, AIAA J. 26 (1988), pp. 690–697.

[97] A. F. GHONIEM AND O. M. KNIO, *Numerical simulation of flame propagation in constant volume chambers*, Twenty-first Symposium (International) on Combustion/The Combustion Institute, 1986, pp. 1313–1320.

[98] A. F. GHONIEM AND K. K. NG, *Numerical study of the dynamics of the forced shear layer*, Phys. Fluids, 30 (1987), pp. 706–721.

[99] A. F. GHONIEM AND J. A. SETHIAN, *Effect of Reynolds number on the structure of recirculating flow*, AIAA J., 25 (1987), pp. 168–171.

[100] A. F. GHONIEM AND F. S. SHERMAN, *Grid-free simulation of diffusion using random walk methods*, J. Comput. Phys., 61 (1985), pp. 1–37.

[101] J. GOODMAN, *Convergence of the random vortex method*, Comm. Pure Appl. Math., 60 (1987), pp. 189–220.

[102] C. GREENGARD, *Convergence of the vortex filament method*, Math. Comp., 47 (1986), pp. 387–398.

[103] ———, *The core spreading vortex method approximates the wrong equation*, J. Comput. Phys., 61 (1985), pp. 345–348.

[104] L. GREENGARD AND V. ROKHLIN, *A fast algorithm for particle simulations*, J. Comput. Phys., 73 (1987), pp. 325–348.

[105] R. W. GRIFFITHS AND P. F. LINDEN, *The stability of vortices in a rotating, stratified fluid*, J. Fluid Mech., 105 (1981), pp. 283–316.

[106] J. P. GUIRAUD AND R. KH. ZEYTOUNIAN, *A double-scaled investigation of the asymptotic structure of the rolled-up vortex sheets*, J. Fluid Mech., 79 (1977), pp. 93–112.

[107] O. HALD, *Convergence of a random method with creation of vorticity*, SIAM J. Sci. Statist. Comput., 7 (1986), pp. 1373–1386.

[108] O. H. HALD, *Convergence of random methods for a reaction-diffusion equation*, SIAM J. Sci. Statist. Comput., 2 (1981), pp. 83–96.

[109] ———, *Convergence of vortex methods for Euler's equations*, II, SIAM J. Numer. Anal., 16 (1979), pp. 726–755.

[110] ———, *Convergence of vortex methods for Euler's equations*, III, SIAM J. Numer. Anal., 24 (1987), pp. 538–582.

[111] O. H. HALD AND V. M. DEL PRETE, *Convergence of vortex methods for Euler's equations*, Math. Comp., 32 (1978), pp. 791–809.

[112] M. G. HALL, *Vortex breakdown*, Ann. Rev. Fluid Mech., 4 (1972), pp. 195–217.

[113] F. H. HARLOW, Report No. LA-2301, Los Alamos National Laboratory, Los Alamos, NM, 1959.

[114] ———, *Numerical methods for fluid mechanics, an annotated bibliography*, Report No. LA-4821, Los Alamos National Laboratory, Los Alamos, NM, 1970.

[115] G. HEIDARINEJAD AND A. F. GHONIEM, *Vortex simulation of the reacting shear layer*;

effects of Reynolds and Damkohler number, AIAA '89, paper AIAA-89-0573, American Institute of Aeronautics and Astronautics, NY, 1989.

[116] D. W. HEWETT AND A. B. LANGDON, *Electromagnetic direct implicit plasma simulation*, J. Comput. Phys., 72 (1987), pp. 121–155.

[117] J. J. L. HIGDON AND C. POZRIKIDIS, *The self-induced motion of vortex sheets*, J. Fluid Mech., 150 (1985), pp. 203–231.

[118] C. M. HO AND L. S. HUANG, *Subharmonics and vortex merging in mixing layers*, J. Fluid Mech., 119 (1982), pp. 443–473.

[119] R. W. HOCKNEY AND J. W. EASTWOOD, *Computer Simulations Using Particles*, McGraw-Hill, New York, 1981.

[120] H. HONJI, *Flow over a backward facing step*, J. Fluid Mech., 69 (1975), pp. 229–240.

[121] O. INOUE, *A new approach to flow problems past a porous plate*, AIAA J., 23 (1985), pp. 1916–1921.

[122] ———, *Vortex simulation of a turbulent mixing layer*, AIAA J., 23 (1985), pp. 367–373.

[123] O. INOUE AND A. LEONARD, *Vortex simulation of forced/unforced mixing layers*, AIAA J., 25 (1987), pp. 1417–1418.

[124] P. A. JACOBS AND D. I. PULLIN, *Coalescence of stretching vortices*, Phys. Fluids, 28 (1985), pp. 1619–1625.

[125] H. KADEN, *Aufwicklung Einer Unstablien Unstetigkeitsflache*, Ing. Arch., 2 (1931), pp. 140.

[126] J. KATZ, *A discrete vortex method for the non-steady separated flow over an airfoil*, J. Fluid Mech., 102 (1981), pp. 315–328.

[127] R. M. KERR, *Simulation of Rayleigh–Taylor using vortex blobs*, J. Comput. Phys., 76 (1988), pp. 48–84.

[128] S. KIDA, *Stabilizing effects of finite core on Karman vortex street*, J. Fluid Mech., 122 (1982), pp. 487–504.

[129] T. KIMURA AND M. TSUTAHARA, *Flows about a rotating circular cylinder by the discrete-vortex method*, AIAA J., 25 (1987), pp. 182–184.

[130] M. KIYA AND M. ARIE, *An inviscid numerical simulation of vortex shedding from an inclined flat plate in shear flow*, J. Fluid Mech., 82 (1977), pp. 241–253.

[131] M. KIYA, K. SASAKI, AND M. ARIE, *Discrete vortex simulation of a turbulent separation bubble*, J. Fluid Mech., 120 (1982), pp. 219–244.

[132] O. M. KNIO AND A. F. GHONIEM, *Three dimensional vortex simulation of entrainment augmentation due to streamwise structures*, AIAA '89, American Institute of Aeronautics and Astronautics, NY, 1989.

[133] R. KRASNY, *A study of singularity formation in a vortex sheet by the point-vortex approximation*, J. Fluid Mech., 167 (1986), pp. 65–93.

[134] ———, *Computation of the vortex sheet roll-up in the Treffitz plane*, J. Fluid Mech., 184 (1987), pp. 123–155.

[135] ———, *Desingularization of periodic vortex sheet roll-up*, J. Comp. 65 (1986), pp. 292–313.

[136] K. KUWAHARA AND H. TAKAMI, *Numerical studies of two-dimensional vortex motion by a system of points*, J. Phys. Soc. Japan, 34 (1973), pp. 247–253.

[137] J. A. LAITONE, *A numerical solution for gas particle flows at high Reynolds numbers*, J. Appl. Mech., 48 (1981), pp. 465–471.

[138] M. J. LANDMAN AND P. G. SAFFMAN, *The three-dimensional instability of strained vortices in a viscous fluid*, Phys. Fluids, 30 (1987), pp. 2339–2342.

[139] A. B. LANGDON, B. I. COHEN, AND A. FRIEDMAN, *Direct implicit large time-step particle simulation of plasmas*, J. Comput. Phys., 51 (1983), pp. 107–138.

[140] J. C. LASHERAS AND H. CHOI, *Three-dimensional instability of a plane free shear layer: an experimental study of the formation and evolution of streamwise vortices*, J. Fluid Mech., 189 (1988), pp. 51–86.

[141] S. LEIBOVICH, *The structure of vortex breakdown*, Ann. Rev. Fluid Mech., 10 (1978), pp. 221–246.

[142] A. LEONARD, *Computing three-dimensional incompressible flows with vortex elements*, Ann. Rev. Fluid Mech., 17 (1985), pp. 523–559.

[143] ———, *Vortex methods for flow simulation*, J. Comput. Phys., 37 (1980), pp. 289–335.

[144] S. J. LIN AND G. M. CORCOS, *The mixing layer: deterministic models of a turbulent flow. Part 3. The effect of plane strain on the dynamics of streamwise vortices*, J. Fluid Mech., 141 (1984), pp. 139–178.

[145] C. H. LIU, J. TAVANTZIS, AND L. TING, *Numerical studies of motion and decay of vortex filaments*, AIAA J., 24 (1986), pp. 1290–1297.

[146] M. LOEWENSTEIN AND W. G. MATHEWS, *Adiabatic particle hydrodynamics in three dimensions*, J. Comput. Phys., 62 (1986), pp. 414–428.

[147] D. G. LONG, *Convergence of the random vortex method in two dimensions*, J. Amer. Math. Soc., 1 (1988), pp. 799–804.

[148] M. F. MCCRACKEN AND C. S. PESKIN, *A vortex method for blood flow through heart valves*, J. Comput. Phys., 35 (1980), pp. 183–205.

[149] A. MAJDA, *Vortex Dynamics: Numerical Analysis, Scientific Computing, and Mathematical Theory*, in ICIAM '87, J. McKenna and R. Temam, eds., Society for Industrial and Applied Mathematics, Philadelphia, 1988, pp. 153–182.

[150] ———, *Vorticity and the mathematical theory of incompressible fluid flow*, Comm. Pure Appl. Math., 19 (1986), Special Volume, pp. S187–S220.

[151] A. MAJDA AND J. A. SETHIAN, *The derivation and numerical solution of the equations for zero mach number combustion*, Combustion Sci. Tech., 42 (1985), pp. 185–205.

[152] K. W. MANGLER AND J. WEBER, *The flow near the center of a rolled-up vortex sheet*, J. Fluid Mech., 30 (1967), pp. 177–193.

[153] C. MARCHIRO AND M. PULVIRENTI, *Hydrodynamics in two dimensions and vortex theory*, Comm. Math. Phys., 84 (1982), pp. 483–503.

[154] T. MAXWORTHY, *Turbulent vortex rings*, J. Fluid Mech., 64 (1974), pp. 227–239.

[155] A. MAYO, *The fast solution of Poisson's and the biharmonic equations on irregular regions*, SIAM J. Numer. Anal., 21 (1984), pp. 285–299.

[156] E. MEIBURG AND J. C. LASHERAS, *Experimental and numerical investigation of the three-dimensional transition in plane wakes*, J. Fluid Mech., 190 (1988), pp. 1–37.

[157] D. I. MEIRON, P. G. SAFFMAN, AND J. C. SCHATZMAN, *The linear two-dimensional stability of inviscid vortex streets of finite-cored vortices*, J. Fluid Mech., 147 (1984), pp. 187–212.

[158] R. MENDEZ, *Numerical study of incompressible flow about immersed elastic boundaries*, SIAM J. Sci. Statist. Comput., 4 (1983), pp. 94–104.

[159] J. C. S. MENG, *The physics of vortex-ring evolution in a stratified and shearing environment*, J. Fluid Mech., 84 (1978), pp. 455–469.

[160] J. C. S. MENG AND J. A. L. THOMSON, *Numerical studies of some nonlinear hydrodynamic problems by discrete vortex element methods*, J. Fluid Mech., 84 (1978), pp. 433–453.

[161] F. MILINAZZO AND P. G. SAFFMAN, *The calculation of large Reynolds number two-dimensional flow using discrete vortices with random walk*, J. Comput. Phys., 23 (1977), pp. 380–392.

[162] P. MOIN, A. LEONARD, AND J. KIM, *Evolution of a curved vortex filament into a vortex ring*, Phys. Fluids, 29 (1986), pp. 955–963.

[163] J. J. MONAGHAN, *Why particle methods work*, SIAM J. Sci. Statist. Comput. 3 (1982), pp. 422–433.

[164] J. J. MONAGHAN AND R. A. GINGOLD, *Shock simulation by the particle method* SPH, J. Comput. Phys., 52 (1983), pp. 374–389.

[165] D. W. MOORE, *A numerical study of the rollup of a finite vortex sheet*, J. Fluid Mech., 63 (1974), pp. 225–235.

[166] ———, *On the point vortex method*, SIAM J. Sci. Statist. Comput., 2 (1981), pp. 65–84.

[167] ———, *The spontaneous appearance of a singularity in the shape of an evolving vortex sheet*, Proc. Roy. Soc. London Ser. A, 365 (1979), pp. 105–119.

[168] D. W. MOORE AND P. G. SAFFMAN, *The density of organized vortices in a turbulent mixing layer*, J. Fluid Mech., 69 (1984), pp. 465–473.

[169] J. C. NEU, *The dynamics of a columnar vortex in an imposed strain*, Phys. Fluids, 27 (1984), pp. 2397–2402.

[170] ———, *The dynamics of stretched vortices*, J. Fluid Mech., 143 (1984), pp. 253–276.

[171] E. A. OVERMAN AND N. J. ZABUSKY, *Coaxial scattering of Euler-equation translating V-states via contour dynamics*, J. Fluid Mech., 125 (1982), pp. 187–202.

[172] E. A. OVERMAN, N. J. ZABUSKY, AND S. L. OSSAKOW, *Ionospheric plasma cloud dynamics via regularized contour dynamics*. I. *Stability and nonlinear evolution of one-contour models*, Phys. Fluids, 26 (1983), pp. 1139–1153.

[173] J. PERIAUX, O. PIRONNEAU, AND F. THOMASSET, *Flow over backward-facing step*, in Notes on Numerical Methods in Fluid Mechanics, Vol. 7, M. Pandolfi and R. Piva, eds., Vieweg, FGR, 1984.

[174] M. PERLMAN, *On the accuracy of vortex methods*, J. Comput. Phys., 59 (1985), pp. 200–223.

[175] R. T. PIERREHUMBERT, *A family of steady, translating vortex pairs with distributed vorticity*, J. Fluid Mech., 99 (1980), pp. 129–144.

[176] ———, *Remarks on a paper by Aref and Flinchem*, J. Fluid Mech., 163 (1986), pp. 21–26.

[177] R. T. PIERREHUMBERT AND S. E. WIDNALL, *The structure of organized vortices in a free shear layer*, J. Fluid Mech., 102 (1981), pp. 301–313.

[178] Y. B. POINTIN AND T. S. LUNDGREN, *Statistical mechanics of two-dimensional vortices in a bounded container*, Phys. Fluids, 19 (1976), pp. 1459–1471.

[179] C. POZRIKIDIS AND J. J. L. HIGDON, *Instability of compound vortex layers and wakes*, Phys. Fluids, 30 (1987), pp. 2965–2975.

[180] E. G. PUCKETT, *A study of the vortex sheet method and its rate of convergence,* SIAM J. Sci. Statist., Comput. 10 (1989), pp. 298–326.

[181] N. K. PUI AND I. GARTSHORNE, *Measurements of the growth rate and structure in plane turbulent mixing layers*, J. Fluid Mech., 91 (1978), pp. 111–130.

[182] D. I. PULLIN, *The large-scale structure of unsteady self-similar rolled-up vortex sheets*, J. Fluid Mech., 88 (1978), pp. 401–430.

[183] ———, *Numerical studies of surface-tension effects in nonlinear Kelvin–Helmholtz and Rayleigh–Taylor instability*, J. Fluid Mech., 119 (1982), pp. 507–532.

[184] D. I. PULLIN AND P. A. JACOBS, *Inviscid evolution of stretched vortex arrays*, J. Fluid Mech., 171 (1986), pp. 377–406.

[185] D. I. PULLIN AND W. R. C. PHILLIPS, *On a generalization of Kaden's problem*, J. Fluid Mech., 104 (1981), pp. 45–53.

[186] A. PUMIR AND E. D. SIGGIA, *Vortex dynamics and the existence of solutions to the Navier–Stokes equations*, Phys. Fluids, 30 (1987), pp. 1606–1626.

[187] S. ROBERTS, *Accuracy of the random vortex method for a problem with non-smooth initial conditions*, J. Comput. Phys., 58 (1985), pp. 29–43.

[188] A. C. ROBINSON AND P. G. SAFFMAN, *Three-dimensional stability of an elliptical vortex in a straining field*, J. Fluid Mech., 142 (1984), pp. 451–466.

[189] ———, *Three-dimensional stability vortex arrays*, J. Fluid Mech., 125 (1982), pp. 411–427.

[190] A. RIZZI AND B. ENGQUIST, *Selected topics in the theory and practice of computational fluid dynamics*, J. Comput. Phys., 72 (1987), pp. 1–69.

[191] L. ROSENHEAD, *The formation of vortices from a surface of discontinuity*, Proc. Roy. Soc. London Ser. A, 134 (1931), pp. 170–192.

[192] J. W. ROTTMAN, J. E. SIMPSON AND P. K. STANSBY, *The motion of a cylinder of fluid released from rest in a cross-flow*, J. Fluid Mech., 177 (1987), pp. 307–337.

[193] P. G. SAFFMAN, *The approach of a vortex pair to a plane surface in an inviscid fluid*, J. Fluid Mech., 92 (1979), pp. 497–503.

[194] ———, *Dynamics of vorticity*, J. Fluid Mech., 106 (1981), pp. 49–58.

[195] P. G. SAFFMAN AND G. R. BAKER, *Vortex interactions*, Ann. Rev. Fluid Mech., 11 (1979), pp. 95–122.

[196] P. G. SAFFMAN AND J. C. SCHATZMAN, *Properties of a vortex street of finite vortices*, SIAM J. Sci. Statist. Comput., 2 (1981), pp. 285–295.

[197] ———, *An inviscid model for the vortex-street wake,* J. Fluid Mech., 122 (1982), pp. 467–486.

[198] ———, *Stability of a vortex street of finite vortices*, J. Fluid Mech., 117 (1982), pp. 171–185.

[199] P. G. SAFFMAN AND R. SZETO, *Equilibrium shapes of a pair of equal uniform vortices*, Phys. Fluids, 23 (1980), pp. 2339–2342.

[200] P. G. SAFFMAN AND S. TANVEER, *The touching pair of equal and opposite uniform vortices*, Phys. Fluids, 25 (1982), pp. 1929–1931.

[201] T. SARPKAYA, *An analytic study of separated flow about circular cylinders,* ASME J. Basic Engrg, 90 (1968), pp. 511–520.
[202] ———, *An inviscid model of two-dimensional vortex shedding for transient and asymptotically steady separated flow over an inclined plate*, J. Fluid Mech., 68 (1975), pp. 109–128.
[203] H. SCHLICTING, *Boundary Layer Theory*, McGraw-Hill, New York, 1968.
[204] L. W. SCHWARTZ, *A semi-analytic approach to the self-induced motion of vortex sheets*, J. Fluid Mech., 111 (1981), pp. 475–490.
[205] J. A. SETHIAN, *Results of a numerical convergence study of the random vortex method*, in Vortex Methods, C. Anderson and C. Greengard, eds., Vol. 1360, Springer-Verlag, New York, 1988, pp. 83–95.
[206] ———, *Turbulent combustion in open and closed vessels*, J. Comput. Phys., 54 (1984), pp. 425–456.
[207] ———, *The wrinkling of a flame due to viscosity*, in Fire Dynamics and Heat Transfer, J. G. Quintiere, R. A. Alpert, and R. A. Altenkirch, eds., Vol. 25, HTD, ASME, New York, 1983, pp. 29–32.
[208] J. A. SETHIAN AND A. F. GHONIEM, *Validation study of vortex methods*, J. Comput. Phys., 74 (1988), pp. 283–317.
[209] J. A. SETHIAN AND J. B. SALEM, *Animation of interactive fluid flow visualization tools on a data parallel machine*, Internat. J. Super. Appl., 3.2 (1988), pp. 10–39.
[210] D. L. SHARP, *An overview of Rayleigh–Taylor instability*, in Fronts, Interfaces and Patterns, Proc. Third International Conference of the Center for Nonlinear Studies, North-Holland, Amsterdam, 1984, pp. 3–10.
[211] A. I. SHESTAKOV, *A hybrid vortex-*ADI *solution for flows of low viscosity*, J. Comput. Phys., 31 (1979), pp. 313–334.
[212] E. D. SIGGIA, *Collapse and amplification of a vortex filament*, Phys. Fluids, 28 (1985), pp. 794–805.
[213] E. D. SIGGIA, AND H. AREF, *Point-vortex simulation of the inverse energy cascade in two-dimensional turbulence*, Phys. Fluids, 24 (1981), pp. 171–173.
[214] J. H. B. SMITH, *Vortex flows in aerodynamics*, Ann. Rev. Fluid Mech., 18 (1986), pp. 221–242.
[215] P. A. SMITH AND P. K. STANSBY, *Generalized discrete vortex method for cylinders without sharp edges*, AIAA J., 25 (1987), pp. 199–200.
[216] ———, *Impulsively started flow around a circular cylinder by the vortex method*, J. Fluid Mech., 194 (1988), pp. 45–77.
[217] ———, *Wave-induced bed flows by a Langrangian vortex scheme*, J. Comput. Phys., 60 (1985), pp. 489–516.
[218] P. K. STANSBY, *A generalized discrete vortex method for sharp-edged cylinders*, AIAA J., 23 (1985), pp. 856–861.
[219] P. K. STANSBY AND A. G. DIXON, *Simulation of flows around cylinders by a Lagrangian vortex scheme*, Appl. Ocean. Res., 5 (1983), pp. 167–178.
[220] M. E. STERN AND L. J. PRATT, *Dynamics of vorticity fronts*, J. Fluid Mech., 161 (1985), pp. 513–532.
[221] D. M. SUMMERS, T. HANSON, AND C. B. WILSON, *A random vortex simulation of wind-flow over a building*, Internat. J. Numer. Methods Fluid Mech., 5 (1985), pp. 849–871.
[222] NAK WON SUNG, J. A. LAITONE, AND D. J. PATTERSON, *Angled jet flow model for a diesel engine intake process—random vortex method*, Internat. J. Numer. Methods Fluid Mech., 3 (1983), pp. 283–293.
[223] J. G. TELSTE, *Inviscid flow about a cylinder rising to a free surface*, J. Fluid Mech., 182 (1987), pp. 149–168.
[224] Z. TENG, *Elliptic-vortex method for incompressible flow at high Reynolds number*, J. Comput. Phys., 46 (1982), pp. 54–68.
[225] ———, *Variable-elliptic-vortex method for incompressible flow simulation*, J. Comput. Math., 4 (1986), pp. 255–263.
[226] G. TRYGGVASON, *Numerical simulations of the Rayleigh–Taylor instability*, J. Comput. Phys., 75 (1988), pp. 253–282.
[227] G. TRYGGVASON AND H. AREF, *Finger-interaction mechanisms in stratified Hele–Shaw flow*, J. Fluid Mech., 154 (1985), pp. 287–301.

[228] G. TRYGGVASON AND H. AREF, *Numerical experiments on Hele–Shaw flow with a sharp interface*, J. Fluid Mech., 136 (1983), pp. 1–30.

[229] C. TSAI AND S. E. WIDNALL, *The stability of short waves on a straight vortex filament in a weak externally imposed strain field*, J. Fluid Mech., 73 (1976), pp. 721–733.

[230] J. J. W. VAN DER VEGT AND W. C. DE BOOM, *Numerical simulation of flow around circular cylinders at high Reynolds number*, in Behaviour of Offshore Structures, Elsevier, Amsterdam, the Netherlands, 1985.

[231] A. I. VAN DER VOOREN, *A numerical investigation of the rolling up of vortex sheets*, Proc. Roy. Soc. London Ser. A, 373 (1980), pp. 67–91.

[232] F. L. WESTWATER, *Rolling up of the surface of discontinuity behind an aerofoil of finite span*, Aero. Res. Counc., R and M, Vol. 1962 (1935).

[233] S. E. WIDNALL, *The stability of a helical vortex filament*, J. Fluid Mech., 54 (1972), pp. 641–663.

[234] ———, *The structure and dynamics of vortex filaments*, Ann. Rev. Fluid Mech., (1975), pp. 141–165.

[235] J. H. WILLIAMSON, *Four cheap improvements to the particle-mesh code*, J. Comput. Phys., 41 (1981), pp. 256–269.

[236] C. D. WINANT AND F. K. BROWAND, *Vortex pairing: the mechanism of turbulent mixing-layer growth at moderate Reynolds number*, J. Fluid Mech., 63 (1974), pp. 237–255.

[237] N. J. ZABUSKY, *Computational synergetics and mathematical innovation*, J. Comput. Phys., 43 (1981), pp. 195–249.

[238] N. J. ZABUSKY, M. H. HUGHES, AND K. V. ROBERTS, *Contour dynamics for the Euler equations in two dimensions*, J. Comput. Phys., 30 (1979), pp. 96–106.

[239] N. J. ZABUSKY AND E. A. OVERMAN, *Regularization and contour dynamical algorithms. I. Tangential Regularization*, J. Comput. Phys., 52 (1983), pp. 351–373.

2

COVERGENCE OF VORTEX METHODS

Ole H. Hald

Introduction

In this paper we discuss the theory of vortex methods. At first, the vortex method was supported by physical considerations and heuristic arguments. By now there are many theoretical studies of various types of vortex methods. However, there is still no convergence proof for the original vortex algorithm applied to the flow of a slightly viscous fluid around a cylinder. It is simply too complicated. This algorithm involves a solution of Euler's equations combined with a random walk solution of the heat equation to simulate the viscous effects. Furthermore, it contains an intricate device for creating vorticity along the boundary. Currently, we have excellent theories for inviscid flows and a good understanding of the viscous flows provided there are no boundaries. It is the creation at the boundary that is not understood from the numerical analysis point of view. The purpose of this paper is to present some of the results that have been obtained and to point out some of the open questions.

The goal of numerical analysis is to get the best answer for the least amount of work. In practice we might ask: "How should we choose our parameters to get an accuracy of three digits?" The typical answer of a numerical analyst is: "If you arrange the calculations in this way, the rate of convergence is so and so." This is not an answer to the question, but in many cases it is the best we can do. One reason is that the error bounds depend on the solution we are trying to approximate. Another is that the error bounds are often wild overestimates. The theory of vortex methods suffers from both difficulties.

This is not the first survey of vortex methods. In his paper, Leonard [57] emphasizes the physical interpretation of two- and three-dimensional vortex methods. Beale and Majda [6] present a concise summary of the theory of inviscid methods. The survey by Anderson and Greengard [2] is really a research paper. They study the connection between various methods, improve consistency arguments, and prove convergence with discretization in time. In his lecture notes, Raviart [71] studies particle methods in general. He divides the theory into parts and seeks optimal results for the parts. The lecture notes by Marchioro and Pulvirenti [62] contain a study of the random vortex method, and explain the connection with diffusion processes and the Navier–Stokes equations. Finally, Majda [60] presents the theory of viscous and inviscid calculations, gives the historical development of the ideas, and outlines a measure theoretic interpretation of numerical experiments with vortex sheets. I am in debt to all.

In this paper we will emphasize those methods for which convergence proofs exist. Other methods may well be faster or more accurate, but we shall only regard this as a stimulus for further research. There will be no proofs, not even an outline of the arguments. Instead, I have tried to take the point of view of a practician. There are two questions: "What's in it for me?" and "Are all those papers really necessary?" In this presentation we will use a slightly heavy notation, which reveals the connection between the numerical methods and the mathematical formulas without a lot of back-tracking. We hope that this will make it easy to program the methods, at the cost of some beauty.

Our presentation is lopsided. There is too much emphasis on two-dimensional flow. However, there is a good reason for this imbalance. There are twice as many papers on two-dimensional flows as there are on three-dimensional flows. Two-dimensional flows are simpler to understand and therefore have been the testing ground for new theories. The focus of our presentation is also quite narrow. For instance, we do not discuss the convergence proof by Cottet for the vortex-in-cell method, nor do we describe the mathematical results obtained by Pulvirenti and his co-workers.

1. Two-dimensional Inviscid Flows

Let us begin with the simplest possible case, namely, a two-dimensional flow. We assume that the flow is incompressible and that the viscosity is zero. As there are no convergence results for problems involving boundaries, we also assume that the flow takes place in R^2. We will first introduce the flow map. At time $t = 0$ we place a particle at the point α. At later time t this particle will be at the point $x(\alpha, t)$. The mapping between α and $x(\alpha, t)$ is called the flow map. To determine the flow map we must solve the ordinary differential equations

$$\frac{d}{dt}x(\alpha,t)=u(x(\alpha,t),t),$$

$$x(\alpha,0)=\alpha.$$

Here $u(x, t)$ is the velocity field for the flow and a solution of Euler's equations. At this point we could derive the Lagrangian equations (see Anderson and Greengard [2], Beale and Majda [4], Hald [45], Long [58] or Chapter 1 of this book). Instead we simply write down the equations we want to solve:

$$\frac{d}{dt}x(\alpha,t)=\int_{R^2}K(x(\alpha,t)-x(\alpha',t))\omega(\alpha',0)\,d\alpha', \tag{1.1}$$

$$u(x,t)=\int_{R^2}K(x-x(\alpha',t))\omega(\alpha',0)\,d\alpha', \tag{1.2}$$

$$\omega(x,t)=\int_{R^2}\delta(x-x(\alpha',t))\omega(\alpha',0)\,d\alpha' \tag{1.3}$$

with $x(\alpha, 0) = \alpha$ as initial conditions. In (1.2) and (1.3), x is a dummy variable, while $x(\alpha', t)$ represents the flowmap. We regard (1.1) as an infinite system of ordinary differential equations. The kernel K is given by $K(x) = (2\pi r^2)^{-1}(-x_2, x_1)^T$ where $r^2 = x_1^2 + x_2^2$ and ^{T} means transpose. The function $\omega(x, t)$ is the vorticity at time t, and we notice that the flow map, the velocity field, and the vorticity distribution depend only on the initial vorticity $\omega(x, 0)$. The function δ in (1.3) denotes the delta function and the equation says, in a complicated way, that the vorticity is constant along the particle paths, i.e., $\omega(x(\alpha, t), t) = \omega(\alpha, 0)$. This will not be true for three-dimensional flows.

The vortex method is a numerical technique for solving (1.1)–(1.3). To get a well-behaved system of equations, we smooth the kernel K near the origin and replace the integral by a sum. We cover the α plane by nonoverlapping squares with side length h and centered at $\alpha_j = jh$ with $j = (j_1, j_2)$. In this way we get the vortex method

$$\frac{d}{dt}\tilde{x}(\alpha_i, t) = \sum_j K_\delta(\tilde{x}(\alpha_i, t) - \tilde{x}(\alpha_j, t))\omega(\alpha_j, 0)h^2, \tag{1.4}$$

$$\tilde{u}(x, t) = \sum_j K_\delta(x - \tilde{x}(\alpha_j, t))\omega(\alpha_j, 0)h^2, \tag{1.5}$$

$$\tilde{\omega}(x, t) = \sum_j \zeta_\delta(x - \tilde{x}(\alpha_j, t))\omega(\alpha_j, 0)h^2 \tag{1.6}$$

with $\tilde{x}(\alpha_j, 0) = \alpha_j$. Here ζ_δ is an approximation to the delta function and K_δ approximates K. Even this is not the complete recipe. We must also solve the ordinary differential equations (1.4) numerically. We will deal with this problem later.

If we are ever going to use the vortex method we must have a finite number of vortices. Otherwise we cannot solve the ordinary differential equations. It is therefore natural to assume that the initial vorticity $\omega(x, 0)$ has compact support. This assumption has been made by Dushane [34], Del Prete and Hald [29], Hald [45], [48], Beale and Majda [4], [5], Anderson and Greengard [2], Beale [9], and Hou [49]. It ensures that the velocity vanishes at infinity. We may, of course, be interested in the movement of particles that start at points α_j outside the support of the initial vorticity. However, the computed velocity field and the computed vorticity distribution are independent of such particles, because $\omega(\alpha_j, 0) = 0$. A different point of view can be found in the work of Cottet [25], [26]. For reasons of generality, he allows the initial vorticity to have unbounded support, and sums over all j in (1.4)–(1.6). To prove convergence for such a problem it becomes necessary to add some regularity at infinity. In Cottet's early work [25] this is done by assuming that all the derivatives of the initial vorticity exist and decay at a certain rate at infinity. In the later work [26], he assumes that a fixed, finite number of derivatives of the initial vorticity are uniformly bounded and absolutely integrable. I have not seen any numerical tests for inviscid flows with unbounded support, although those by Perlman [68] come close.

To get good error bounds we must look closely at the smoothing K_δ of K. The natural approach is to modify the behavior of K near the origin, so that the velocity is bounded, and make a continuous transition to the flow outside a disk with radius δ. The first such cutoff is due to Chorin [17]. It is

$$K_\delta(x) = \begin{Bmatrix} (2\pi\delta r)^{-1} & 0 < r \leqq \delta \\ (2\pi r^2)^{-1} & \delta < r < \infty \end{Bmatrix} \begin{pmatrix} -x_2 \\ x_1 \end{pmatrix}. \tag{1.7}$$

In this case the distance δ from the origin really deserves the name "cutoff length."

In more advanced theories K_δ is determined differently and it is not clear what the cutoff length is geometrically. We observe that Chorin's cutoff is not continuous at the origin. The velocities at a pair of points $(\pm\varepsilon, 0)$ are opposite and large. The value at the origin is undefined, but it is natural to set $K_\delta(0) = 0$ and hope that it will be used sparingly in the numerical calculations. Another early cutoff is due to Milinazzo and Saffman [64]. They used

$$K_\delta(x) = \begin{Bmatrix} (2\pi\delta^2)^{-1} & 0 \leqq r \leqq \delta \\ (2\pi r^2)^{-1} & \delta < r < \infty \end{Bmatrix} \begin{pmatrix} -x_2 \\ x_1 \end{pmatrix}. \tag{1.8}$$

Instead of looking at K_δ as an approximation to the kernel K we can regard K_δ as a velocity field. The corresponding vorticity distribution is then found by $\zeta_\delta = \text{curl } K_\delta$. For Chorin's cutoff (1.7) $\zeta_\delta = (2\pi\delta r)^{-1}$ for $0 < r < \delta$ and 0 for $r > \delta$. The vorticity distribution corresponding to (1.8) is $(\pi\delta^2)^{-1}$ for $r \leqq \delta$ and 0 for $r > \delta$. Note that Chorin's ζ_δ is singular at the origin. This makes the convergence proof difficult, but not impossible (see Hald [48]). Otherwise the properties of ζ_δ are

(i) Discontinuous at $r = \delta$,
(ii) Compact support,
(iii) Positive,
(iv) $\zeta_\delta(x) = \delta^{-2}\zeta(\delta^{-1}x) = \delta^{-2}\zeta(\delta^{-1}|x|, 0)$,
(v) $\int_{R^2} \zeta_\delta(x)\, dx = 1$.

In the search for higher order cutoffs, the first property is harmful and the second is irrelevant. The third property restricts the rate of convergence to second order at best. The fourth and the fifth properties are fundamental. It follows from (iv) that ζ_δ satisfies a scaling law and is radially symmetric. It is therefore convenient to use the same symbol ζ for both the function $\zeta(x)$, which depends on two variables (x_1, x_2), and for the function $\zeta(r)$, which depends on one variable $r = |x|$. As $\delta \to 0$ the mass is concentrated near the origin, and (v) says that the total mass is independent of δ. We may therefore regard ζ_δ as an approximation to the delta function. If ζ is given, then we know ζ_δ. The velocity field corresponding to the vorticity distribution ζ_δ is $K_\delta(x) = K * \zeta_\delta(x)$ where $*$ means convolution. To obtain this result we have used (1.2) and (1.3) for $t = 0$. Since ζ_δ is radially symmetric it follows that

$$\begin{aligned} K_\delta(x) &= K(x) \int_{|y| \leqq r} \zeta_\delta(y)\, dy \\ &= \frac{(-x_2, x_1)^T}{2\pi r^2} f\left(\frac{r}{\delta}\right). \end{aligned}$$

The function $f(r) = 2\pi \int_0^r s\zeta(s)\, ds$ is called the shape factor or (somewhat confusingly) the cutoff function. Once ζ is given it is easy to evaluate K_δ either analytically or numerically. In this formulation δ is just a scaling parameter and since $\zeta(s)$ may be positive for all s, the name cutoff length is no longer appropriate. In the rest of this paper, we will therefore call δ the "cutoff parameter."

We now turn to the crux of the matter, namely, the moment conditions. In their search for higher order methods, Beale and Majda [4], [5] formulated three conditions that lead to rapid convergence. They begin by assuming that the function ζ is twice differentiable and radially symmetric, i.e., $\zeta(x) = \zeta(|x|)$. The three conditions are

(1.9) $$\int_{R^2} \zeta(x)\,dx = 1,$$

(1.10) $$\int_{R^2} x^\alpha \zeta(x)\,dx = 0, \qquad 1 \leqq |\alpha| \leqq m-1,$$

(1.11) ζ is smooth and rapidly decreasing.

It is (1.10) that is called the moment condition. The left hand side is similar to the definition of moments in probability theory, and the equation says that a certain number of moments vanish. Here $\alpha = (\alpha_1, \alpha_2)$ and $x^\alpha = x_1^{\alpha_1} x_2^{\alpha_2}$. Equation (1.9) is the same normalization requirement as (v). Condition (1.11) has had a checkered fate. It is really a technical requirement and ensures that the remainder in certain integrals are easy to estimate. The original version by Beale and Majda [5] was formulated in terms of the Fourier transform of ζ. They assumed that there exists a positive constant γ such that

(1.12) $$(1+|y|)^{\gamma+|\beta|}\,|(\partial^\beta \hat{\zeta})(y)| \leqq C$$

for any multi-index $\beta = (\beta_1, \beta_2)$. Here y is the dual variable to x. Subsequently, Beale and Majda [7] changed this condition to

(1.13) $$(1+|x|)^\gamma |\partial^\beta \zeta(x)| \leqq C_{\beta\gamma}$$

for any multi-index β and for every integer γ. Note that these cutoff functions are infinitely differentiable and decay rapidly at infinity. This may be convenient from a theoretical point of view, but it is not necessary. Our question must therefore be: "How should we enlarge the class of cutoff functions and still preserve the convergence theory?" The first, and in some sense the most natural choice, is to assume that the cutoff function ζ has compact support and a finite number of derivatives, say k. Such cutoff functions have been considered by Hald [45] with $k = 3$, Cottet [25] with $k \geqq 2$, and Raviart [71] with $k \geqq 1$. The next step is to allow exponential decay. Thus Cottet [25] investigated cutoff functions ζ that satisfy (1.13) for all γ and all $|\beta| \leqq k$, where k is fixed and $k \geqq 3$. The final step has been to permit a slower decay at infinity. Thus Anderson and Greengard [2] assume that the cutoff function ζ is k times continuously differentiable with $k \geq 3$ and satisfies

(1.14) $$|x|^{4+m}|\zeta(x)| \leqq C,$$

(1.15) $$|x|^{2+|\beta|}|\partial^\beta \zeta(x)| \leqq C, \qquad |\beta| \leqq k,$$

(1.16) $$\int_{R^2} |x|^m |\zeta(x)|\,dx \leqq C.$$

If the cutoff function satisfies (1.9), (1.10), (1.14)–(1.16), then Anderson and Greengard [2] say that it belongs to $M^{k,m}$. The letter M means mollifier. A slightly larger class has been studied by Raviart [71]. However, instead of comparing the different classes of cutoff functions, let us look at some actual examples. We will present both ζ_δ and K_δ because ζ_δ is used to calculate the vorticity distribution, and K_δ is used to compute the velocity field. The most elegant class is due to Beale and Majda [7]. It consists of

(1.17) $$\zeta_\delta(x) = \frac{1}{\pi\delta^2} P_n\left(\frac{r^2}{\delta^2}\right) e^{-r^2/\delta^2},$$

$$K_\delta(x)=\frac{(-x_2,x_1)^T}{2\pi r^2}\left[1-Q_n\left(\frac{r^2}{\delta^2}\right)e^{-r^2/\delta^2}\right], \tag{1.18}$$

where $P_n = Q_n - Q_n'$ and Q_n are the Laguerre polynomials normalized so that the constant term is 1. For practical calculations the information in Table 2.1 should be sufficient. Note that the order $m = 2n + 2$. When $n = 0$, ζ_δ is positive and we have a second order cutoff. Observe also that there is no cutoff length. We could try to define the cutoff length as the radius of the disk that contain 99 percent of the mass. Even this is unsatisfactory because the oscillations of ζ_δ for $n = 2$ leads to three different solutions, namely, 0.757δ, 1.925δ, and 2.662δ. Since $P(r^2)\exp(-r^2)$ is smooth at the origin and decays rapidly at infinity, we find that the cutoff functions (1.17) by Beale and Majda satisfy (1.11) with any reasonable interpretation. The cutoffs (1.18) are therefore included in the theories of Anderson and Greengard [2], Beale and Majda [5], Cottet [25], [26], Hald [45], [48], and Raviart [71]. The numerical efficiency of the cutoffs (1.18) has been investigated by Beale and Majda [7], and the performance of a similar class of cutoffs, also suggested by Beale and Majda [7], has been studied by Perlman [68]. The conclusion is, roughly speaking, that the rate of convergence suggested by the theory can be observed initially, but deteriorates as time increases.

So far we have considered cutoffs with high, but finite order. We now ask: "Can we find an infinite order cutoff?" The surprising answer is yes! Beale and Majda [4] noted the existence of such cutoffs, but did not give explicit examples. The following one was proposed by Hald [48]:

$$\zeta_\delta(x)=\frac{4}{45\pi\delta^2(r/\delta)^3}\left[16J_3\left(\frac{4r}{\delta}\right)-10J_3\left(\frac{2r}{\delta}\right)+J_3\left(\frac{r}{\delta}\right)\right], \tag{1.19}$$

$$K_\delta(x)=\frac{(-x_2,x_1)^T}{2\pi r^2}\left[1-\frac{8}{45(r/\delta)^2}\left(4J_2\left(\frac{4r}{\delta}\right)-5J_2\left(\frac{2r}{\delta}\right)+J_2\left(\frac{r}{\delta}\right)\right)\right]. \tag{1.20}$$

Here J_2 and J_3 are the Bessel functions of order 2 and 3. The infinite order cutoff function does not satisfy the Beale–Majda condition (1.12) because the Fourier transform of (1.19) is only once continuously differentiable. Neither does it satisfy the moment conditions for $m = 4$ or the Anderson and Greengard's conditions (1.14)–(1.16) for $k = m = 2$. The cutoff function simply decays too slowly. Thus new arguments were required to establish the high order convergence for these cutoffs (see Hald [48]). The rapid convergence with the infinite order cutoffs has been observed by Nordmark [66]. His experiments show that the rate of convergence basically depends on the smoothness of the initial vorticity ω. There is no saturation as for the finite order cutoffs. Equations (1.19), (1.20) are the simplest of a whole class of infinite order cutoffs, all expressed in terms of Bessel functions (see Nordmark [66]). The disadvantages of these cutoffs are obvious. The first drawback, which they share with the Beale–Majda cutoffs, is the infinite support. Thus the interpretation of the computed solution becomes unclear when we discuss problems with boundaries. The second

Table 2.1. The coefficients for (1.17) and (1.18), m = order.

n	$P_n(s)$	$Q_n(s)$	m
0	1	1	2
1	$2 - s$	$1 - s$	4
2	$3 - 3s + s^2/2$	$1 - 2s + s^2/2$	6
3	$4 - 6s + 2s^2 - s^3/6$	$1 - 3s + 3s^2/2 - s^3/6$	8

difficulty is the presence of the Bessel functions in the formulas, because the Bessel functions are expensive to evaluate. One solution is to partition the interval $(0, \infty)$ into smaller intervals, and in each interval replace J_n by its best polynomial approximation in the maximum norm. In the last interval we use a short asymptotic expansion of J_n. The cost of this preprocessing is small compared to a typical vortex calculation (see Nordmark [66]), but the resulting method is no longer simple. Hence we return to cutoff functions with finite support.

Cutoff functions with finite support have become important after the discovery by Greengard and Rokhlin [44] of fast methods for vortex calculations. We observe first that there are no infinite order cutoff functions with compact support (see Hald [48]). Thus we seek simple high order cutoffs. One of the nicest is the eighth order cutoff due to Nordmark [66]. It is

$$(1.21)\quad \zeta_\delta(x)=\begin{cases}\dfrac{52}{\pi\delta^2}\left(1-21\dfrac{r^2}{\delta^2}+105\dfrac{r^4}{\delta^4}-140\dfrac{r^6}{\delta^6}\right)\left(1-\dfrac{r^2}{\delta^2}\right)^9 & \text{for } r\leqq\delta,\\ 0 \quad \text{for } r>\delta,\end{cases}$$

$$(1.22)\quad K_\delta(x)=\begin{cases}\dfrac{(-x_2,x_1)^T}{2\pi r^2}\left[1+\left[286-1092\left(1-\dfrac{r^2}{\delta^2}\right)+1365\left(1-\dfrac{r^2}{\delta^2}\right)^2\right.\right.\\ \qquad\left.\left.-560\left(1-\dfrac{r^2}{\delta^2}\right)^3\right]\left(1-\dfrac{r^2}{\delta^2}\right)^{10}\right] & \text{for } r\leqq\delta,\\ \dfrac{(-x_2,x_1)^T}{2\pi r^2} \quad \text{for } r>\delta.\end{cases}$$

The behavior of this cutoff has been investigated on three stationary, radially symmetric flows with $\omega(x,t)=(1-r^2)^3, (1-r^2)^7, (1-r^2)^{14}$ for $|x|\leqq 1$ and $\omega=0$ for $|x|>1$. At time $t=50$ the rate of convergence for the three velocity fields were 2.8, 4.5, 4.0. Here s is the rate of convergence if the error goes to zero like const.$\cdot h^s$. In these experiments $\delta=1.7\sqrt{h}$ and Nordmark used a rezoning technique, which will be discussed later. The corresponding results for the infinite order cutoff (1.20) with $\delta=0.3\sqrt{h}$ were 3.0, 4.1, 8.3 (see [65]). Even though the infinite order cutoff is more accurate for smooth flows, I think that the eighth order cutoff is perfectly adequate in practice. Figure 2.1 shows that the cutoff functions (1.17), (1.19), and (1.21) are surprisingly similar. To carry out this comparison it is important to normalize the cutoff functions, say $\zeta_\delta(0)=1$. Hence we have chosen $\delta=\sqrt{4/\pi}$ for $n=3$ in (1.17), $\delta=\sqrt{7/(4\pi)}$ in (1.19), and $\delta=\sqrt{52/\pi}$ in (1.21). Suppose we have made an experiment and found a good choice of h and δ. Assume also that we want to run it again, but this time with a different cutoff. Nordmark's [66] calculations indicate that we should keep h and choose the new value of δ such that the two cutoff functions have the same height at the origin.

We now come to one of the most difficult questions: How do we choose h and δ? Here we must distinguish between the actual behavior of the numerical method and our mathematical analysis of the method. In practice we take the mesh length h as small as we can afford, and choose δ so that the solution looks right, or so that the error is small on test problems. The truth is that we do not know how to find the optimal value. The theoretical approach is much simpler. Here we choose δ such that the error bounds are minimized. Thus Del Prete and Hald [29] use $\delta=h^q$, where q depends on the smoothness of the flow. This choice was also used by Beale and Majda [5], but here q depends on the decay of the Fourier transform of the cutoff func-

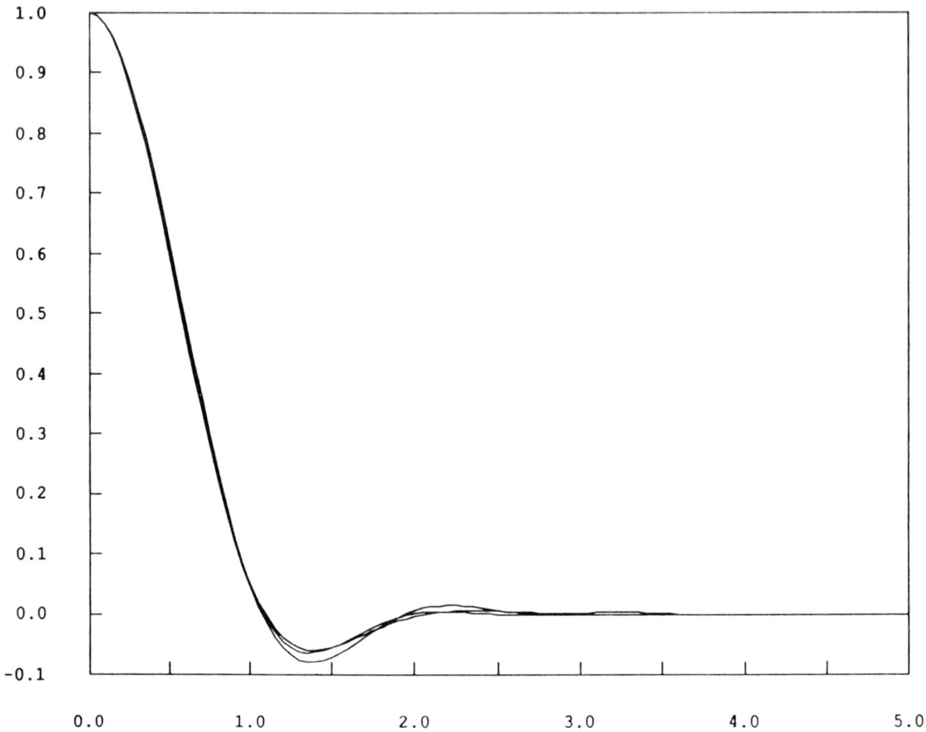

Fig. 2.1. The cutoff functions (1.17), (1.19), and (1.21).

tion and the number of vanishing moments. Cottet [25] uses the condition $\delta = ch^q$, where q depends on the smoothness of the cutoff function and on the number of moments that vanish. Any choice of h and δ can be accommodated by choosing c properly, but the proof requires that h and δ are sufficiently small. In Beale and Majda [5] and Cottet [25], $q = 1 - \varepsilon$ gives rapid convergence. On the other hand, Hald [45] finds that $\delta = h^{1/2}$ is nearly optimal. The requirement by Anderson and Greengard [2] is that $C[\delta^m + (h/\delta)^k \delta] \leqq \frac{1}{2} h\delta$ for cutoff functions $\zeta \in M^{k,m}$. Raviart [71] uses const.$\cdot\delta^{1+\varepsilon'} \leqq h \leqq$ const.$\cdot\delta^{1+\varepsilon}$, while Cottet [26] uses $h \leqq c\delta^{1+\varepsilon}$. In all cases the cutoff parameter tends to zero slower than the mesh length and this fact is used in the proofs. The theoretical choice of δ will lead to an effective numerical method only if our theory captures the essence of the problem and the assumptions are satisfied. The experiments by Del Prete and Hald [29], Perlman [68], Beale and Majda [7] and Nordmark [66] have not settled the question. I believe that there is a reason for the conflicting recommendations for the choice of δ. The question is this: Is the flow smoother or more rough than the order of the cutoff function? For smooth flows the finite order cutoffs are fully exploited and $\delta = h^{1-\varepsilon}$ gives small error bounds (see Beale and Majda [5], Cottet [25], [26], Anderson and Greengard [2], Raviart [71]). On the other hand, if the flow has only a few derivatives and we use a higher order cutoff, then $\delta = h^{1/2}$ seems optimal (see Hald [45]). This holds in particular for the infinite order cutoffs (see Hald [48]). The explanation fails for rough flows computed with singular cutoff functions and it has not been checked by numerical experiments.

To measure the error in the vortex method we use both a discrete p-norm and a continuous p-norm. The discrete norm is $\|g\|_p = (\sum_j |g(\alpha_j)|^p h^2)^{1/p}$ where we sum over all α_j in the initial vorticity. The continuous norm is $\|g\|_{L^p(B)} = (\int_B |g(x)|^p\,dx)^{1/p}$ where B is a bounded set in R^2. The following result, with some modifications, can be found in Beale and Majda [5], Cottet [25], Anderson and Greengard [2], and Raviart [71].

THEOREM 1. *Assume that the velocity field is smooth enough and that the initial vorticity has compact support. Let $\zeta \in M^{k,m}$ with k sufficiently large and $m \geqq 2$. Set $\delta = ch^q$ with $0 < q < 1$. There exist a constant C such that for all h sufficiently small we have:*

(1) *Convergence of the particle paths*

$$\max_{0 \leqq t \leqq T} \|\tilde{x}(\alpha_j, t) - x(\alpha_j, t)\|_p \leqq C[\delta^m + (h/\delta)^k \delta],$$

(2) *Convergence of the particle velocities*

$$\max_{0 \leqq t \leqq T} \|\tilde{u}(\tilde{x}(\alpha_j, t), t) - u(x(\alpha_j, t), t)\|_p \leqq C[\delta^m + (h/\delta)^k \delta],$$

(3) *Convergence of the velocity field*

$$\max_{0 \leqq t \leqq T} \|\tilde{u}(x, t) - u(x, t)\|_{L^p(B)} \leqq C[\delta^m + (h/\delta)^k \delta].$$

2. Bells and Whistles

So far we have emphasized the basic theory. In this section we will discuss some of the refinements. We begin with the choice of time step. After the initial discretization, the vortex method (1.4) is just a system of ordinary differential equations. To solve these we can, for example, use a Runge–Kutta method or a multistep scheme. As the time step Δt tends to zero the numerical solution converges to the exact solution of (1.4), which in turn approximates the solution of the original equation (1.1). The problem occurs when the time step, the mesh length, and the cutoff length tend to zero simultaneously, because we are then dealing with larger systems of equations and the equations are more singular. Anderson and Greengard were the first to study the convergence of the full problem. In [2] they consider a class of explicit multistep schemes that includes the Adams–Bashforth methods. Suppose that the local truncation error for the ordinary differential equation solver is of order $(\Delta t)^{r+1}$. If the flow is sufficiently smooth and the assumptions in Theorem 1 are satisfied, then

$$\|\text{comp } x_j - \text{exact } x_j\|_p \leqq C[\delta^m + (h/\delta)^k \delta + \Delta t^r], \tag{2.1}$$

provided the right hand side of (2.1) is less than $\frac{1}{2} h^{2/p} \delta$. Thus we get the expected rate of convergence if $\Delta t = h$, $\delta = h^{1-\varepsilon}$, and p is sufficiently large. In practice, the first, second, and fourth order Runge–Kutta methods are the most popular ones, because they are easy to program (see Chorin [17], Milinazzo and Saffman [64], Del Prete and Hald [29], Beale and Majda [7], Nakamura, Leonard, and Spalart [65], Anderson [1], Sethian and Ghoniem [76]). Since Euler's method is of Adams–Bashforth type, its convergence is guaranteed. In addition, Anderson and Greengard [2] prove that the modified Euler method converges quadratically. Their technique can be extended to other second order Runge–Kutta methods. The treatment of the higher order methods is more difficult, but the convergence of the classical fourth order Runge–Kutta method was obtained by Hald [48] for vortex methods with infinite order cutoffs and $\Delta t \leqq \sqrt{h}$. The result is, of course, special. However, there is no reason to believe that similar results cannot be obtained for high order cutoffs and smooth flows.

Just because we can prove the convergence for large time intervals does not mean that the computed solutions are accurate or that the expected rate of convergence can be observed (see Perlman [68]). To achieve accurate solutions we can use a rezoning technique proposed by Beale and Majda [7]. Their idea is to redistribute the vorticity on a rectangular grid after a fixed number of steps and then restart the calculation. Nordmark [66] has added two refinements. The first change is to represent the vorticity distribution by four times as many particles as needed to calculate the velocity field. The second change is to check if it is necessary to rezone. The test uses a special property of two-dimensional flows, namely, that the vorticity is constant along the particle paths. Using only a quarter of the particles, Nordmark compares the computed vorticity at the particle positions with the expected value, and rezones when the difference has increased more than 25 percent. For stationary, radially symmetric flows, where the solution is known, Nordmark's approach leads to smaller errors than Beale and Majda's technique, and gives a more regular growth of the error. The algorithm has also been applied to the merger of two vortex blobs. Here the initial vorticity is $\omega = \max(0, (1 - 4(|x| - \frac{1}{2})^2 - 4y^2)^7)$ and the exact solution is not known. For times $0 \leqq t \leqq 100$ the computed rate of convergence for the infinite order cutoff (1.20) increases steadily from 3.7 to 4.9 (see [66]). This increase may indicate a smoothing of the computed solution. Indeed, Beale and Majda [7] do not recommend their own method without further testing and warns against diffusive errors due to the rezoning. A local regridding technique has been presented by Hou, Lowengrub, and Shelley [53]. The idea is to replace a single vortex by two vortices, when the original grid becomes distorted. This still has to be implemented, but the authors prove that their method converges quadratically.

Beale [8] has suggested another way of improving the accuracy, which we will now discuss. Consider (1.6) and set $c_j = \omega(\alpha_j, 0)h^2$. Since the vorticity is conserved along particle paths we should have $c_j = \tilde{\omega}(\tilde{x}(\alpha_j, t), t)h^2$. However, this will not hold, even at $t = 0$. Thus Beale lets c_j change in time and determines $c_j(t)$ by solving the equations

$$\omega(\alpha_i, 0) = \sum_j \zeta_\delta(\tilde{x}(\alpha_i, t) - \tilde{x}(\alpha_j, t))c_j(t)$$

iteratively. The coefficients $c_j(t)$ are then used in (1.4) and (1.5). It improves the accuracy significantly, but Nordmark [66] has found that it is not competitive with the method of rezoning for large time integrations.

Other ways of improving the accuracy have been proposed by Hou [52] and Hou, Lowengrub, and Shelley [53]. Let us rewrite (1.1) as

$$\frac{d}{dt}x(\alpha,t) = \int_{R^2} K(x(\alpha,t) - x(\alpha',t))[\omega(\alpha',0) - \omega(\alpha,0)g(x(\alpha,t) - x(\alpha',t))]\,d\alpha'$$

$$+ \omega(\alpha,0)\int_{R^2} K(x(\alpha,t) - x(\alpha',t))g(x(\alpha,t) - x(\alpha',t))\,d\alpha'. \tag{2.2}$$

Suppose that g is a radially symmetric function with compact support and that $g(0) = 1$. The last integral in (2.2) will then vanish and we can approximate (2.2) by

$$\frac{d}{dt}\tilde{x}(\alpha_i,t) = \sum_j K_\delta(\tilde{x}(\alpha_i,t) - \tilde{x}(\alpha_j,t))[\omega(\alpha_j,0) - \omega(\alpha_i,0)g(\tilde{x}(\alpha_i,t) - \tilde{x}(\alpha_j,t))]h^2.$$

To use this method we must follow the movements of particles placed outside the support of the initial vorticity. What is the advantage of this approach? It increases the stability of the method and decreases the growth of the errors in the calculations. A convergence proof for the maximum norm can be found in [53]. A precursor of this algorithm, which is more complicated and not as accurate, was investigated by Hou [52].

All the theories in § 1 share one assumption, namely, that $h/\delta \to 0$. Is this necessary? Working with the vortex method in three dimensions, Beale [9] noted that the kernel $K_\delta(x)$ is an odd function of x and that this property can be used to decrease the error and improves the error bound from $C[\delta^m + (h/\delta)^k\delta]$ to $C[\delta^m + (h/\delta)^k\delta^2]$. Using Beale's technique, Del Prete [31] proved quadratic convergence for positive cutoffs with $\delta = ch$, provided the vorticity is three times continuously differentiable. The constant c depends on the derivatives of ω. It is large if the derivatives are large. Subsequently, Del Prete [32] obtained almost quadratic convergence with $\delta = ch$ if the second derivatives of ω are Hölder continuous. The ultimate result in this direction is due to Goodman, Hou, and Lowengrub [40]. They prove that the point vortex method converges for all times if the flow is smooth, and provide numerical experiments that support the claim for short times. The stability argument has been shortened by Cottet [28]. Does this mean that we can forget all about cutoffs? I think not. Using cutoff functions is more than a mathematical trick. It is also a practical device. It

(1) Gives smaller errors for long time,
(2) Gives higher order of convergence for smooth flows,
(3) Allows pointwise evaluation of velocity and vorticity,
(4) Allows rezoning.

The recommendations gathered from different studies are equivocal. Thus Conte [24] has shown that three point vortices can coalesce in finite time. This is not likely to happen in calculations. Indeed, Marchioro and Pulvirenti have looked at the initial conditions that give rise to a singularity, and have shown that they have Lebesgue measure zero (see [62, p. 26]). Against this stands the experience of Chorin and Bernard [18]. They studied the roll up of a vortex sheet and found that without a cutoff the numerical results looked chaotic. To get accurate results one must balance the time step, the mesh length, the cutoff parameter, and the level of roundoff errors in the calculation (see Krasny [56]). In practice the vortex method is used together with random walks to simulate viscosity, and here the computed velocity may become unbounded if we do not use a cutoff. The running times for the higher order methods are seldom more than twice the running time for the point vortex method or for the vortex method with cutoff (1.7). Finally, the theories of Beale [9], Del Prete [31]–[33], and Goodman, Hou, and Lowengrub [40] all require that the flow be smooth, but it is not clear if this assumption is necessary. Hald [48] has shown that with a cutoff, the vortex method will converge if the initial vorticity is just Hölder continuous. This is the natural class of initial data for the mathematical existence and uniqueness theory for Euler's equations (see Wolibner [78], Kato [54], McGrath [63]). It is therefore of interest to prove the convergence of the point vortex method for this class of initial data.

There is a class of methods which we have not yet discussed, namely, vortex methods with integral evaluation of the vorticity. This means that the coefficients $\omega(\alpha_j, 0)$ in (1.4) are replaced by the averages of the initial vorticity ω over squares centered at α_j and with side length h. These methods converge if the initial vorticity is Hölder continuous and we use a cutoff (see Hald [48]). In his thesis Cottet [25] has shown that methods with integral evaluation of the vorticity cannot converge faster than quadratically.

So far we have assumed that the initial vorticity is fixed, and our error bounds depend upon the estimates of the derivatives of ω. What happens if the initial vorticity is highly oscillatory, say $\omega(x, 0) = \max [0, (1 - r^2)^3][1 + \sin(2\pi r/\varepsilon)]$ with $r = |x|$ and $\varepsilon \to 0$. Using techniques from homogenization theory, Hou [49] has shown that the computed solution (flow map and velocity field) converges essentially to the solution generated by the average of the initial vorticity. In our example it would be $\bar{\omega}(x, 0) = \max [0, (1 - r^2)^3]$. No statement is made about the computed vorticity distribution.

Finally, we mention four methods that fall outside our framework. Throughout this paper we use an equidistant partition of the initial vorticity. Variable partitions have been considered by Anderson and Greengard [2], and a general convergence theory has been developed by Chiu and Nicolaides [16]. Second, we have only considered radially symmetric cutoff functions. This is not appropriate for periodic flows. Here Fishelov [36] has used a cutoff function that is expressed in terms of a truncated Fourier series. It should also be mentioned that Hou [52] has proved the convergence of a vortex method where the shape of the cutoff function depends on its position. Third, we assume that the flow is incompressible. A numerical method for flows with slight density variations can be found in Anderson [1] together with a convergence proof. Fourth, we have restricted our attention to smooth flows. Buttke [12] has suggested a special vortex method for flows that can be represented by patches of constant vorticity. Here it is enough to move the points on the boundary of the patches. In each step the vortex blobs are squares of different sizes and fit inside the patches. The convergence of this method has not been established.

3. Two-dimensional Viscous Flows

In describing the flow of a fluid, we should not forget the viscosity. Hence we add a diffusion term to the Euler's equations and get the Navier–Stokes equations. To simulate the diffusion we can use random walks. This was suggested by Chorin [17]. Assume we know a solution $u(x, t)$ of the Navier–Stokes equations with viscosity ν. Let W be the standard Brownian motion in R^2 (a Wiener process) and consider the stochastic differential equation

$$\begin{aligned} dX(\alpha,t) &= u(X(\alpha,t),t)\,dt + \sqrt{2\nu}\,dW(t), \\ X(\alpha,0) &= \alpha. \end{aligned}$$

What does this mean? Every Brownian motion path $W(t)$ is a highly irregular continuous function with $W(0) = 0$. For each α and each path of $W(t)$ the solution of the stochastic differential equation is given by

$$X(\alpha,t) = \alpha + \int_0^t u(X(\alpha,\tau),\tau)\,d\tau + \sqrt{2\nu}\,W(t).$$

This integral equation can be solved by Picard iteration if u is smooth. The solution $X(\alpha, t)$ is a continuous function that is just as irregular as $W(t)$. We interpret $X(\alpha, t)$ as the position at time t of a particle that is at α at time $t = 0$. Since the Brownian motion is random, X is also random. Let $G(x, t, \alpha, s)$ be the transition probability. Then $\int_A G\,dx$ is the probability that a particle starting in α at time s

reaches the set A at time t. We can now write down the equations that replace the Navier–Stokes equations:

$$(3.1)\quad dX(\alpha,t)=\left\{\int_{R^2} E'[K(X(\alpha,t)-X(\alpha',t))\omega(\alpha',0)]\,d\alpha'\right\}dt+\sqrt{2\nu}\,dW(\alpha,t),$$

$$(3.2)\quad u(x,t)=\int_{R^2} E'[K(x-X(\alpha',t))\omega(\alpha',0)]\,d\alpha',$$

$$(3.3)\quad \omega(x,t)=\int_{R^2} E'[\delta(x-X(\alpha',t))\omega(\alpha',0)]\,d\alpha'$$

with $X(\alpha,\ 0)=\alpha$. Here E' means the expected value and $E'[g(X(\alpha',t))]=\int g(x)G(x,\ t,\ \alpha',\ 0)\ dx$. Equation (3.1) represents an infinite system of stochastic equations, one for each α in R^2. To get a finite number of equations we drop the expected value and assume that the vorticity has compact support. For each point $\alpha_j=jh$ we select a Brownian motion $W(\alpha_j,\ t)$. The processes should be independent. Finally, we solve the equations

$$(3.4)\quad d\tilde{X}(\alpha_i,t)-\left\{\sum_j K_\delta(\tilde{X}(\alpha_i,t)-\tilde{X}(\alpha_j,t))\omega(\alpha_j,0)h^2\right\}dt+\sqrt{2\nu}\,dW(\alpha_i,t),$$

$$(3.5)\quad \tilde{u}(x,t)=\sum_j K_\delta(x-\tilde{X}(\alpha_j,t))\omega(\alpha_j,0)h^2,$$

$$(3.6)\quad \tilde{\omega}(x,t)=\sum_j \zeta_\delta(x-\tilde{X}(\alpha_j,t))\omega(\alpha_j,0)h^2$$

with $\tilde{X}(\alpha_j,\ 0)=\alpha_j$. If $\nu=0$ these equations reduce to (1.4)–(1.6), as they should. In this presentation we have followed Long [58], but there is a different approach. Suppose we distribute the particles $\tilde{X}_j(0)$ randomly such that the probability of landing in a set A is $\int_A p(x)\ dx$. Set $\omega(x,0)=c(x)p(x)$. We can then replace $\omega(\alpha_j,\ 0)\cdot h^2$ by $c(\tilde{X}_j(0))\cdot$(the total number of particles)$^{-1}$. This technique has been studied by Marchioro and Pulvirenti [61] and Goodman [39]. The numerical experiments by Roberts [72] indicate that the first approach gives smaller errors.

To solve (3.4) numerically we must discretize in time. The simplest choice is Euler's method. Since each component of $W(t+\Delta t)-W(t)$ is a Gaussian random variable with zero mean and variance Δt we can approximate (3.4) by

$$(3.7)\quad \tilde{X}_i(t+\Delta t)=\tilde{X}_i(t)+\Delta t\sum_j K_\delta(\tilde{X}_i(t)-\tilde{X}_j(t))\omega_j h^2+\sqrt{2\nu\Delta t}\underline{N_i},$$

where $\omega_j=\omega(\alpha_j,0)$. The two components in $\underline{N}=(N^{(1)},N^{(2)})$ are Gaussian distributed random variables with zero mean and unit variance. Here is an algorithm for finding $N^{(1)}$ and $N^{(2)}$. Let $\{U^{(1)},\ U^{(2)}\}$ be a collection of independent random variables, uniformly distributed in [0, 1]. For each pair $U^{(1)}$, $U^{(2)}$ we get a pair of Gaussian random variables by

$$N^{(1)}=\cos(2\pi U^{(1)})(-2\log(U^{(2)}))^{1/2},$$

$$N^{(2)}=\sin(2\pi U^{(1)})(-2\log(U^{(2)}))^{1/2}.$$

From a practical point of view, it is hard to create uniformly distributed independent random numbers on the computer, but it is seldom obvious from the numerical calculations that the numbers are correlated. Instead of using Euler's method, we could solve the stochastic differential equation (3.4) by a Taylor series method or a higher order Runge–Kutta method. This does not give the expected improvement. Let us assume, for the moment, that the only available information about the Brownian motion is the increments $W(t+\Delta t)-W(t)$. It then follows from the work of Clark and Cameron [23] that the global error for any Taylor series method is at best $O(\Delta t)$ (see Pardoux and Talay [67]). Rümelin [74] has studied a large class of Runge–Kutta methods and has shown that the best local error is $O(\Delta t^{3/2})$. To get higher order methods we must use more information about the Wiener process. Such a method has been presented by Chang [13]. If we use Chang's method to solve (3.4) we get the following scheme:

$$P_i=\tilde{X}_i(t)+\frac{1}{2}\Delta t\sum_j K_\delta(\tilde{X}_i(t)-\tilde{X}_j(t))\omega_j h^2,$$

$$Q_i=\tilde{X}_i(t)+\frac{1}{2}\Delta t\sum_j K_\delta(\tilde{X}_i(t)-\tilde{X}_j(t))\omega_j h^2+\sqrt{2\nu\Delta t}\left(\frac{1}{4}\underline{N_i}+\frac{\sqrt{3}}{4}\underline{N_i'}\right),$$

$$\tilde{X}_i(t+\Delta t)=\tilde{X}_i(t)+\frac{1}{3}\Delta t\sum_j K_\delta(P_i-P_j)\omega_j h^2 \tag{3.8}$$

$$+\frac{2}{3}\Delta t\sum_j K_\delta(Q_i-Q_j)\omega_j h^2+\sqrt{2\nu\Delta t}\,\underline{N_i}.$$

The components of $\underline{N}$ and $\underline{N'}$ should be independent Gaussian distributed random variables with zero mean and variance 1. Note that $\underline{N_i}$ is used twice. If $\nu=0$, then $P_i=Q_i$ and we have a conventional second order Runge–Kutta method. If $\nu\neq 0$, then we solve a stochastic differential equation and the global error is $O(\Delta t^{3/2})$. Chang [14] has compared the performance of (3.7) and (3.8). His initial vorticity is $\omega(x,0)=4/\pi$ for $|x|\leqq\frac{1}{2}$ and zero elsewhere. The constant $\nu=0.002$. The conclusion is that Chang's method is more accurate than Euler's method, but also that you need a variance reduction technique to see the difference.

It has been difficult to prove convergence for the random vortex method, because it is a mixture of a deterministic method and a probabilistic method. It was therefore natural to attack one-dimensional problems first. Here, Hald [46] investigated a reaction-diffusion equation, Roberts [73] investigated Burger's equation, and Puckett [69] investigated the Kolmogorov equation. The convergence for the random vortex method was obtained by Marchioro and Pulvirenti [61], Goodman [39], and Long [58]. The most precise result is due to Long. To make it easier for the reader, he avoids discretization in time and concentrates on (3.4). The idea is to compare the solution of the whole system (3.4) with the solutions of the single stochastic differential equations (3.1), but generated by the same paths of the Brownian motion, i.e.,

$$dX(\alpha_i,t)=u(X(\alpha_i,t),t)dt+\sqrt{2\nu}\,dW(\alpha_i,t),$$

$$X(\alpha_i,0)=\alpha_i.$$

The following result is due to Long [58].

THEOREM 2. *Assume that the velocity field is smooth enough and that the initial vorticity has compact support. Let $\zeta\in M^{k,m}$ with k sufficiently large and $m\geqq 2$. Set $\delta=ch^q$ with $0<q<1$. Let B be a bounded set in R^2. There exists a constant C such that for all h sufficiently small we have:*

(1) *Convergence of the particle paths*

$$\max_{0 \leqq t \leqq T} \|\tilde{X}(\alpha_j, t) - X(\alpha_j, t)\|_p \leqq C[\delta^m + (h/\delta)^k \delta + h|\log(h)|],$$

(2) *Convergence of the particle velocities*

$$\max_{0 \leqq t \leqq T} \|\tilde{u}(\tilde{X}(\alpha_j, t), t) - u(X(\alpha_j, t), t)\|_p \leqq C[\delta^m + (h/\delta)^k \delta + h|\log(h)|],$$

(3) *Convergence of the velocity field*

$$\max_{0 \leqq t \leqq T} \|\tilde{u}(x, t) - u(x, t)\|_{L^p(B)} \leqq C[\delta^m + (h/\delta)^k \delta + h|\log(h)|]$$

with probability $1 - o(h)$.

Now, the computed velocity field $\tilde{u}$ is a random quantity. The exact solution u is a deterministic quantity. Theorem 2 says that $\|\tilde{u} - u\|$ is small with high probability. Thus if we run the problem just once, with h sufficiently small, we should expect an accurate solution. This is our main conclusion. In this presentation we have suppressed two observations by Long [58]. The first is that $o(h)$ can be replaced by h^r for any r. To sharpen the result in this way we must increase the constant C. The second observation is that the error bound holds for all $\nu \leqq 1$. We would have liked to exhibit this dependence as a factor multiplying $h|\log(h)|$. Such a result is not yet available for the random vortex method, but similar results have been obtained for one-dimensional model problems (see Hald [46], [47], Puckett [69]). Finally, Hou [51] has extended Theorem 2 to a case, where the shape of the cutoff function depends upon its position.

Is it necessary to use random walks to simulate effects of the viscosity? The answer is no. However, not everything that looks reasonable will work. The natural approach is to use the Gaussian cutoff (1.17) for $n = 1$ with $\delta = (\delta_0^2 + 4\nu t)^{1/2}$. With this definition the cutoff function ζ_δ will be an exact solution of the heat equation. We can interpret the method as a fractional step algorithm. In the first step we approximate Euler's equations. In the second step we approximate the heat equation. Nevertheless, Greengard [42] has shown that the numerical solution of this algorithm will not converge to the solution of the Navier–Stokes equations. Numerical experiments with radially symmetric flows do not reveal the flaw, and it is doubtful whether the regridding technique used by Chang and Chern [15] will bring about convergence for other flows. A different, and successful, deterministic algorithm has been studied by Cottet and Mas-Gallic [27]. They choose $\delta = \sqrt{4\nu\Delta t}$ and assume that $h \leqq c\delta^{1+\varepsilon}$. Instead of changing the cutoff function they change the strengths of the vortices. This involves solving the heat equation after each step, and they prove that the method converges. The error is $O(\nu\Delta t)$, but in this algorithm we cannot let $\nu \rightarrow 0$ without changing h and δ. A related technique has been suggested by Fishelov [36]. Once again the idea is to change the strengths of the vortices by solving the equation $(d/dt)\omega(x(\alpha, t), t) = \Delta\omega(x, t)$, where $x = x(\alpha, t)$ and $\Delta = \nabla^2$ is the Laplacian. This is done by differentiating the approximate vorticity field explicitly and using a second order Runge–Kutta method in time. Numerical experiments for radially symmetric flows indicate that the algorithm converges, but it has not been proved.

4. Three-dimensional Inviscid Flows

The difference between two-dimensional flows and three-dimensional flows is that the vorticity is no longer constant along the particle paths. For two-dimensional flows we can prove existence for all times, but for three-dimensional flows we can only show existence for a short time. It depends on the initial data and the viscosity (see Kato [55], Temam [77]). This does not matter to the numerical analyst, who studies convergence as long as the solution is smooth. In practice the solution becomes highly irregular (see Chorin [21], [22] or Greengard [41]). We can then insert new points, but we quickly run out of money or storage.

For three-dimensional flows, we can present the Lagrangian equations in several ways. Each of these suggests a different vortex method. We begin with a formulation that can be extended to the viscous case. Let $u(x, t)$ be the velocity field and let $\omega(x, t)$ be the vorticity field. We get the particle trajectory $x(\alpha, t)$ by solving the equations

$$(4.1)\quad \frac{d}{dt}x(\alpha,t)=\int_{R^3} K(x(\alpha,t)-x(\alpha',t))\omega(x(\alpha',t),t)\,d\alpha',$$

$$(4.2)\quad u(x,t)=\int_{R^3} K(x-x(\alpha',t))\omega(x(\alpha',t),t)\,d\alpha',$$

$$(4.3)\quad \frac{d}{dt}\omega(x(\alpha,t),t)=(\omega(x(\alpha,t),t)\cdot\nabla_x)\int_{R^3} K(x(\alpha,t)-x(\alpha',t))\omega(x(\alpha',t),t)\,d\alpha'$$

with $x(\alpha, 0) = \alpha$ and $\omega(x(\alpha, 0), 0) = \omega(\alpha, 0)$ as initial conditions. To have a consistent set of equations we need that $\operatorname{div}\omega(x, 0) = 0$. The derivation of (4.1)–(4.3) can be found in this book in the article by Sethian. All the vectors u, ω, x, α have three components and ∇_x is the gradient. The right hand side of (4.3) is $(\omega(x, t)\cdot\nabla_x)u(x, t)$ evaluated at $x = x(\alpha, t)$. If $\omega(\alpha, 0) = 0$, then the vorticity vanishes on the whole trajectory $x(\alpha, t)$. Thus if the vorticity has compact support initially it will have compact support at later times. The kernel K is given by

$$K(x)=\frac{1}{4\pi}\begin{bmatrix} 0 & \frac{x_3}{|x|^3} & \frac{-x_2}{|x|^3} \\ \frac{-x_3}{|x|^3} & 0 & \frac{x_1}{|x|^3} \\ \frac{x_2}{|x|^3} & \frac{-x_1}{|x|^3} & 0 \end{bmatrix}$$

In [2] the matrix has the wrong sign.

To solve (4.1)–(4.3) we smooth the kernel and replace the integral by a sum. Let $\alpha_j = jh$ with $j = (j_1, j_2, j_3)$ be a grid in R^3. The following vortex method was presented by Anderson and Greengard [2]:

$$(4.4)\quad \frac{d}{dt}\tilde{x}(\alpha_i,t)=\sum_j K_\delta(\tilde{x}(\alpha_i,t)-\tilde{x}(\alpha_j,t))\tilde{\omega}(\tilde{x}(\alpha_j,t),t)h^3,$$

$$(4.5)\quad \tilde{u}(x,t)=\sum_j K_\delta(x-\tilde{x}(\alpha_j,t))\tilde{\omega}(\tilde{x}(\alpha_j,t),t)h^3,$$

$$(4.6)\quad \frac{d}{dt}\tilde{\omega}(\tilde{x}(\alpha_i,t),t)=(\tilde{\omega}(\tilde{x}(\alpha_i,t),t)\cdot\nabla_x)\sum_j K_\delta(\tilde{x}(\alpha_i,t)-\tilde{x}(\alpha_j,t))\tilde{\omega}(\tilde{x}(\alpha_j,t),t)h^3$$

with $\tilde{x}(\alpha_j, 0) = \alpha_j$ and $\tilde{\omega}(\tilde{x}(\alpha_j, 0), 0) = \omega(\alpha_j, 0)$. Thus we have a finite number of ordinary differential equations if the initial vorticity has compact support. The convergence of this scheme with time discretization has not been established and will not be discussed. Note also that it is illegal to differentiate under the integral sign in (4.3) because $K(x)$ is too singular near $x = 0$. Yet, this is in essence what we do in (4.6). It is therefore surprising that the algorithm converges.

We construct the approximate kernel as in two dimensions, i.e., $K_\delta(x) = K * \zeta_\delta(x)$. Here $\zeta_\delta(x) = \delta^{-3}\zeta(\delta^{-1}x)$ and ζ is a smooth radially symmetric function that vanishes at infinity. We can therefore factor the kernel like this

$$K_\delta(x) = K(x) \int_{|y| \leqq r} \zeta_\delta(y)\, dy = K(x) f\left(\frac{r}{\delta}\right),$$

where $r = |x|$ and $f(r) = 4\pi \int_0^r s^2 \zeta(s)\, ds$ is the shape factor. There are no explicit classes of higher order cutoffs for three-dimensional flows and no real tests of infinite order cutoffs. Instead, Beale [9] has experimented with

$$f(r) = r^3(1 + r^6)^{-1/2}, \tag{4.7}$$

$$f(r) = r^3\left\{(1 + r^6)^{-1/2} + \frac{3}{2}(1 + r^6)^{-3/2}\right\}, \tag{4.8}$$

$$f(r) = r^3\left\{(1 + r^6)^{-1/2} + \frac{27}{8}(1 + r^6)^{-5/2}\right\} \tag{4.9}$$

We get the corresponding cutoff functions by $\zeta(x) = (4\pi r^2)^{-1} f'(r)$. Beale's shape factors are of the form $\sum_n c_n r^n f^{(n)}(r)$ where $c_0 = 1$ and f is given by (4.7). This construction was suggested by Beale and Majda [7]. The coefficients c_n are determined by the moment conditions. These are

$$\int_{R^3} \zeta(x)\, dx = 1, \tag{4.10}$$

$$\int_{R^3} x^\alpha \zeta(x)\, dx = 0, \qquad 1 \leqq |\alpha| \leqq m - 1, \tag{4.11}$$

(4.12) ζ is smooth and decays at infinity.

The cutoff functions corresponding to (4.7)–(4.9) satisfy (4.11) with $m = 2, 4, 6$ and there is no eighth order cutoff in this particular family. An eighth order cutoff similar to (1.21)–(1.22) has been derived by Nordmark (personal communication). Other finite order cutoffs can be found in Beale and Majda [7], Beale [9] and Greengard [41]. It is also possible to construct infinite order methods. While the cutoff functions can be expressed in terms of elementary functions, the shape factors cannot and it is laborious to implement the methods (Nordmark, personal communication). Preliminary numerical experiments indicate that the behavior of infinite order methods in two and three dimensions is similar. To apply the moment conditions we must make (4.12) more precise. For this we can use the shape factor. Thus Beale [9] assumes that $r^{-3}f(r)$ is a smooth function of r^2 and that

$$r^{m+1+j}|D^j f(r)| \leqq C, \tag{4.13}$$

where $m \geqq 4$ and D means differentiation with respect to r. We say that the cutoff function $\zeta \in M^{k,m}$ if the moment conditions are satisfied and (4.13) holds for $j = 1$,

$2, \cdots, k+1$. Similar requirements have been used by Anderson and Greengard [2] and Greengard [43] while Cottet [26] assumes that all derivatives of ζ are bounded and integrable.

There is an unexpected complication for three-dimensional flows: How should we define the computed vorticity $\tilde{\omega}$? A natural candidate is $\sum \zeta_\delta(x-\tilde{x}_j)\tilde{\omega}_j h^3$. However, the divergence of this vector field is not zero. Hence we choose $\tilde{\omega} = \operatorname{curl} \tilde{u}$. More precisely,

$$\tilde{\omega}(x,t) = \sum_j Z_\delta(x-\tilde{x}(\alpha_j,t))\tilde{\omega}(\tilde{x}(\alpha_j,t),t)h^3,$$

where $Z_\delta(x) = \delta^{-3}Z(\delta^{-1}x)$ and

$$Z(x) = \frac{f(|x|)}{4\pi|x|^5}\begin{bmatrix} 2x_1^2-x_2^2-x_3^2 & 3x_1x_2 & 3x_1x_3 \\ 3x_1x_2 & 2x_2^2-x_1^2-x_3^2 & 3x_2x_3 \\ 3x_1x_3 & 3x_2x_3 & 2x_3^2-x_1^2-x_2^2 \end{bmatrix}$$

$$+\frac{\zeta(x)}{|x|^2}\begin{bmatrix} x_2^2+x_3^2 & -x_1x_2 & -x_1x_3 \\ -x_1x_2 & x_1^2+x_3^2 & -x_2x_3 \\ -x_1x_3 & -x_2x_3 & x_1^2+x_2^2 \end{bmatrix}$$

The first term dominates if x is large while $Z(0) = \frac{2}{3}\zeta(0)I$. Since $\operatorname{div}\tilde{\omega} = 0$ it follows that $\tilde{u}$ is the velocity field corresponding to $\tilde{\omega}$ (see Beale [9]).

To measure the error in the vortex method we use both a discrete p-norm and a continuous p-norm. The discrete norm is $\|g\|_p = (\sum_j |g(\alpha_j)|^p h^3)^{1/p}$, where we sum over all α_j in the initial support of the vorticity. The continuous norm is $\|g\|_{L^p(B)} = (\int_B |g(x)|^p\,dx)^{1/p}$, where B is a bounded set in R^3. The following very complete result is due to Beale [9].

THEOREM 3. *Assume that the velocity field is smooth enough for $0 \leqq t \leqq T$ and that the initial vorticity has compact support. Let $\zeta \in M^{k,m}$ with k sufficiently large and $m \geqq 4$. Set $\delta = ch^q$ with $\frac{1}{3} < q < 1$. There exists a constant C such that for all h sufficiently small we have*:

(1) *Convergence of the particle paths*

$$\max_{0\leqq t\leqq T} \|\tilde{x}(\alpha_j,t) - x(\alpha_j,t)\|_p \leqq Ch^{mq},$$

(2) *Convergence of the particle velocities*

$$\max_{0\leqq t\leqq T} \|\tilde{u}(\tilde{x}(\alpha_j,t),t) - u(x(\alpha_j,t),t)\|_p \leqq Ch^{mq},$$

(3) *Convergence of the particle vorticities*

$$\max_{0\leqq t\leqq T} \|\tilde{\omega}(\tilde{x}(\alpha_j,t),t) - \omega(x(\alpha_j,t),t)\|_p \leqq Ch^{mq-1},$$

(4) *Convergence of the velocity field*

$$\max_{0\leqq t\leqq T} \|\tilde{u}(x,t) - u(x,t)\|_{L^p(B)} \leqq Ch^{mq},$$

(5) *Convergence of the vorticity field*

$$\max_{0\leqq t\leqq T} \|\tilde{\omega}(x,t) - \omega(x,t)\|_{L^p(B)} \leqq Ch^{mq-1}.$$

The convergence of (4.4)–(4.6) has also been studied by Cottet [26]. Using a different technique he obtains an estimate similar to (4), but for a wider class of methods. Thus he allows $m = 2$ and $h \leqq c\delta^{1+\epsilon}$. However, there can be no error bound similar to (5), because Cottet defines the computed vorticity as a sum of delta functions. This formulation can also be found in Raviart [71].

We have emphasized the method by Anderson and Greengard [2] because it is a building block for the random walk method, which we will present in the next section. However, the first convergence proof for a three-dimensional vortex method was by Beale and Majda [4] for a method of their own design. Instead of (4.3) they use

$$\frac{d}{dt}\omega(x(\alpha,t),t)=(\omega(\alpha,0)\cdot\nabla_\alpha)\int_{R^3}K(x(\alpha,t)-x(\alpha',t))\omega(x(\alpha',t),t)\,d\alpha'. \tag{4.14}$$

To approximate (4.14) we smooth the kernel K and replace the integral by a sum. This leads to

$$\frac{d}{dt}\tilde{\omega}(\tilde{x}(\alpha_i,t),t)=(\omega(\alpha_i,0)\cdot\nabla_\alpha^h)\sum_j K_\delta(\tilde{x}(\alpha_j,t)-\tilde{x}(\alpha_j,t))\tilde{\omega}(\tilde{x}(\alpha_j,t),t)h^3, \tag{4.15}$$

where ∇_α^h is a finite difference approximation to the gradient ∇_α. There is a problem here. Assume that the initial vorticity ω has compact support and that we approximate the derivatives in the gradient by second order finite difference formulas. To use (4.15) we need the position of particles that start outside the initial support. Fortunately the vorticity vectors for these particles remain zero and do not contribute to the sums in (4.4)–(4.6). Since the velocity field is independent of the extra particles we can follow their movement without any difficulty. If we use a higher order approximation to the gradient we must include many particles outside the initial support.

We can solve (4.1), (4.2), (4.14) by using (4.4), (4.5), (4.15). This method was introduced by Beale and Majda [4]. However, it is simpler, in practice, to replace (4.14) and (4.15) by an equivalent pair of equations:

$$\omega(x(\alpha,t),t)=(\omega(\alpha,0)\cdot\nabla_\alpha)x(\alpha,t), \tag{4.16}$$

$$\tilde{\omega}(\tilde{x}(\alpha_i,t),t)=(\omega(\alpha_i,0)\cdot\nabla_\alpha^h)\tilde{x}(\alpha_i,t). \tag{4.17}$$

We get these formulas by integrating both sides of (4.14), (4.15) with respect to t using (4.1), (4.4). The convergence of these two methods has been studied by Beale and Majda [4] and Anderson and Greengard [2]. In each case the vorticity has compact support. Beale and Majda [4] assume that the flow and the cutoff function are smooth, while Anderson and Greengard [2] allow a finite number of derivatives. The estimates are similar to those in Theorem 3, but contain an extra term $O(h^s)$ when the approximation to the gradient is of order s. In addition, Anderson and Greengard [2] consider the discretization in time and prove convergence for a large class of explicit multistep schemes and for first and second order Runge–Kutta methods. Finally, we mention that Hou and Lowengrub [51] have proved almost quadratic convergence for the point vortex method, i.e., (4.4), (4.5), (4.15) with $\delta = 0$, provided the flow map and the vorticity field are three times continuously differentiable.

The last method to be discussed is the vortex filament method. Thus we are, in effect, reversing the historical development, as the filament methods were invented before the blob methods (see Chorin [21], [22], Del Prete [30], Leonard [57]). We will present a version due to Greengard [43]. It is designed to study the interaction of vortex rings and exploit the special structure of the initial data (see Fig. 2.2). Let

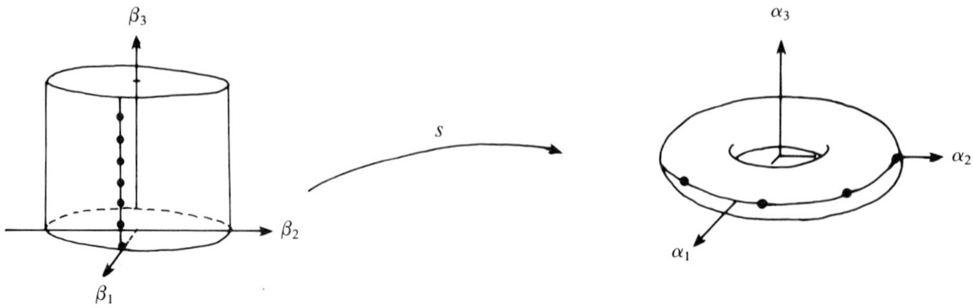

Fig. 2.2. Parametrization $\alpha = s(\beta)$ of a vortex ring.

$\alpha = s(\beta)$ where α, β have three components and s is periodic in the third coordinate with period 1. We assume that the initial vorticity satisfies

$$\omega(s(\beta),0)=c(\beta)(\partial_3 s)(\beta).$$

Here $c(\beta)$ should vanish if $s(\beta)$ is outside the support of $\omega(\cdot, 0)$. Let $x(\alpha, t)$ be the flow map and define $\zeta(\beta, t) = x(s(\beta), t)$. It follows from (4.1) and (4.16) and a change of variables ($\alpha = s(\beta)$) that

$$\frac{d}{dt}\chi(\beta,t)=\int_{R^2\times[0,1]} K(\chi(\beta,t)-\chi(\beta',t))\omega(\chi(\beta',t),t)J(\beta')\,d\beta',$$

$$\omega(\chi(\beta,t),t)=c(\beta)\partial_3\chi(\beta,t).$$

Here $J(\beta) = |\det(\partial_i s_j)(\beta)|$ and we assume that the function s, restricted to one period, is one to one and onto the support of the initial vorticity. To discretize the equations we replace the integral by a sum, smooth the kernel K, and set $\beta_j = jh$ where h^{-1} is an integer. Thus we arrive at

$$\frac{d}{dt}\tilde{\chi}(\beta_i,t)=\sum_j K_\delta(\tilde{\chi}(\beta_i,t)-\tilde{\chi}(\beta_j,t))\tilde{\omega}(\tilde{\chi}(\beta_j,t),t)J(\beta_j)h^3,$$

$$\tilde{\omega}(\tilde{\chi}(\beta_i,t),t)=c(\beta_i)\partial_3^h\tilde{\chi}(\beta_i,t),$$

where ∂_3^h is a finite difference approximation to ∂_3 and $j_3 = 1, 2, \cdots, h^{-1}$. Thus instead of a variable partition in the physical space with coordinates α, we have an equidistant partition in the coordinate space β. Greengard [43] solves the ordinary differential equations by Euler's method and proves the convergence of the particle paths. He assumes that the moment condition (4.11) is satisfied with $m \geqq 4$ and that ∂_3^h is a centered difference operator of order four or higher.

5. Three-dimensional Viscous Flows

Our description of the three-dimensional vortex method will be based mainly on a paper by Long [59]. The algorithm is (4.4)–(4.6) by Anderson and Greengard [2] suitably randomized. Once again we do not follow the historical development, as there is no convergence proof for the filament method applied to viscous flows or for the method with vortex segments and random walks (see Leonard [57], Del Prete [30],

Chorin [20]). Let u be the solution of the Navier–Stokes equations with viscosity ν and consider the equations

$$dX(\alpha,t)=u(X(\alpha,t),t)\,dt+\sqrt{2\nu}\,dW(t),$$

$$\frac{d}{dt}\xi(\alpha,t)=(\xi(\alpha,t)\cdot\nabla_x)u(X(\alpha,t),t)$$

with $X(\alpha,0)=\alpha$ and $\xi(\alpha,0)=\omega(\alpha,0)$. Here $\omega=\operatorname{curl} u$. For each Brownian motion path $W(t)$ we get a particle path $X(\alpha,t)$ by solving the stochastic differential equation. Using the particle path and solving the ordinary differential equation we obtain the function $\xi(\alpha,t)$. It can be shown that the velocity field u can be expressed in terms of X and ξ. This leads to

$$(5.1)\quad dX(\alpha,t)=\left\{\int_{R^3}E'[K(X(\alpha,t)-X(\alpha',t))\xi(\alpha',t)]\,d\alpha'\right\}dt+\sqrt{2\nu}\,dW(\alpha,t),$$

$$(5.2)\quad u(x,t)=\int_{R^3}E'[K(x-X(\alpha',t))\xi(\alpha',t)]\,d\alpha',$$

$$(5.3)\quad \frac{d}{dt}\xi(\alpha,t)=(\xi(\alpha,t)\cdot\nabla_x)\int_{R^3}E'[K(X(\alpha,t)-X(\alpha',t))\xi(\alpha',t)]\,d\alpha'$$

with $X(\alpha,0)=\alpha$ and $\xi(\alpha,0)=\omega(\alpha,0)$ as initial conditions. The letter E' stands for expectation. The derivation of (5.2) is far from obvious and will be presented in Long [59]. Equations (5.3) and (4.3) are clearly similar. However, $\xi(\alpha,t)$ and $\omega(X(\alpha,t),t)$ are not equal, partly because the chain rule fails for stochastic processes. The vectors u, ξ, X, W, x, α have three components, and the kernel K is the same as in § 4.

To solve (5.1)–(5.3) we smooth the kernel, drop the expectation, and replace the integral by a sum. For each point $\alpha_j=jh$ with $j=(j_1,j_2,j_3)$ we select a standard Brownian motion $W(\alpha_j,t)$. We can now present the numerical approximation of equations (5.1)–(5.3):

$$(5.4)\quad d\tilde{X}(\alpha_i,t)=\left\{\sum_j K_\delta(\tilde{X}(\alpha_i,t)-\tilde{X}(\alpha_j,t))\tilde{\xi}(\alpha_j,t)h^3\right\}dt+\sqrt{2\nu}\,dW(\alpha_i,t),$$

$$(5.5)\quad \tilde{u}(x,t)=\sum_j K_\delta(x-\tilde{X}(\alpha_j,t))\tilde{\xi}(\alpha_j,t)h^3,$$

$$(5.6)\quad \frac{d}{dt}\tilde{\xi}(\alpha_i,t)=(\tilde{\xi}(\alpha_i,t)\cdot\nabla_x)\sum_j K_\delta(\tilde{X}(\alpha_i,t)-\tilde{X}(\alpha_j,t))\tilde{\xi}(\alpha_j,t)h^3$$

with $\tilde{X}(\alpha_j,0)=\alpha_j$ and $\tilde{\xi}(\alpha_j,0)=\omega(\alpha_j,0)$. If $\nu=0$ the equations reduce to (4.4)–(4.6) as they should. To solve (5.4)–(5.6) we need a discretization in time. The simplest choice is Euler's method as in (3.7). If we want a higher order method we can modify the method by Chang [13]. We regard (5.4), (5.6) as a system of stochastic differential equations where the diffusion coefficient is zero for half of the equations. The appropriate generalization of (3.8) is therefore

$$P_i=\tilde{X}_i(t)+\frac{1}{2}\Delta t\sum_j K_\delta(\tilde{X}_i(t)-\tilde{X}_j(t))\tilde{\xi}_j(t)h^3,$$

$$Q_i = \tilde{X}_i(t) + \frac{1}{2}\Delta t \sum_j K_\delta(\tilde{X}_i(t) - \tilde{X}_j(t))\tilde{\xi}_j(t)h^3 + \sqrt{2\nu\Delta t}\left(\frac{1}{4}\underline{N_i} + \frac{\sqrt{3}}{4}\underline{N_i'}\right),$$

$$R_i = \tilde{\xi}_i(t) + \frac{1}{2}\Delta t(\tilde{\xi}_i(t)\cdot\nabla_x)\sum_j K_\delta(\tilde{X}_i(t) - \tilde{X}_j(t))\tilde{\xi}_j(t)h^3,$$

$$\tilde{X}_i(t+\Delta t) = \tilde{X}_i(t) + \frac{1}{3}\Delta t \sum_j K_\delta(P_i - P_j)R_j h^3 + \frac{2}{3}\Delta t \sum_j K_\delta(Q_i - Q_j)R_j h^3 + \sqrt{2\nu\Delta t}\underline{N_i},$$

$$\tilde{\xi}_i(t+\Delta t) = \tilde{\xi}_i(t) + \frac{1}{3}\Delta t(R_i\cdot\nabla_x)\sum_j K_\delta(P_i - P_j)R_j h^3 + \frac{2}{3}\Delta t(R_i\cdot\nabla_x)\sum_j K_\delta(Q_i - Q_j)R_j h^3.$$

To get a finite number of equations we assume that the initial vorticity has compact support. The components of $\underline{N}$ and $\underline{N'}$ should be independent Gaussian distributed random variables with zero mean and variance one. The rate of convergence for this scheme is $O(\Delta t^{3/2})$. For simplicity, Long [59] does not take time discretization into account. Instead, he compares the solution of (5.4)–(5.6) with the solution of (5.1) which is generated by the same Brownian paths, i.e.,

$$dX(\alpha_i, t) = u(X(\alpha_i, t), t)dt + \sqrt{2\nu}dW(\alpha_i, t),$$

$$\frac{d}{dt}\xi(\alpha_i, t) = (\xi(\alpha_i, t)\cdot\nabla_x)u(X(\alpha_i, t), t)$$

with $X(\alpha_i, 0) = \alpha_i$ and $\xi(\alpha_i, 0) = \omega(\alpha_i, 0)$. The following statement can be found in Long [58].

THEOREM 4. *Assume that the velocity field is smooth enough for $0 \leqq t \leqq T$ and that the initial vorticity has compact support. Let $\zeta \in M^{k,m}$ with k sufficiently large and $m \geqq 4$. Set $\delta = ch^q$ with $0 < q < \frac{3}{5}$. Let B be a bounded set in R^3. There exists a constant C such that for all h sufficiently small we have*:

(1) *Convergence of the particle paths*

$$\max_{0\leqq t\leqq T} \|\tilde{X}(\alpha_j, t) - X(\alpha_j, t)\|_p \leqq C[\delta^m + h(h/\delta)^{1/2}|\log(h)|],$$

(2) *Convergence of the particle velocities*

$$\max_{0\leqq t\leqq T} \|\tilde{u}(\tilde{X}(\alpha_j, t), t) - u(X(\alpha_j, t), t)\|_p \leqq C[\delta^m + h(h/\delta)^{1/2}|\log(h)|],$$

(3) *Convergence of the velocity field*

$$\max_{0\leqq t\leqq T} \|\tilde{u}(x, t) - u(x, t)\|_{L^p(B)} \leqq C[\delta^m + h(h/\delta)^{1/2}|\log(h)|]$$

with probability $1 - o(h)$.

As for three-dimensional inviscid flows, we get convergence only as long as the solution exists and is smooth. As for two-dimensional viscous flows, we observe that $\|\text{comp } u - \text{exact } u\|$ is small with high probability. Thus, when h is sufficiently small, one experiment should give a reliable picture of the flow. The convergence of (5.4)–(5.6) has also been proved by Esposito and Pulvirenti [35]. They use a different technique and do not give error bounds in terms of h and δ.

6. Open Problems

We have now gone through the main results for viscous and inviscid flows in two and three dimensions. Let me end with five open problems. We have seen that to get good accuracy for long time intervals we must use the rezoning technique by Beale and Majda [7] or its modification by Nordmark [66]. We get the best results if the decision to rezone is made by studying the computed solution itself, and not a priori. This suggests:

(1) *Prove the convergence of the automatic rezoning technique.*

The automatic procedure in § 2 uses a special property of two-dimensional flows, namely, that the vorticity is constant along the particle paths. For three-dimensional flows a different technique is used. Here one inserts new vortices, when the distances between the particles become too big or the vorticity has increased too much (see Chorin [21], [22] or Anderson and Greengard [2]). The decision to create new vortices depends on the computed flow. This suggests:

(2) *Prove the convergence when we add new particles.*

It is irritating that the optimal choice of the cutoff parameter δ in our theory depends on the smoothness of the flow. Instead we want a method that gives high order convergence for smooth flows, as in Beale [9], and linear convergence for Hölder continuous initial data, as in Hald [48]. This suggests:

(3) *Find a foolproof choice of h, δ, and Δt.*

One of the most interesting aspects of the random methods is the creation of vorticity at the boundary. Near the boundary the Navier–Stokes equations can be approximated by the Prandtl equations, and the vorticity approximated by vortex sheets. This can be done in both two and three dimensions (see Chorin [19], [20], Ghoniem, Chorin, and Oppenheim [37], Sethian [75]). The convergence of the numerical methods has been attacked from various directions. Thus Benfatto and Pulvirenti [10] have studied a diffusion process associated with the Prandtl boundary layer equations. Hald [47] has studied a one-dimensional model of free convection near a heated wall. The model was suggested by Ghoniem and Sherman [38]. Puckett [70] has studied the consistency of the vortex sheet method. Finally, Anderson [3] has formulated the boundary conditions in a new way. Still the basic problem has not been solved and this suggests:

(4) *Prove the convergence of the vortex sheet algorithm.*

In practice we use vortex sheets near the boundary, and vortex blobs in the interior of the domain. Benfatto and Pulvirenti [11] have shown that the creation of vorticity at the boundary leads to the correct solution of the Navier–Stokes equations. Their model is discrete in time and continuous in space. Thus we end with the hardest problem of them all:

(5) *Prove the convergence for the random vortex method with creation at the boundary.*

Acknowledgments

This paper was written while the author was visiting Aarhus Universitet. He appreciates the hospitality he received at the Matematisk Institut and thanks Kirsti Andersen and Svend Eric Graversen for many helpful discussions. In addition, the author thanks Alexandre Chorin, Dalia Fishelov, Ding-Gwo Long, Henrik Nordmark, and James Sethian for final criticism.

The work was supported in part by Det Naturvidenskabelige Forskningsraad and NSF grant No. DMS-8701585.

References

[1] C. R. ANDERSON, *A vortex method for flows with slight density variations*, J. Comput. Phys., 61 (1985), pp. 417–444.

[2] C. ANDERSON AND C. GREENGARD, *On vortex methods*, SIAM J. Numer. Anal., 22 (1985), pp. 413–440.

[3] C. R. ANDERSON, *Vorticity boundary conditions and boundary vorticity generation for two-dimensional viscous incompressible flows*, J. Comput. Phys., 80 (1989), pp. 72–97.

[4] J. T. BEALE AND A. MAJDA, *Vortex methods*. I: *convergence in three dimensions*, Math. Comp., 39 (1982), pp. 1–27.

[5] ———, *Vortex methods*. II: *higher order accuracy in two and three dimensions*, Math. Comp., 39 (1982), pp. 29–52.

[6] ———, *Vortex methods for fluid flow in two or three dimensions*, Contemp. Math., 28 (1984), pp. 221–229.

[7] ———, *Higher order accurate vortex methods with explicit velocity kernels*, J. Comput. Phys., 58 (1985), pp. 188–208.

[8] J. T. BEALE, *On the accuracy of vortex methods at large times*, in Computational Fluid Dynamics and Reacting Gas Flow, B. Engquist, M. Luskin, and A. Majda, eds., Springer-Verlag, New York, 1988, pp. 19–33.

[9] ———, *A convergent* 3-D *vortex method with grid-free stretching*, Math. Comp., 46 (1986), pp. 401–424, and supplement S15–S20.

[10] G. BENFATTO AND M. PULVIRENTI, *A diffusion process associated to the Prandtl equation*, J. Funct. Anal., 52 (1983), pp. 330–343.

[11] ———, *Convergence of Chorin–Marsden product formula in the half-plane*, Comm. Math. Phys., 106 (1986), pp. 427–458.

[12] T. F. BUTTKE, *A fast adaptive vortex method for patches of constant vorticity in two dimensions*, J. Comput. Phys., 89 (1990), pp. 161–186.

[13] C. C. CHANG, *Numerical solution of stochastic differential equations with constant coefficients*, Math. Comp., 49 (1987), pp. 523–542.

[14] ———, *Random vortex methods for the Navier–Stokes equations*, J. Comput. Phys., 76 (1988), pp. 281–300.

[15] C.-C. CHANG AND R.-L. CHERN, *The effect of viscous diffusion in discrete vortex dynamics for slightly viscous flows*, manuscript 32 pages, National Taiwan University, Taipei, Taiwan, 1989.

[16] C. CHIU AND R. A. NICOLAIDES, *Convergence of a higher-order vortex method for two-dimensional Euler equations*, Math. Comp., 51 (1988), pp. 507–534.

[17] A. J. CHORIN, *Numerical study of slightly viscous flow*, J. Fluid Mech., 57 (1973), pp. 785–796.

[18] A. J. CHORIN AND P. S. BERNARD, *Discretization of a vortex sheet, with an example of roll-up*, J. Comput. Phys., 13 (1973), pp. 423–429.

[19] A. J. CHORIN, *Vortex sheet approximation of boundary layers*, J. Comput. Phys., 27 (1978), pp. 428–442.

[20] ———, *Vortex models and boundary layer instability*, SIAM J. Sci. Statist. Comput., 1 (1980), pp. 1–21.

[21] ———, *Estimates of intermittency, spectra, and blow-up in developed turbulence*, Comm. Pure Appl. Math., 34 (1981), pp. 853–866.

[22] ———, *The evolution of a turbulent vortex*, Comm. Math. Phys., 83 (1982), pp. 517–535.

[23] J. M. C. CLARK AND R. J. CAMERON, *The maximum rate of convergence of discrete approximations for stochastic differential equations*, in Stochastic Differential Systems, B. Grigelionis, ed., Proc. IFIP-WG 7/1 Working Conference, Wilnius, 1978, Lecture Notes in Control and Information Sciences, Vol. 25, Springer-Verlag, Berlin, 1980, pp. 162–171.

[24] M. R. CONTE, *Etude d'ecoulements bidimensionnels tourbillonnaires*: *mouvement de trois ou quatre tourbillons ponctuels et de nappes tourbillonnaires*, These de doctorat d'etat, l'Université Pierre et Marie Curie, Paris 6, France, 1979.

[25] G. H. COTTET, *Methodes particulaires pour l'e'quation d'Euler dans le plan*, these de 3eme cycle, Université Pierre et Marie Curie, Paris 6, France, 1982.

[26] G. H. COTTET, *A new approach for the analysis of vortex methods in two and three dimensions*, Ann. Inst. H. Poincaré Anal. Non Linéaire, 5 (1988), pp. 227–285.

[27] G. H. COTTET AND S. MAS-GALLIC, *A particle method to solve transport-diffusion equations. Part* II: *the Navier–Stokes equations*, Numer. Math., 57 (1990), pp. 805–827.

[28] G. H. COTTET, *A remark on the stability of the point vortex method*, manuscript 3 pages, University of California, Los Angeles, CA, 1989.

[29] V. M. DEL PRETE AND O. H. HALD, *Convergence of vortex methods for Euler's equations*, Math. Comp., 32 (1978), pp. 791–809.

[30] V. DEL PRETE, *Numerical simulation of vortex breakdown*, Boll. Un. Mat. Ital. B (6), 2 (1983), pp. 241–253.

[31] ———, *Quadratic convergence of vortex methods*, Math. Comp., 52 (1989), pp. 457–470.

[32] ———, *Almost quadratic convergence of vortex methods*, manuscript 40 pages, University of Genoa, Genoa, Italy, 1987.

[33] ———, *Convergence of vortex methods with pointwise evaluation*, manuscript 33 pages, University of Genoa, Genoa, Italy, 1990.

[34] T. E. DUSHANE, *Convergence for a vortex method for solving Euler's equation*, Math. Comp., 27 (1973), pp. 719–728.

[35] R. ESPOSITO AND M. PULVIRENTI, *Three-dimensional stochastic vortex flows*, Math. Methods Appl. Sci., 11 (1989), pp. 431–445.

[36] D. FISHELOV, *A new vortex scheme for viscous flows*, J. Comput. Phys., 86 (1990), pp. 211–224.

[37] A. F. GHONIEM, A. J. CHORIN, AND A. K. OPPENHEIM, *Numerical modeling of turbulent flow in a combustion tunnel*, Philos. Trans. Roy. Soc. London Ser. A., 304 (1982), pp. 303–325.

[38] A. F. GHONIEM AND F. SHERMAN, *Grid-free simulation of diffusion using random walk methods*, J. Comput. Phys., 61 (1985), pp. 1–37.

[39] J. GOODMAN, *Convergence of the random vortex method*, Comm. Pure Appl. Math., 60 (1987), pp. 189–220.

[40] J. GOODMAN, T. Y. HOU, AND J. LOWENGRUB, *Convergence of the point vortex method for the* 2-D *Euler equations*, Comm. Pure and Appl. Math., 43 (1990), pp. 415–430.

[41] C. GREENGARD, *Three dimensional vortex methods*, Ph.D. thesis, University of California, Berkeley, CA, 1984.

[42] ———, *The core spreading vortex method approximates the wrong equation*, J. Comput. Phys., 61 (1985), pp. 345–348.

[43] ———, *Convergence of the vortex filament method*, Math. Comp., 47 (1986), pp. 387–398.

[44] L. GREENGARD AND V. ROKHLIN, *A fast algorithm for particle simulations*, J. Comput. Phys., 73 (1987), pp. 325–348.

[45] O. H. HALD, *Convergence of vortex methods for Euler's equations*, II, SIAM J. Numer. Anal., 16 (1979), pp. 726–755.

[46] ———, *Convergence of random methods for a reaction-diffusion equation*, SIAM J. Sci. Statist. Comput., 2 (1981), pp. 85–94.

[47] ———, *Convergence of a random method with creation of vorticity*, SIAM J. Sci. Statist. Comput., 7 (1986), pp. 1373–1386.

[48] ———, *Convergence of vortex methods for Euler's equations*, III, SIAM J. Numer. Anal., 24 (1987), pp. 538–582.

[49] T. Y. HOU, *Convergence of vortex methods for highly oscillatory vorticity*, Ph.D. thesis, University of California, Los Angeles, CA, 1988.

[50] T. Y. HOU AND J. LOWENGRUB, *Convergence of the point vortex method for the* 3-D *Euler equations*, Comm. Pure and Appl. Math., to appear.

[51] T. Y. HOU, *Convergence of a variable blob vortex method for Euler and Navier–Stokes equations*, manuscript 22 pages, New York University, New York, 1988.

[52] ———, *A new desingularization for vortex methods*, Math. Comp., to appear.

[53] T. Y. HOU, J. LOWENGRUB, AND M. J. SHELLEY, *The convergence of an exact desingularization and local regridding for vortex methods*, SIAM J. Sci. Statist. Comput., submitted.

[54] T. KATO, *On classical solutions of the two-dimensional non-stationary Euler equation*, Arch. Rational Mech. Anal., 25 (1967), pp. 188–200.

[55] T. KATO, *Nonstationary flows of viscous and ideal fluids in R^3*, J. Funct. Anal., 9 (1972), pp. 296–305.

[56] R. KRASNY, *Computation of vortex sheet roll-up in the Trefftz plane*, J. Fluid Mech., 184 (1987), pp. 123–155.

[57] A. LEONARD, *Vortex methods for flow simulation*, J. Comput. Phys., 37 (1980), pp. 289–335.

[58] D.-G. LONG, *Convergence of the random method in two dimensions*, J. Amer. Math. Soc., 1 (1988), pp. 779–804.

[59] ———, *Convergence of random vortex method in three dimensions*, manuscript 31 pages, Princeton University, Princeton, NJ, 1988.

[60] A. MAJDA, *Vortex Dynamics: Numerical Analysis, Scientific Computing, and Mathematical Theory*, in ICIAM '87, J. McKenna and R. Temam, eds., Society for Industrial and Applied Mathematics, Philadelphia, 1988, pp. 153–182.

[61] C. MARCHIORO AND M. PULVIRENTI, *Hydrodynamics in two dimensions and vortex theory*, Comm. Math. Phys., 84 (1982), pp. 483–503.

[62] ———, *Vortex Methods in Two-Dimensional Fluid Dynamics*, Lecture Notes in Physics, Vol. 203, Springer-Verlag, New York, 1984.

[63] F. J. MCGRATH, *Nonstationary plane flow of viscous and ideal fluids*, Arch. Rational Mech. Anal., 27 (1967/68), pp. 329–348.

[64] F. MILINAZZO AND P. G. SAFFMAN, *The calculation of large Reynolds number two-dimensional flow using discrete vortices with random walk*, J. Comput. Phys., 23 (1977), pp. 380–392.

[65] Y. NAKAMURA, A. LEONARD, AND P. SPALART, *Vortex simulation of an inviscid shear layer*, AIAA/ASME 3rd Joint Thermophysics, Fluids, Plasma and Heat Transfer Conference, St. Louis, Missouri, AIAA-82-0948, 1982.

[66] H. O. NORDMARK, *Higher order vortex methods with rezoning*, Ph.D. thesis, University of California, Berkeley, CA, 1988.

[67] E. PARDOUX AND D. TALAY, *Discretization and simulation of stochastic differential equations*, Acta Appl. Math., 3 (1985), pp. 23–47.

[68] M. B. PERLMAN, *On the accuracy of vortex methods*, J. Comput. Phys., 59 (1985), pp. 200–223.

[69] G. PUCKETT, *Convergence of a random particle method to solutions of the Kolmogorov equation $u_t = \nu u_{xx} + u(1 - u)$*, Math. Comp., 52 (1989), pp. 615–645.

[70] ———, *A study of the vortex sheet method and its rate of convergence*, SIAM J. Sci. Statist. Comput., 10 (1989), pp. 298–326.

[71] P. A. RAVIART, *An analysis of particle methods*, Lecture Notes in Mathematics, Vol. 1127, Springer-Verlag, Berlin, 1985, pp. 243–324.

[72] S. ROBERTS, *Accuracy of the random vortex method for a problem with non-smooth initial data*, J. Comput. Phys., 58 (1985), pp. 29–43.

[73] ———, *Convergence of a random walk method for the Burgers equation*, Math. Comp., 52 (1989), pp. 647–673.

[74] W. RÜMELIN, *Numerical treatment of stochastic differential equations*, SIAM J. Numer. Anal., 19 (1982), pp. 604–613.

[75] J. SETHIAN, *Turbulent combustion in open and closed vessels*, J. Comput. Phys., 54 (1984), pp. 425–456.

[76] J. A. SETHIAN AND A. F. GHONIEM, *The validation of vortex methods*, J. Comput. Phys., 74 (1988), pp. 283–317.

[77] R. TEMAM, *Navier–Stokes Equations, Theory and Numerical Analysis*, 2nd ed., North Holland, Amsterdam, 1979.

[78] W. WOLIBNER, *Un theoreme sur l'existence du mouvement plan d'un fluide parfait, homogene, incompressible, pendent un temps infiniment long*, Math. Z., 37 (1933), pp. 698–726.

A GALLERY OF FLUID MOTION USING VORTEX METHODS

3

James A. Sethian

Introduction

In this paper, we provide some examples of vortex simulations of two-dimensional, viscous, incompressible flow in a variety of flow regimes and confinement geometries. The goal of these calculations is to analyze the generation, development, and evolution of large fluid structures in the flow. The basic algorithm underlying these numerical simulations is the random vortex method, described in Chapter 1.

The results of these numerical simulations are visualized using real-time interactive flow tools built on a massively parallel processor. Since most fluid visualization techniques are highly parallel in nature, a massively parallel processor with parallel output to a frame buffer allows rapid animation of fluid motion. The flow tools mimic those in the experimental laboratory, and include display of moving color contours for scalar fields, smoke/dye injection of passive particles to identify coherent flow structures, and bubble wire tracers for velocity profiles, as well as three-dimensional interactive rotation and zoom and pan. These provide an effective and interactive way to analyze the underlying physical mechanisms, and to compare results with laboratory experiment.

We study several flow configurations, including flow in an expanding/contracting channel, backward-facing step, closed cavity, piston/valve, and enclosed obstacles. The results reveal a variety of fascinating fluid structures, including eddy formation, merger, propagation and decay, mixing and intertwining of fluid structures, and pairing of counteroscillatory vortical structures.

1. The Vortex Method

All of the calculations in this gallery use the random vortex method described in some detail in Chapter 1. Briefly, the velocity is approximated by a set of discrete vortex elements, or "blobs," whose relative positions and strengths determine the overall velocity field. The motion of the fluid is approximated by updating the position of the individual vortex elements according to the influence of the aggregate velocity field, plus a random step to model diffusion processes. Boundary conditions are met through the addition of potential flow to satisfy the normal boundary condition and a vorticity creation algorithm to satisfy the tangential no-slip condition on solid walls. Full details of the algorithm may be found in Sethian [5].

2. Flow Visualization on a Massively Parallel Machine

The data from the vortex simulations are visualized using a new graphics environment for essentially real-time interactive visualization of computational fluid mechanics (see Sethian and Salem [8]). Within this environment, the researcher may interactively examine fluid data on a graphics display using animated flow visualization diagnostics which mimic those in the experimental laboratory. The fluid diagnostic tools include display of moving color contours for scalar fields, smoke/dye injection of passive particles to identify coherent flow structures, and bubble wire tracers for velocity profiles, as well as three-dimensional interactive rotation and zoom and pan. These provide an effective and interactive way to analyze the underlying physical mechanisms, and to compare results with laboratory experiments.

The system is implemented on a data parallel supercomputer with parallel input/output to a framebuffer. Since most fluid visualization techniques are highly parallel in nature, this allows rapid animation of the fluid motion. The key components of this interactive graphics environment are a Connection Machine CM-2™ data parallel supercomputer and a CM-2™ framebuffer. Images are updated on the framebuffer at nine frames per second, providing essentially real-time motion.

As in a laboratory experiment, this graphics environment allows the researcher to study fluid velocity by injecting color-coded dye that is passively advected as the animation unfolds. Because the animation is both interactive and in real-time, moving eddy structures can be tagged and tracked with dye as they form. Different colored dyes can be used to watch regions merge or diffuse, and mixing mechanisms may be studied. Time-dependent scalar fields, such as temperature and density, can be color-tagged, and the range of displayed values may be changed during the animation. For three-dimensional flows, interactive rotation is possible as the animation proceeds, as well as zoom and pan. In this environment, the researcher typically performs dozens of realistic flow visualization experiments in rapid succession.

The environment assumes a stored data field, obtained either through numerical simulation or physical experiment. The data is preprocessed so that the relevant physical quantities, such as velocity, vorticity, density, pressure, temperature, etc., are given at each time step on a fixed discrete grid in the computational domain. In our case, we use the stored velocity computed on a fixed grid from the vortex method as input to the graphics environment. At each time step, the appropriate numerical calculations are performed in parallel on the CM-2™, and the images are displayed on the framebuffer.

The key visualization tools are summarized below.

Vector Quantities. The tools for both two- and three-dimensional vector quantities, such as velocity, are

- Injection of passively advected tracer particles.
- Mouse-driven location of injection points.

- Menu-specified frequency and number of injections.
- Menu-driven color-tagging according to location, frequency, time, and number of injections.
- Menu-driven injection of tracer particles,
  - along a line ("hydrogen bubble" tracers),
  - concentrated in a disk or ball ("smoke/dye" injection),
  - on a ring or sphere, or
  - continuous injection at a point.

Scalar Quantities. For scalar functions of two space variables, such as temperature, density, and pressure, the current tools are

- Motion of regions of selected contoured scalar values, generated by appropriate coloring of all points in selected range; and
- Mouse- or menu-driven selection of range of scalar values and color map to be displayed.

Display Attributes. As the animation sequence proceeds, the user may interactively

- Rotate the view point for three-dimensional flows.
- Change the color display map.
- Zoom and pan.

On a 8,192 processor CM-2™, images are updated on the framebuffer at nine frames per second, producing essentially real-time motion.

Details about the interactive flow visualization environment may be found in Sethian and Salem [8].

3. Implementations of the Vortex Method

All of the simulations shown rely on the basic random vortex method. However, there are significant differences. The code used to compute the flow over a backward-facing step is described in Ghoniem, Chorin, and Oppenheim [4] and is a direct implementation of the original method described in Chapter 1.

The code used to produce flow in a square cavity, described in Baden and Puckett [3], makes use of a fast algorithm to compute the particle-particle interaction. In a straightforward implementation, the N body problem is solved directly, costing $O(N^2)$ operations to evaluate all the pairwise interactions. The fast algorithm capitalizes on the idea that interactions between particles far apart can be effectively "lumped" or averaged with a relatively inexpensive calculation. Only the interactions between vortices that are quite close need be computed directly (see Baden and Puckett [3]). Thus, the vortex-vortex interactions are first calculated using a discrete Poisson equation on a grid, similar to a "vortex-in-cell" method. The computed velocities between particles close together are then "corrected" by solving the direct force law expression for the induced velocity. An intricate set of bins is used to determine which particles are close together and hence require velocity correction. The speedup in computer time using this technique is significant. For details, see Anderson [1], Baden [2], and Baden and Puckett [3].

In the majority of calculations presented here (flow in an expanding/contracting channel, flow around a piston/valve, from around interior islands, flow in a cross section), significant speedup is obtained through the implementation of the random vortex method on a massively parallel processor. Here, a general code has been developed for computing two-dimensional viscous, incompressible flow in arbitrary rectilinear geometries using vortex methods (see Sethian, Brunet, Greenberg, and Mesirov [7]). As input, the user may draw an arbitrary rectilinear geometry, including interior islands, on the screen of a frame buffer, and prescribe arbitrary location of inlets and outlets. The Connection Machine CM-2(TM), a 65,000 processor data parallel machine, is then used to approximate the evolving flow using the random vortex algorithm.

The implementation of the random vortex method on a massively parallel machine requires the efficient solution of both an **N**-body problem and Poisson's equation at each time step. The **N**-body problem requires the evaluation of a particle-particle interaction between all pairs of particles. Two parallel techniques have been developed to evaluate this mutual interaction. The first relies on a digital "orrery" to cycle through all possible interactions. The second relies on an all-to-all broadcast between the nodes of the hypercube by simultaneously sending information along many cube wires at once. The elliptic Poisson solver is solved using an iterative method, and can be made efficient through the use of slicewise stencils in the nearest neighbor grid updates. Finally, dynamic reconfiguration of data in the processors is used as the number of Lagrangian elements increases and decreases. For details on this parallel implementation of the vortex method, see Sethian, Brunet, Greenberg, and Mesirov [7].

4. Analysis of Flows

Expanding/Contracting Channel. We begin by analyzing flow in an expanding/contracting channel, calculated using the parallel implementation described in [7]. Flow enters through an opening on the left with a uniform entrance profile, and immediately encounters a sudden expansion. Near the exit, a sudden contraction occurs, and the flow then exits through a narrow opening on the right.

We show the results of this calculation over a sequence of photographs (see insert after p. 64). The first three photographs visualize the flow by stretching a simulated bubble wire across the vertical width of the channel, just past the large expansion. Starting from potential flow (Fig. 3.1(a)), the no-slip condition is instantaneously imposed at $t = 0$. As vorticity created on the boundary in response to the no-slip condition is shed into the main part of the flow, a pair of large eddies develop on the left sides of the two "arms" of the domain, entrapping fluid (Fig. 3.1(b)). Later still (Fig. 3.1(c)), these two eddies have broken away and moved along the top and bottom sides of the arms.

In Fig. 3.2(a), we illustrate the flow at a later time using both the bubble wire and multicolored dyes injected into the flow. The clockwise rotating yellow eddy in the bottom arm and the counterclockwise rotating blue eddy in the top arm now direct fluid away from the arms, shunting off the fluid, and change the effective geometry of the domain. This illustrates an important flow feature, namely, that the development of large fluid structures in the domain can drastically alter the flow portrait. Finally, in Fig. 3.2(b), the shed eddies in the top and bottom arms have moved to the right. At the left side of each arm an additional counterrotating eddy has appeared. Once again, the yellow eddies are spinning clockwise, and the blue eddies are spinning counterclockwise.

Flow Over a Backward-facing Step. Next, we illustrate flow over a backward-facing step at various Reynolds numbers. The data for this calculation comes from a detailed study of flow over a backward-facing step (see Sethian and Ghoniem [6]). The code used for this calculation was developed in Ghoniem, Chorin, and Oppenheim

[4]. The domain geometry consists of a narrow inlet on the left through which flow enters. A sudden step expansion occurs, and the flow exits 20 step heights downstream on the right. The behavior of the flow depends critically on R, the Reynolds number. In the laminar regime ($R < 250$), the flow is smooth, with a single, steady, and stable recirculation zone in the lower left-hand corner, and with the expected parabolic profile. In the transitional regime ($R \approx 500$–2,000), the downstream edge of the recirculation bubble tears off and is shed, only to be replaced by another growing recirculation bubble. Finally, as the Reynolds number is increased still further, the flow enters a turbulent phase, and eddies shed from the recirculation bubble near the step are interspersed with counterrotating eddies shed off the top wall.

In Fig. 3.3(a), we illustrate these three flow regimes using the simulated hydrogen bubble wire placed vertically across the channel. The flows shown are fully developed. To begin, the laminar flow (R = 125, on the bottom) shows the large recirculation zone in the lower left at the base of the step, and the parabolic flow downstream. The location of the reattachment point can be clearly seen. The transitional flow (R = 500, in the middle) exhibits markedly different behavior. Bubbles released along the wire are now caught up in the large eddies that develop at the base of the step and move downstream as coherent fluid structures. Here, one sees the tremendous mixing that results from this periodic shedding of large eddies. Finally, the turbulent flow (R = 5,000, on the top) exhibits the pairing of eddies shed from the recirculation bubble with counterrotating eddies shed off the top wall, and the thin band of rapidly moving fluid trapped in between.

In Fig. 3.3(b), we illustrate the same set of flows using injected purple and yellow dye. The dyes are injected in exactly the same points and at the same frequency in all three flows (laminar, transitional, and turbulent). Here, the dramatic changes in the flow portraits as a function of Reynolds number are evident. The recirculation zone is clearly shown by the purple dye in the laminar flow, with no mixing with the yellow dye tracing out the main parabolic profile. In contrast, the yellow and purple dyes are considerably mixed in both the transitional (middle) and turbulent (top) regimes.

Flow in a Closed Domain. Next, we study the evolution of a large circular patch of blue dye continuously injected into a counterclockwise circulating flow (Fig. 3.4). The data comes from a calculation of flow in a square cavity by Baden and Puckett [3]. The Reynolds number is R = 1,000 in this calculation. The domain is a closed rectangle, and a large central vortex is placed at the center, whose strength remains constant in time. We inject two different colored dyes at opposite sides of the box. In the upper left, we see the injected dyes shortly after the no-slip condition is turned on. In the upper right, we see the swirling fluid developing rolls and oscillations as it spins around the large central vortex, due to the developing counterrotating eddies in the corners. This can be seen in the "wrinkling" of the dyes. In the lower left, the fluid has made several rotations around the vessel. Flow in the corners is caught up in the counterrotating eddies and outlines the large vortex structures. In the last figure (lower right), the injected dyes have stretched out considerably and mixed, concentrated in some regions while stretched into thin strands in others.

Flow in a Piston/Valve. Next, we show flow in a piston, calculated using the parallel vortex implementation described in [7]. The Reynolds number is R = 5,000 in this calculation. Flow enters through channels on the left, and exits through a single opening on the right. We begin by following a time sequence visualized using a vertical bubble wire. Starting as potential flow (Fig. 3.5(a)), the no-slip condition is instantaneously imposed at $t = 0$. In Fig. 3.5(b), a pair of counterrotating eddies build in the main chamber. These eddies grow into a pair of large eddies in the center of the chamber that remain as essentially stable objects (Fig. 3.5(c)). Counterrotating eddies can also been seen in the upper left part of the main chamber, which further mixes the flow. In Fig. 3.5(d), we move the placement of the bubble wire slightly to the left, revealing further recirculation zones within the flow. Finally, in Fig. 3.6(a), we show the late time development of the same flow, visualized using colored blue dye.

Additional Flows. Finally, we show two more flows, both at Reynolds number $R = 5{,}000$, calculated using the general vortex flow package described in [7]. In Fig. 3.6(b), we show flow entering through a narrow inlet on the left, which passes around an interior island before exiting on the right. Once again, the counterrotating recirculation eddies are clearly seen. Finally, in Fig. 3.6(c), we model an idealized two-dimensional cross section of a Connection Machine CM-2™ to study the cooling patterns set up by air currents. Here, flow enters through narrow inlets on the left and right ends of the bottom wall. The flow then enters a set of narrow vertical channels which represent planes containing components, and then exits near the bottom of the left and right walls. We visualize the flow by stretching a bubble immediately above the bottom wall. The incoming flows meet in the center, and induce a pair of counterrotating eddies which destabilize the flow as it enters the narrow channels.

Acknowledgments

We wish to acknowledge the contributions of Jean-Phillipe Brunet, Adam Greenberg, and Jill P. Mesirov in the parallel implementation of the vortex method, used to calculate the flow through the expanding/contracting channel, piston/valve, confined islands and cross section shown in these photographs. We also acknowledge the work of A. F. Ghoniem in developing the stepcode used to produce the raw data shown within. S. Baden and E. G. Puckett generously provided the raw data for the flow in the closed cavity. Finally, all of the above data was visualized using the interactive flow visualization tools developed in conjunction with James B. Salem.

This work is supported in part by the Applied Mathematics Subprogram of the Office of Energy Research under contract DE-AC03-76SF00098. The author also acknowledges the support of the National Science Foundation.

References

[1] C. ANDERSON, *A method of local corrections for computing the velocity field due to a distribution of vortex blobs*, J. Comput. Phys., 62 (1986), pp. 111–123.

[2] S. B. BADEN, *Very large vortex calculations in two dimensions*, in Vortex Methods, C. Anderson and C. Greengard, eds., Vol. 1360, Springer-Verlag, New York, 1988, pp. 96–120.

[3] S. B. BADEN AND E. G. PUCKETT, *A fast vortex code for computing 2-D flow in a box*, J. Comput. Phys., in press.

[4] A. F. GHONIEM, A. J. CHORIN, AND A. K. OPPENHEIM, *Numerical modeling of turbulent flow in a combustion tunnel*, Philos. Trans. Roy. Soc. London Ser. A., 304 (1982), pp. 303–325.

[5] J. A. SETHIAN, *Turbulent combustion in open and closed vessels*, J. Comput. Phys., 54 (1984), pp. 425–456.

[6] J. A. SETHIAN AND A. F. GHONIEM, *Validation study of vortex methods*, J. Comput. Phys., 74 (1988), pp. 283–317.

[7] J. A. SETHIAN, J. P. BRUNET, A. GREENBERG, AND J. P. MESIROV, *Two-dimensional, viscous, incompressible flow in arbitrary geometries on a massively parallel processor*, J. Comp. Phys., submitted.

[8] J. A. SETHIAN AND J. B. SALEM, *Animation of interactive fluid flow visualization tools on a data parallel machine*, Internat. J. Super. Appl., 3.2 (1988), pp. 10–39.

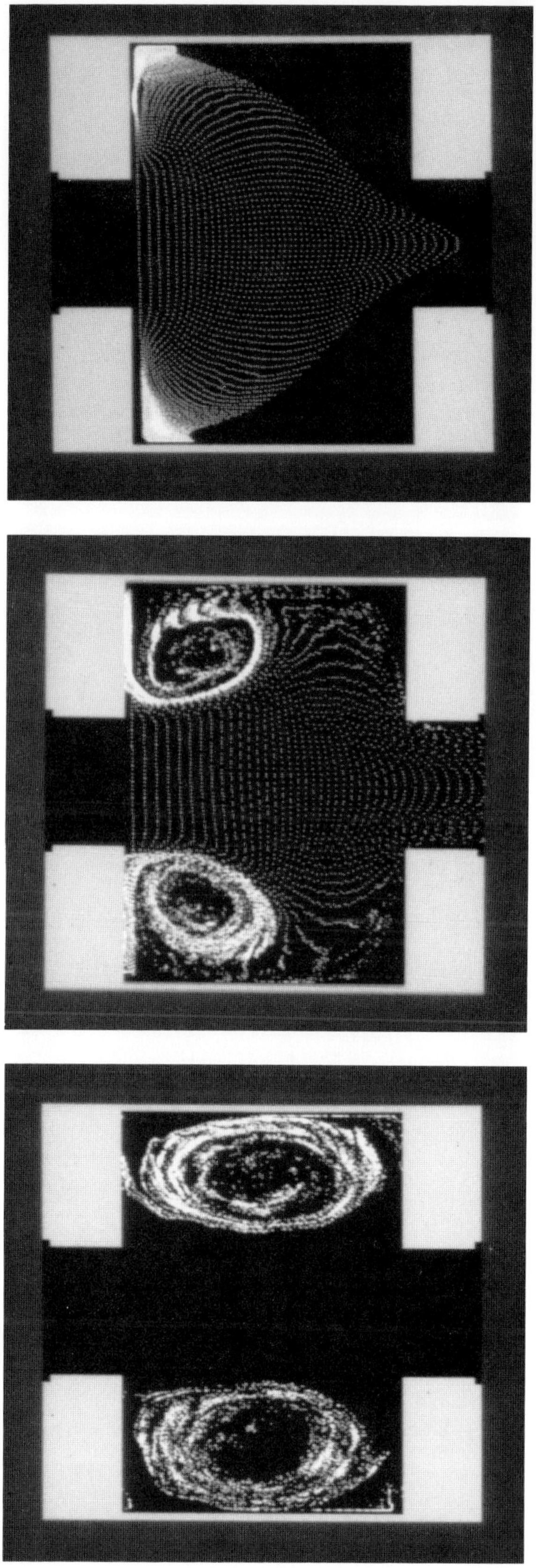

(a)

(b)

(c)

Fig. 3.1. Expanding/contracting channel.

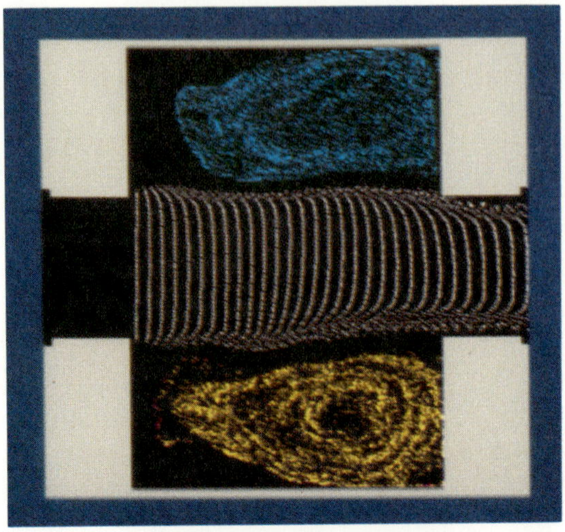

(a)

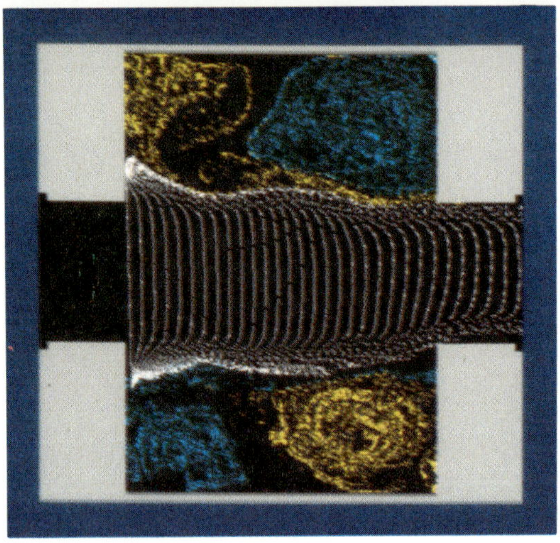

(b)

Fig. 3.2. Expanding/contracting channel.

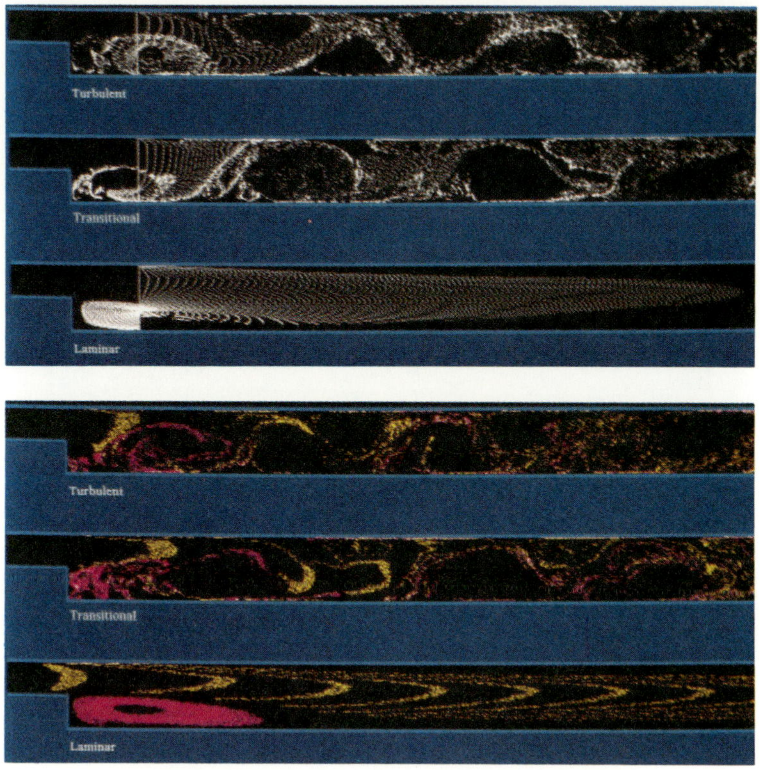

Fig. 3.3. Flow over a backward-facing step

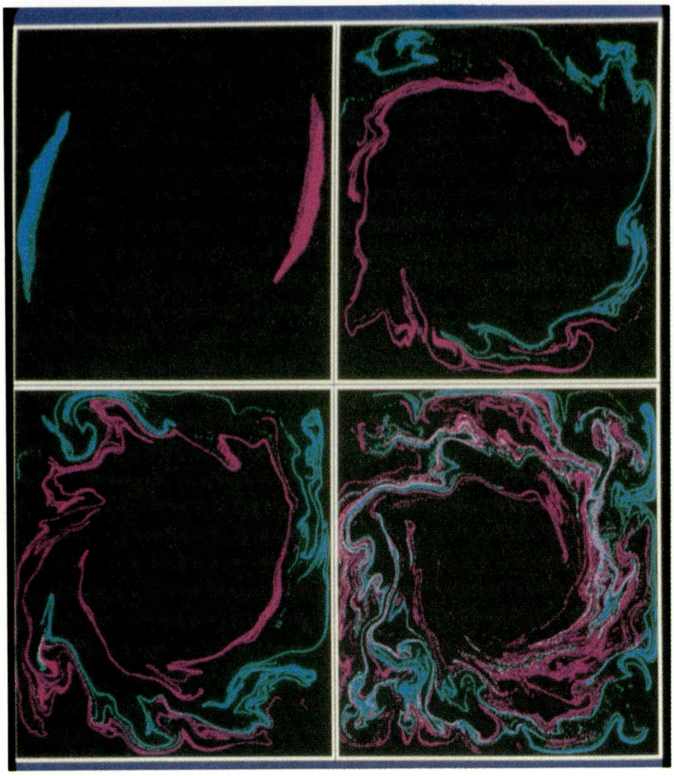

Fig. 3.4. Flow in a closed cavity. Data provided by S. Baden and E. G. Puckett.

(a)

Fig. 3.5. Flow in a piston/valve.

(b)

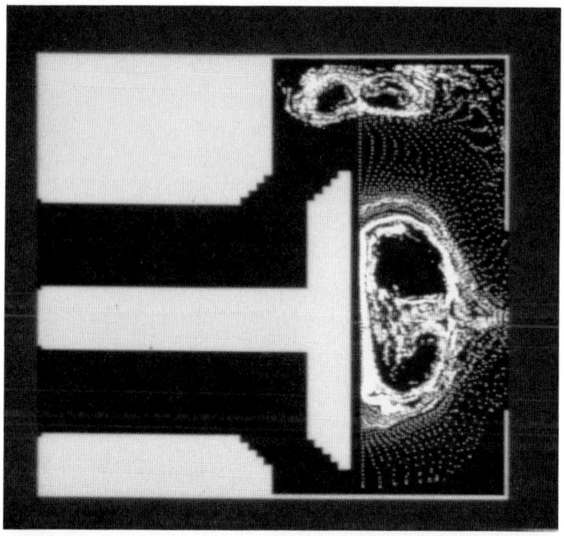

(c)

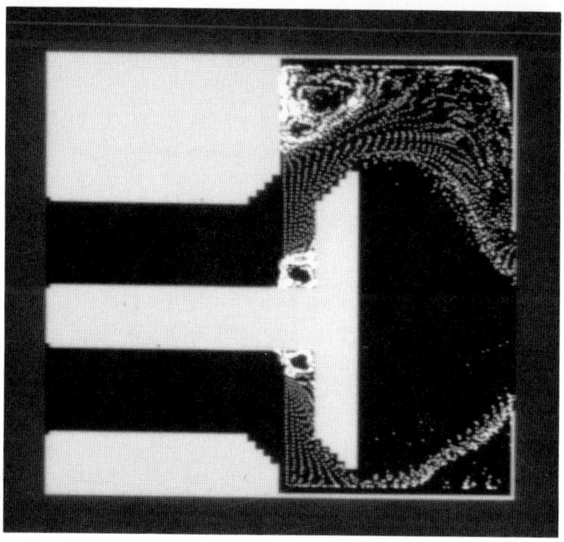

(d)

Fig. 3.5 (continued).

(a) (b) (c)

Fig. 3.6. Additional flows.

4

PHYSICAL VORTEX VISUALIZATION AS A REFERENCE FOR COMPUTER SIMULATION

Peter Freymuth

Introduction

The traditional art and science of flow visualization [1]–[8] is at a crossroads with the new art and science of computer visualization [9]–[14]. Physical flow visualization is increasingly capable of describing the shape of two-dimensional vortices and the connectivity of three-dimensional vortex systems in complex time-dependent flow fields. The challenge is to simulate these vortex patterns by computer at a convincing quality for a large variety of flow parameters and flow configurations. Behind a high quality simulation stands the requisite computer power to produce and control desired aspects of the flow field at the vorticity, velocity, and pressure levels. Combining this power with the cognitive power of computer graphics to visualize from different angles and distances in false color and possibly stereoscopic, promises to "outvisualize" physical flow visualization. Quantitative and therefore more limited physical anemometric and barometric measurements can be easily outperformed by simulation. The ultimate goal is to reach a level of computational power and cognitive accomplishment which can be labeled "understanding" of complex, nonlinear dynamic flow problems.

Currently, physical flow visualization serves as a tool for qualitative exploration, as a physical reference and challenge for computer modeling and as a training ground for developing the descriptive language appropriate for complex vortical flows. Flow visualization also serves as a tool for qualitative computer code validation. A quantitative validation by anemometry is often sought after at great expense and manpower, where internal policing of code and computer quality would suffice.

While flow visualization is capable of resolving vortex patterns in complex time-dependent flows it stops far short of resolving the finer scales of turbulent flows. Turbulent diffusion even blurs or "poisons" the visualization at a scale that has been resolvable prior to the onset of turbulence. As a matter of fact, both physical and numerical visualization are currently ill-equipped to visualize the fine details of turbulence. While turbulence is unavoidable in many flow configurations its visualization will not be emphasized in this paper. This paper documents mathematically challenging two- and three-dimensional vortex patterns in unsteady separating flows and describes their development.

Vortex patterns have been visualized in a variety of unsteady, separating flows. These physical references are used to outline some of the challenges and some of the promises of computer simulation, from the point of view of an experimentalist.

1. Limited Mathematical Approaches

Many computer visualizations do not have ready counterparts of physical visualization. Typically, vortices are "initially" placed into infinite space and their inviscid development is simulated [10]–[14]. In physical experiments vorticity has to be generated or "grown" on solid surfaces and from there diffused and finally convected into the flow. No other production mechanism for vorticity is available in homogeneous, constant density flows. A comparison of computer and physical experiments then appears tenuous and computations backward in time still could not create meaningful boundary conditions. Boundary conditions simply need boundaries. Mathematical calculations of this type nevertheless are useful since they can identify vortical interactions which at some occasions may occur in real flows. Limited computer visualizations may well take a lead in identifying flow processes and in training fluid dynamicists to "see," in the presence of seemingly overwhelming complexity.

Inviscid discrete vortex methods [15]–[17] allow for separation from solid surfaces with sharp corners (airfoils, flat plates) but do not model secondary and higher order vortices near solid surfaces. Furthermore, discreteness is linked to coarseness in simulation.

The streamfunction approach [18]–[22] in two-dimensional viscous flows around cylinders (including airfoils) can be very successful and include some visualizations at the vortical level.

The finest vortical details have been computed by using Navier–Stokes codes on supercomputers by Visbal and Shang [23], [24] for pitching airfoils, and by Tamura, Tsuboi, and Kuwahara [25] for an oscillating circular cylinder.

Simulation of three-dimensional unsteady vortex flows near solid surfaces is still terra incognita while physical visualizations for flows over finite wings are emerging [8], [26]–[29].

2. Limited Experimental Approaches

Computer modelers may be introduced to the enormous variety of interesting unsteady separating flows by physical visualization and their description should greatly influence mathematical modeling. Since physical visualization has many limitations of its own which do not necessarily overlap with limitations of computer visualization, these different visualizations represent a synergistic environment.

Despite its long tradition, physical flow visualization is still a rapidly developing field [4]–[8], [30]. Our subsequent examples are therefore not intended to represent ultimate references but to highlight some of the challenges they pose for computer simulation. Our physical visualization is mostly qualitative. In computer modeling the aim therefore should not be so much a quantitative verification of physical visualization but to capture its essence and to profit from its description. Furthermore, computer visualization should strive for the quality of physical visualization and to exceed it in many ways. Once this level has been accomplished and quantitative field descriptions are at hand, physical visualization will remain as a reference and a link to physical reality and as an aesthetic manifestation of nature.

Most of our physical visualizations are based on a method of vortex tagging in air [31]–[33], with certain limitations painfully present. Liquid titanium tetrachloride ($TiCl_4$) is used as a smoke-generating agent. With application to appropriate body surfaces, smoke gets produced mainly where the vorticity is produced. The smoke then tags or stains the vorticity and makes the vortex development visible. Inadvertently, some smoke gets introduced in areas without vorticity production. This problem, as well as inhomogeneities in smoke introduction, too thick or too thin smoke, and differences in smoke and vorticity diffusion, limit the effectiveness of the method, although the results are impressive. Furthermore, smoke mainly allows the visualization of vortex shapes, without much information on vortex strength.

Streamline techniques where particle paths are traced out, usually in water flows, have very good potential for obtaining more quantitative information [34]. Such methods are just emerging, are highly complex, do not show vortical detail, and are not ready for use in three-dimensional flows. Qualitative streamline visualizations have proven very successful for velocity field visualizations in two-dimensional flows [35].

3. Two-dimensional Vortex Visualization

Flow Characterization by Single Frame Visualization. Even in unsteady separating flows a single photograph can convey a large amount of information about the character of the flow. A large collection of such photographs along with a few photographic sequences are available in the literature [3].

Figure 4.1 represents a collection of four relatively simple vortex streets, none of which have been adequately simulated by computer (adequate simulation could happen any day, since the numerical and computational tools are available). Flow is from left to right in all cases and this will be true for subsequent movie sequences as well.

The top frame [36] shows the classical vortex street behind a circular cylinder at a Reynolds number Re = 120. Vortices in the top row turn clockwise, in the bottom row counterclockwise. By the Biot–Savart law this arrangement of vortices induces an upstream velocity component which is characteristic for a drag indicating wake.

The frame below the top frame [37] shows a vortex street generated behind the trailing edge of a periodically pitching airfoil. The Reynolds number based on the chord length in this case is Re = 12,000, the pitch amplitude is 10° around a 0° mean angle of attack and around the quarter chord axis. The reduced frequency $k = \pi fc/U_0$ in this case is $k = 3.7$, where $f = 2$ Hz is the pitching frequency, $U_0 = 61$ cm/sec is the free stream velocity and $c = 35.6$ cm is the chord length. In this case the top row of vortices rotates counterclockwise, and the bottom row clockwise, which by the Biot–Savart law is indicative of a thrusting jet. The propulsive power generation of a pitching wing should be similar to power generation by a pitching fish tail.

The third frame from the top [38] shows the free shear layer of a jet issuing from a nozzle into still air. The shear layer decays into small vortices due to the Kelvin–Helmholtz instability. Some vortices are in a state of pairing, others have completed the pairing process. Close to the right frame edge a vortex pair has decayed turbulently as the irregularity and diffuseness of the smoke attest to.

Similarly to the inviscid Kelvin–Helmholtz instability of a free shear layer, there exists a viscous Tolmien–Schlichting instability for the shear layer over a solid surface. The resulting vortex development is shown for an instant in accelerated starting flow over an airfoil at a 20° angle of attack in the bottom frame of Fig. 4.1.

All of the vortex streets of Fig. 4.1 and many others still pose a challenge for high quality computer simulation.

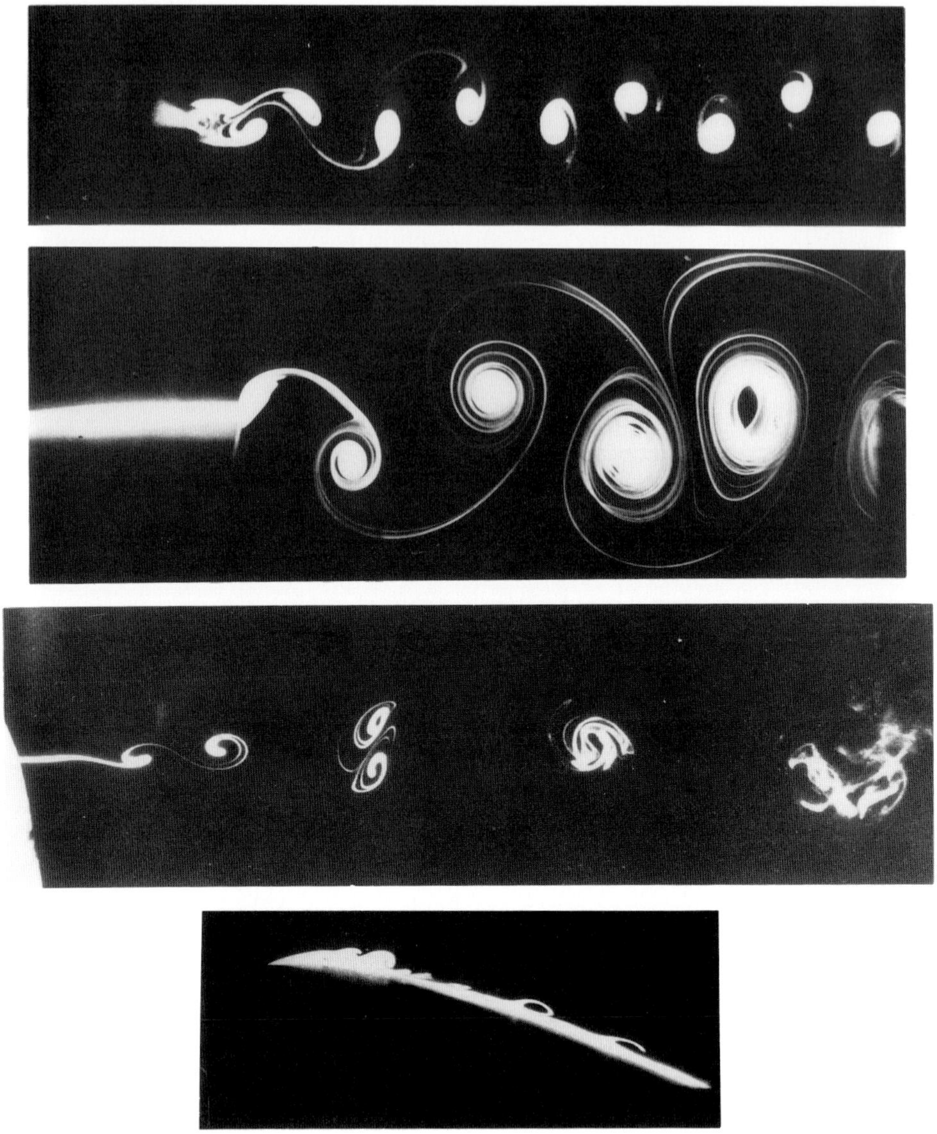

Fig. 4.1. An assortment of vortex streets. Top: vortex street behind a circular cylinder. Second row: vortex street behind a pitching airfoil. Third row: vortex street formed in the boundary layer of a jet issuing from a nozzle (Kelvin–Helmholtz instability). Bottom: vortex street formed by the boundary layer over an airfoil surface (Tolmien–Schlichting instability).

Movie Sequences. A more thorough understanding of vortex development requires movie sequences of visualization, at a sufficient frame rate.

Figure 4.2 shows vortex development over the leading edge of an airfoil at an 80° angle of attack, in accelerated starting flow. The movie frames are ordered into columns from top to bottom and across columns from left to right. Time between consecutive frames is $\Delta t = \frac{1}{16}$ sec. The Reynolds number based on chord length c and acceleration a is $\mathrm{Re} = c^{3/2}a^{1/2}/\nu = 5{,}200$, where ν is the kinematic viscosity in air. The decay of the separating vortex filament into small vortices due to the Kelvin–Helmholtz insta-

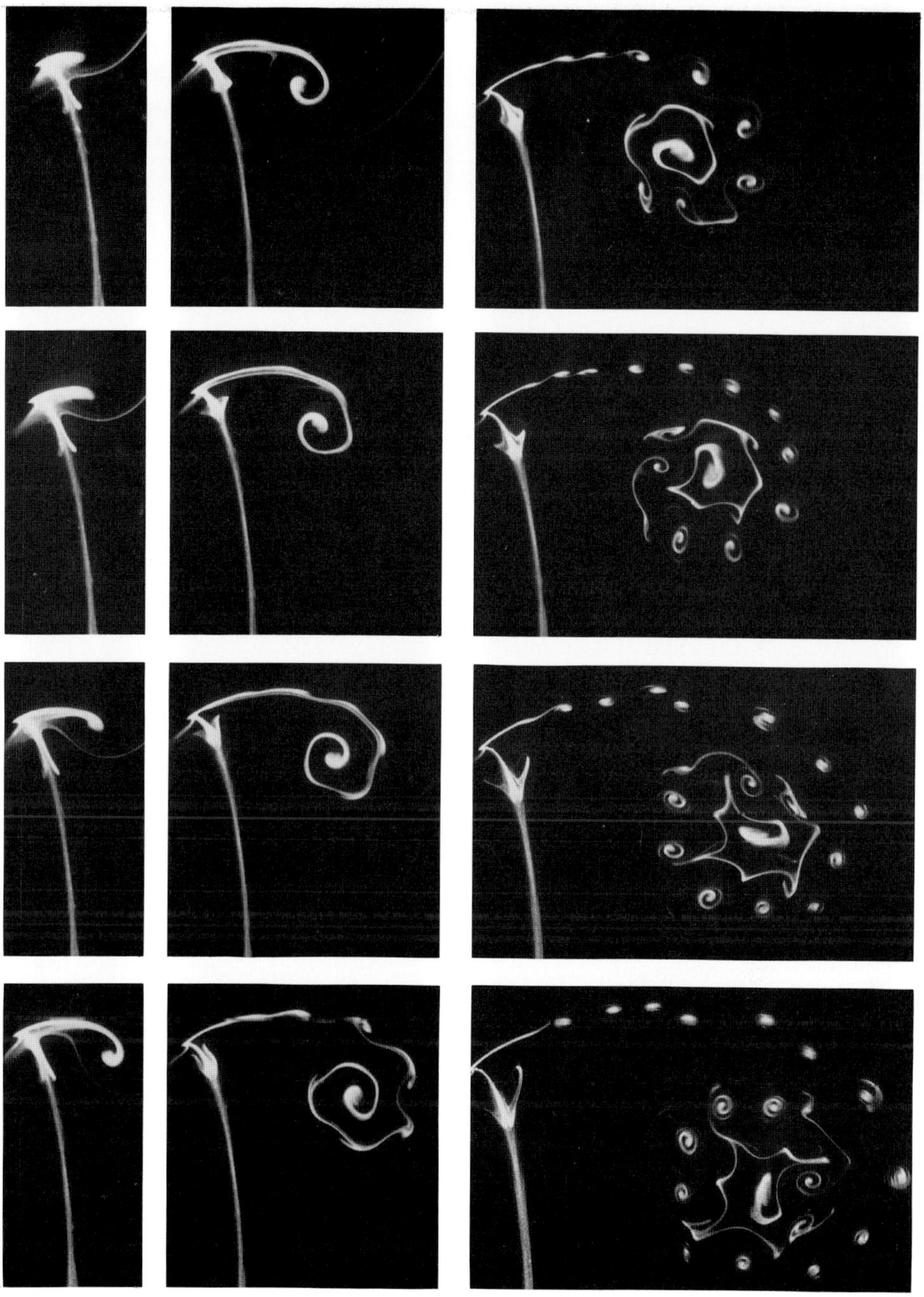

Fig. 4.2. Development of an ornamental leading edge starting vortex for an airfoil in accelerated starting flow at an angle of attack $\alpha = 80°$. $a = 2.4$ m/sec^2, $c = 15.2$ cm, $\Delta t = \frac{1}{16}$ sec, Re = 5,200.

bility and their interaction to form a starting vortex group are crisply shown. We have labeled this combined process of instability and starting vortex formation as ornamentation [39]. This vortex interaction of ornamental appearance awaits computer simulation.

The upper six frames of Fig. 4.3 show formation of two secondary vortices close to the airfoil surface as interactions of two primary vortices with the solid surface. Persistence of a secondary vortex into the turbulent regime can be seen in frames 5 and 6. The airfoil is at a 26° angle of attack in an accelerated starting flow [40]. Both Kelvin–Helmholtz and Tolmien-Schlichting instabilities are likely to contribute to the formation of the vortices.

Under some conditions a tertiary vortex nested inside a secondary vortex can be seen, as the single frame of Fig. 4.3, bottom, shows at a 40° angle of attack. Faint indication of a quartary vortex nested inside the tertiary vortex seems to exist. Secondary vortex formation in computer simulations have become familiar by now [24], [25] but higher-order vortices await simulation.

From the description of Figs. 4.2 and 4.3 it has become clear that vortices interact with each other as well as with solid surfaces. An entire vocabulary to efficiently describe such interactions has been developed [39], [32] and the reader is referred to these sources. Let us focus on a few simple examples which illustrate the usefulness of a descriptive vocabulary in comparative studies. In particular, let us compare in the next three figures some possible interactions between a primary and its secondary vortex.

Figure 4.4 visualizes accelerated starting flow over an airfoil (NACA 0015) at a 40° angle of attack. The first column shows, among other things, development of a leading edge primary vortex and of a secondary vortex. In column 2 the secondary vortex is drawn into the deformed primary spiral and in this way removed from the airfoil surface. This kind of process has been labeled vortex shredding [39].

In Fig. 4.5 the same airfoil is shown in an impulsively started flow. Again primary and secondary vortices form in column 1. In column 2 the trailing edge vortex jumps over the trailing edge of the airfoil, thus coming close enough to the secondary vortex to draw the secondary vortex into it. This process of secondary vortex slipdown and merger with the trailing edge vortex is very different from the process of secondary vortex shredding in Fig. 4.4. The process was first seen in streamline visualizations by Monnet and Coutanceau [35] and in vorticity tagging visualizations by Finaish and Freymuth [41], [42].

Another interaction of a secondary vortex is shown in Fig. 4.6. The airfoil is pitching periodically between 0° and 60° in a steady flow at Re = 12,000 and at a reduced frequency $k = \pi fc/U_0 = 3.1$. The first column again shows primary and secondary leading edge vortex formation. The secondary vortex gets strong enough in column 2 that primary and secondary vortices from a pair which lifts off the airfoil surface, reminiscent of the fireball of an atomic bomb lifting off the ground.

Comparative studies of this kind train the eye to recognize processes in seemingly complex flows and this should be of value to computer modelers if they wish to describe their results. Additional comparative studies can be found in [32].

Figure 4.7 shows vortex development over an airfoil in periodic pitch between 0° and 60° at Re = 1,700 and $k = 0.52$. The sequence is notable for the complexity of sequential and concurrent vortex interactions. Among the interactions we can identify [32], [39] secondary vortex formation, slipdown and merger, splitting, and pair formation of opposite vortices. More time is needed to adequately model complex interactions as shown.

(Text continues on p. 77)

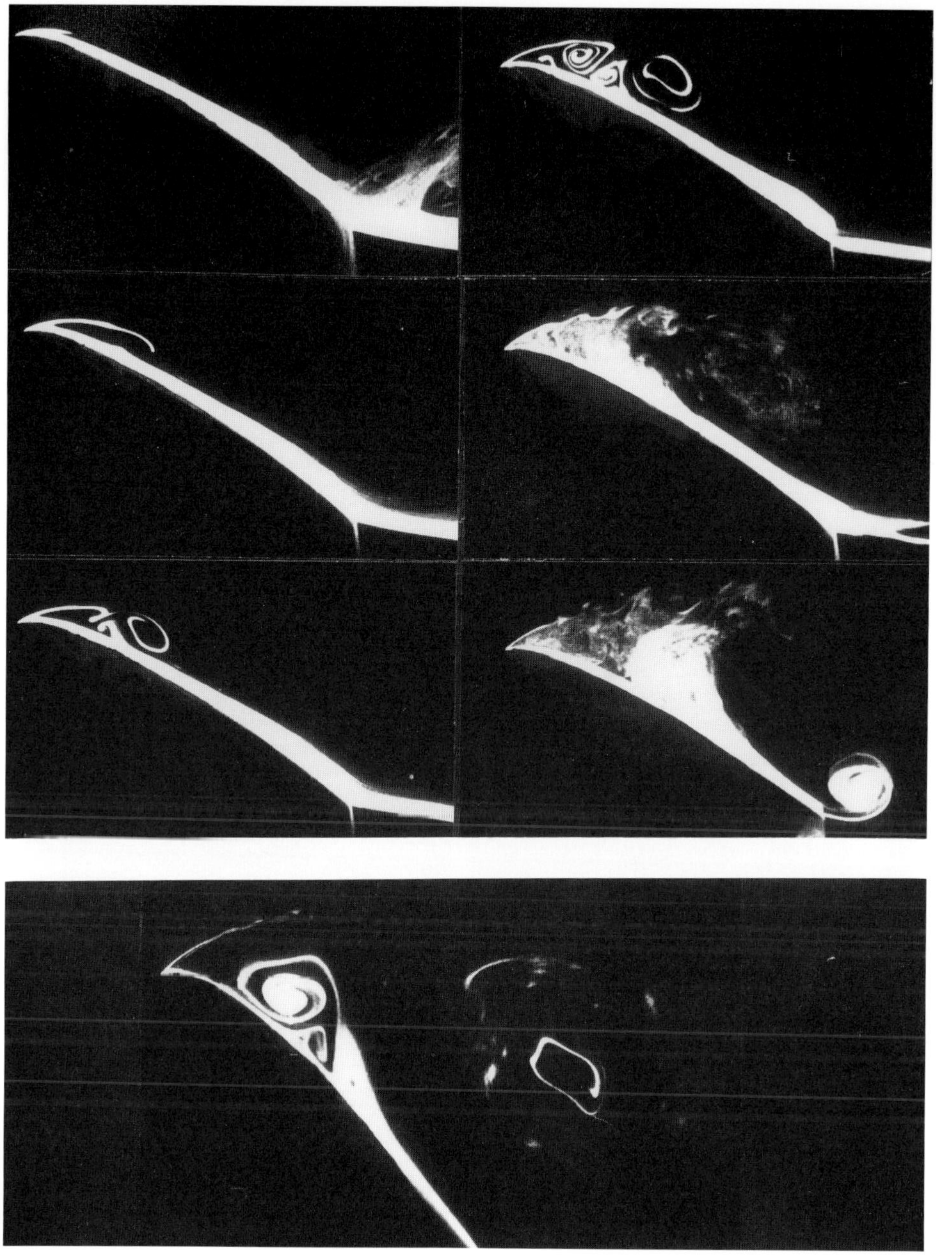

Fig. 4.3. Frames 1–6: accelerated starting flow over an airfoil at $\alpha = 26°$, Re = 5,200. Bottom: single frame at $\alpha = 40°$, Re = 5,200.

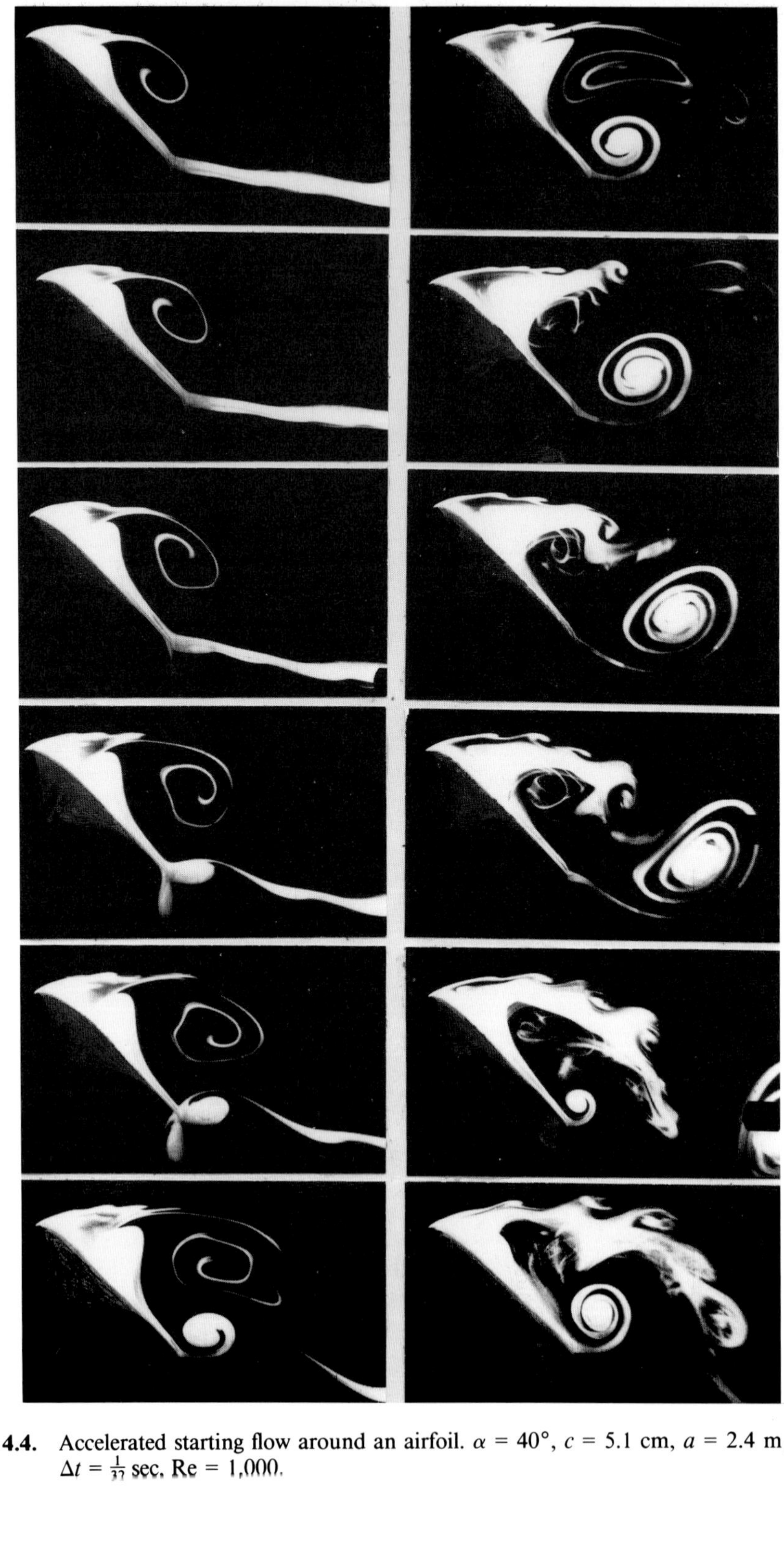

Fig. 4.4. Accelerated starting flow around an airfoil. $\alpha = 40°$, $c = 5.1$ cm, $a = 2.4$ m/sec^2, $\Delta t = \frac{1}{32}$ sec, Re = 1,000.

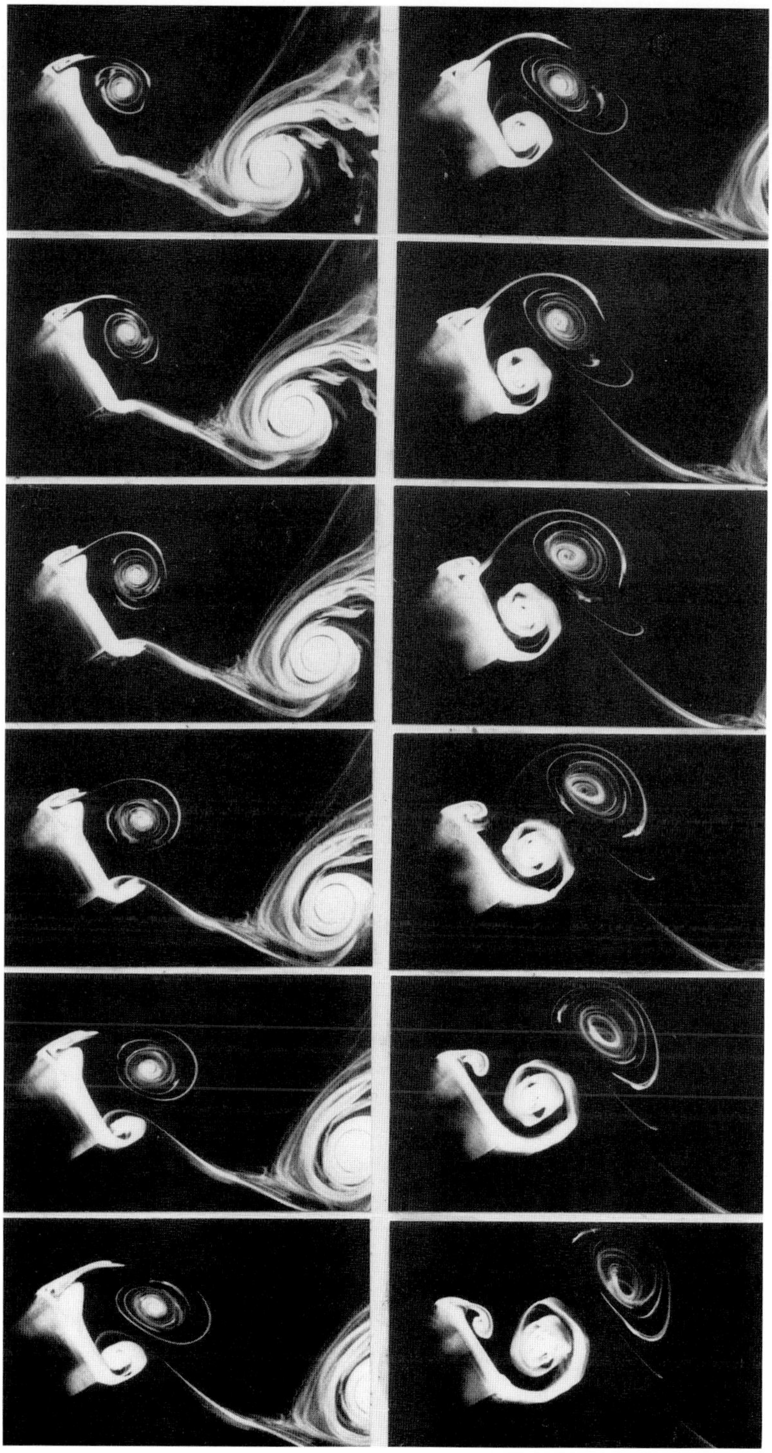

Fig. 4.5. Impulsive starting flow around an airfoil. $\alpha = 50°$, $c = 5.1$ cm, $U_0 = 37$ cm/sec, $\Delta t = \frac{1}{32}$ sec, Re = 1,000.

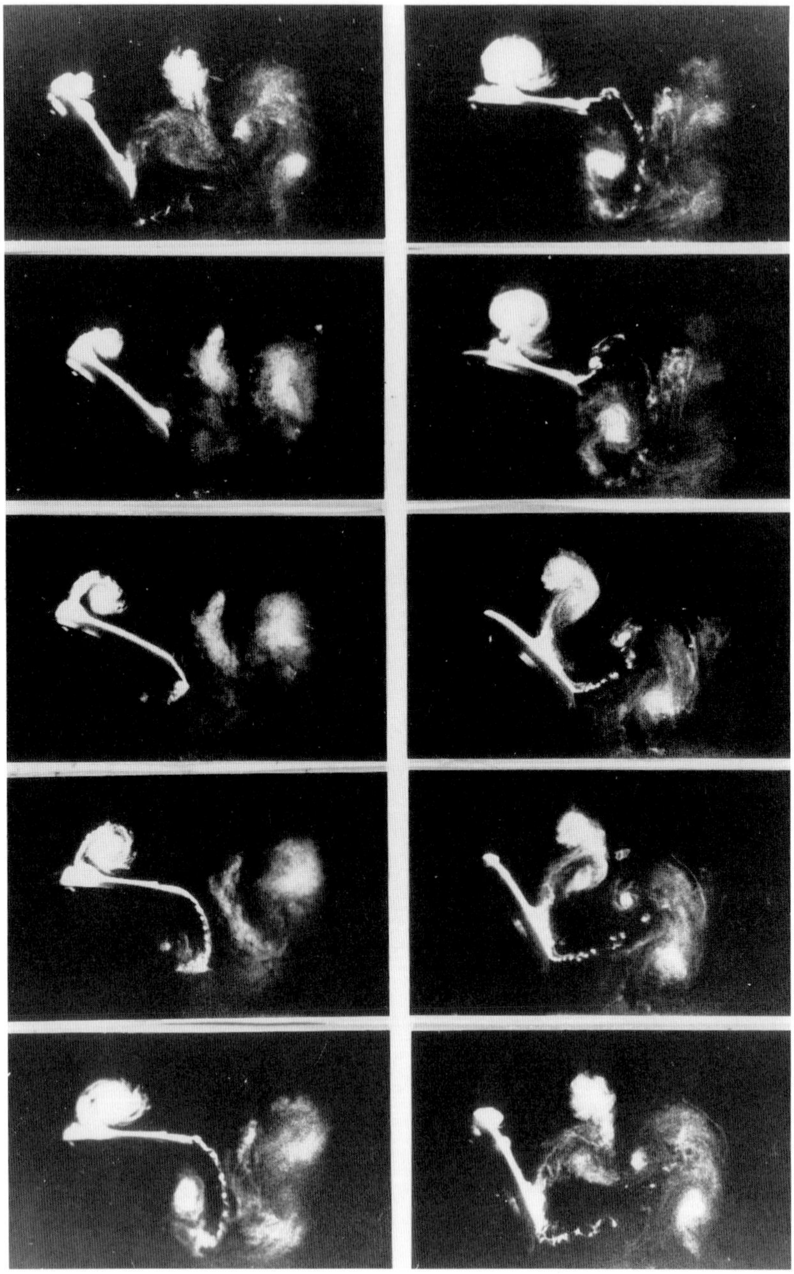

Fig. 4.6. Airfoil in periodic pitch between 0° and 60°. $c = 35.6$ cm, $U_0 = 61$ cm/sec, $\Delta t = \frac{1}{16}$ sec, $k = 3.1$, Re = 12,000.

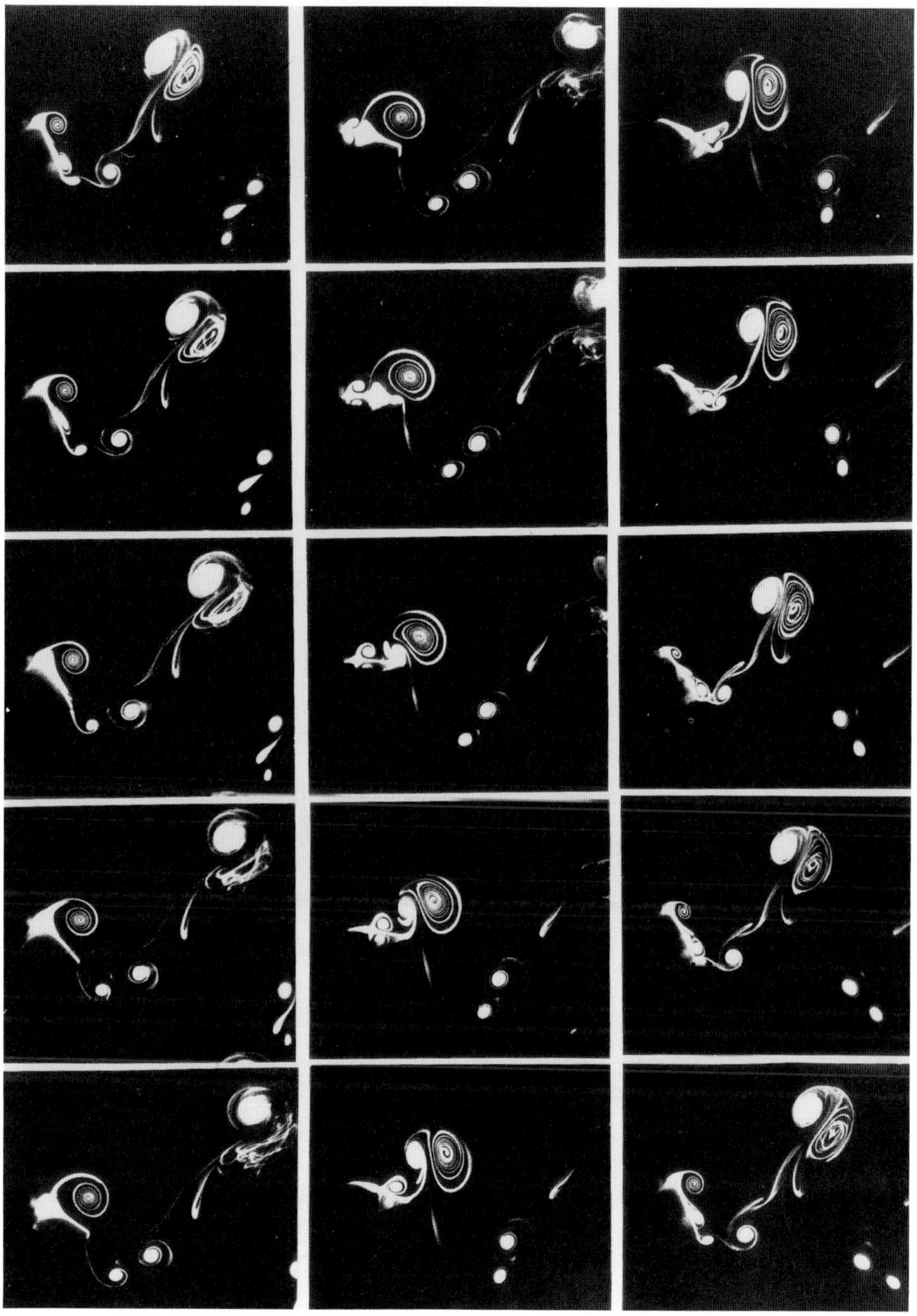

Fig. 4.7. Airfoil in periodic pitch between 0° and 60°. $c = 5.1$ cm, $U_0 = 61$ cm/sec, $\Delta t = \frac{1}{32}$ sec, $k = 0.52$, Re = 1,700.

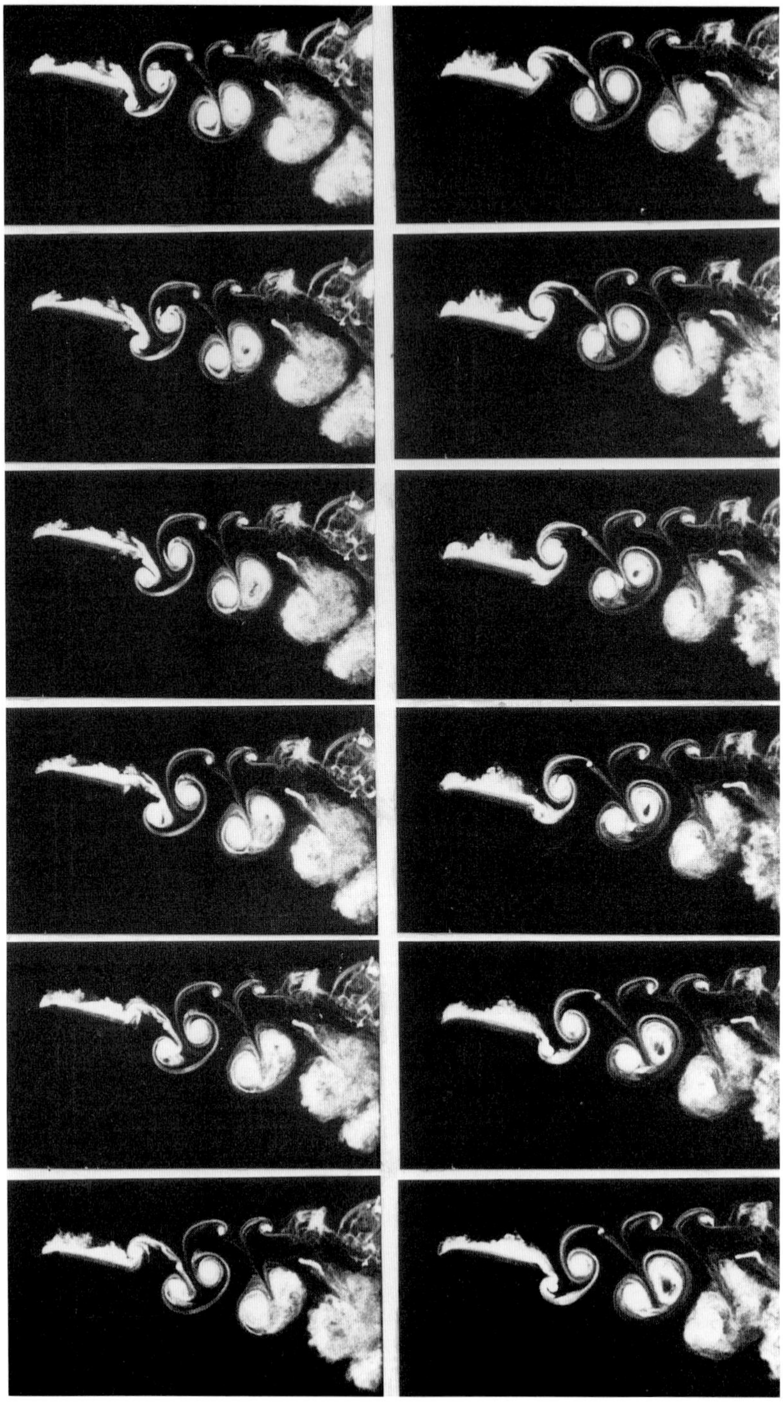

Fig. 4.8. Airfoil in periodic plunge. c = 15.2 cm, α = 5°, f = 6 Hz, U_0 = 61 cm/sec, Δt = $\frac{1}{64}$ sec, k = 4.7, Re = 5,200.

Figure 4.8 sequences an airfoil in periodic plunge at a 5° angle of attack in a steady flow. Plunge frequency is 6 Hz and plunge amplitude is 10 percent of the chord length $c = 15.2$ cm of the NACA 0015 airfoil. Of main interest is the shedding of counter-rotating vortex pairs backward and somewhat downward and simultaneous formation of small clockwise vortices forming a top row. Separation occurs from the trailing edge. We previously [37] found separation of the top row vortices from the leading edge under slightly different conditions. This demonstrates the variability of formation conditions for a particular type of vortex street.

Vortex visualization is not limited to airfoil flows. Figure 4.9 sequences accelerated starting flow over a 5.1 cm high spoiler mounted on a flat plate. Notable is the quick

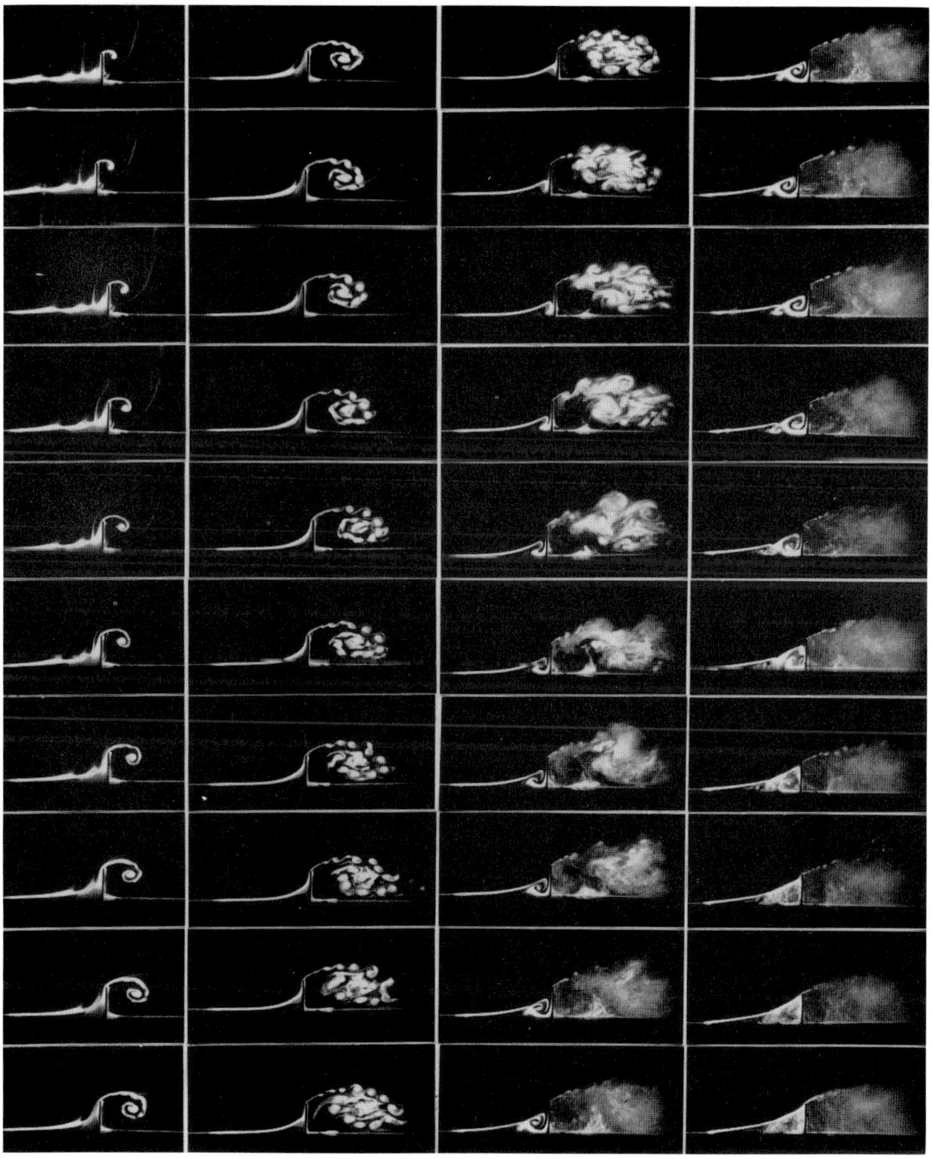

Fig. 4.9. Accelerated starting flow over a spoiler of height $h = 5.1$ cm. $a = 2.4$ m/sec^2, $\Delta t = \frac{1}{32}$ sec (from [43]).

development of an ornamental vortex behind the top of the spoiler and vortex development in front of the spoiler at a vastly slower timescale. Figure 4.10 focuses on the interactions between the starting vortex from a 2.7 cm spoiler and the flat plate, again in accelerated starting flow. Additional visualizations of flow over spoilers, double steps, and cavities can be found in [43].

Undoubtedly, the continued simulation of two-dimensional vortex flow will capture the spirit of vortex visualization and make its own decisive contributions.

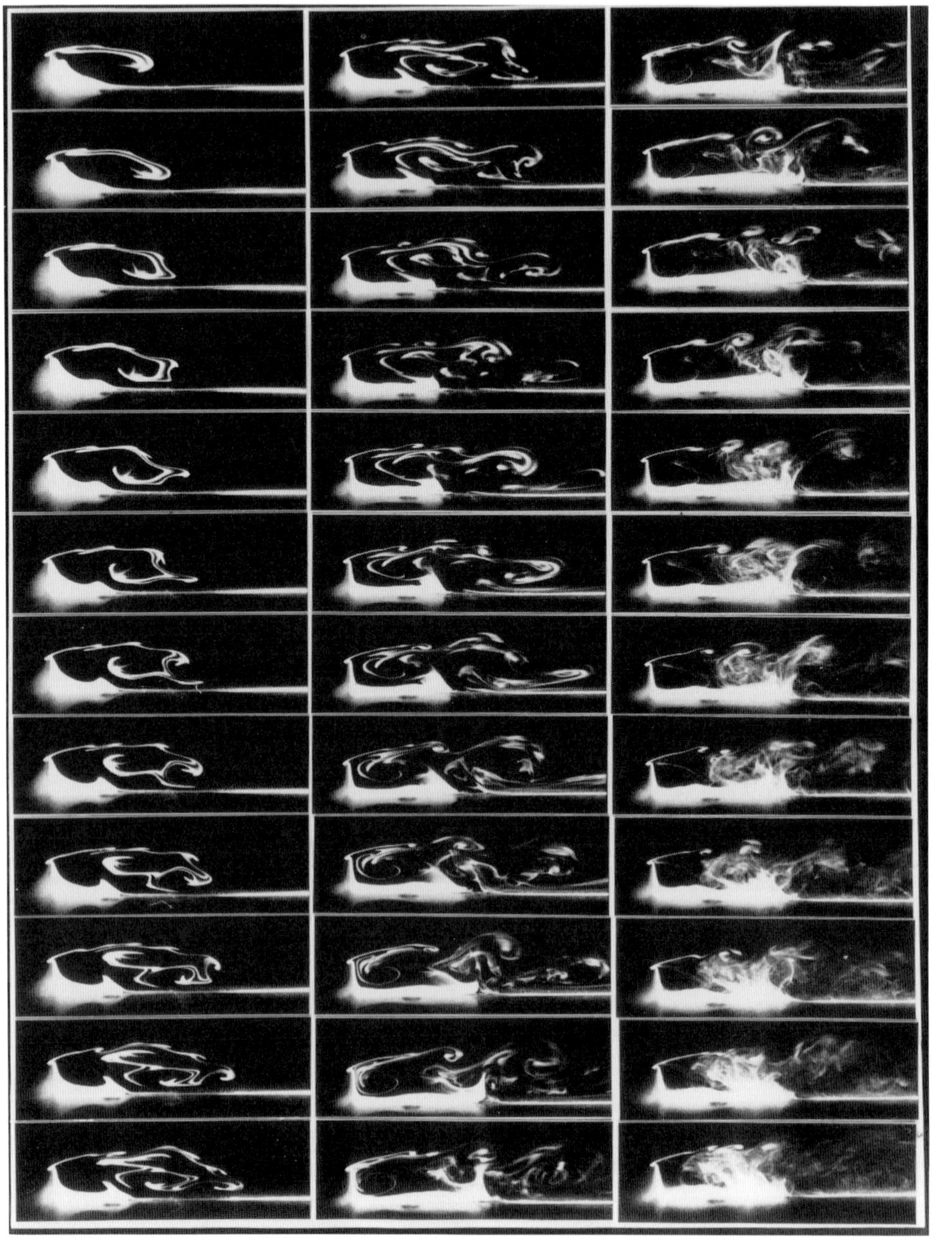

Fig. 4.10. Accelerated starting flow over a spoiler of height h = 1.3 cm, a = 2.4 m/sec^2.

4. Three-dimensional Vortex Visualization

In three-dimensional, unsteady, separated flows one can pursue the transverse structure of the vortex strands by appropriate cuts of visualization, analogous to the cuts in two-dimensional flows shown in the last section. This approach has been taken by Adler and Luttges [44], Gad-el-Hak and Ho [45], and Payne, Ng, and Nelson [46]. The unique challenge in three-dimensional vortex visualization is however to visualize the longitudinal structure of the vortex strands, their entanglement and connectivity. The topological vortex law by Helmholtz requires that finite vortices need to be connected. They cannot just end. In the absence of surface rotation the no slip-condition at walls further requires that the vortices cannot touch down on solid surfaces, they only can be tangent to them. While vortices need to be connected, viscous and combined turbulent-viscous diffusion allows reconnection of strands. The quality of three-dimensional vortex visualization can therefore be judged by how successful the connectivity of the strands is visualized. Our examples for this section are intended to capture the spirit of three-dimensional vortex visualization and to provide qualitative references for computer modelers.

Let us commence with an example of laminar to turbulent transition taken for an airfoil in accelerated starting flow. Figure 4.11, taken from [47], depicts a global view or top view of an NACA 0015 airfoil placed vertically into the wind tunnel, at a 20° angle of attack. Liquid $TiCl_4$ has been allowed to run down the entire span of the airfoil by gravity action, near the leading edge. A rather homogeneous smoke sheet then covers the suction side of the airfoil and allows visualization of the three-dimensional structure of developing vortices. In column 1, frame 1, several two-dimensional vortices have already developed, visible as light, vertical bands. In subsequent frames a waviness develops in spanwise direction. These wavy vortices then wrap around the more dominant two-dimensional vortices giving the mostly two-dimensional vortices a highly corrugated appearance in the second column. Some of the three-dimensional vortices farther behind the leading edge assume hairpin or horseshoe shapes for some time. Such hairpin vortices were also found clustered into a patch which grows into a turbulent spot at a lower angle of 10° in [47], where additional examples of three-dimensional vortex interactions can be found. At higher angles partial pairings of two-dimensional vortices and cross-linkings were common occurrences.

Figure 4.12, taken from [48], visualized the vortex system which develops in the tip region of a half wing in accelerated starting flow. The half wing protrudes from the top wall of the wind tunnel into its center, at an angle of attack $\alpha = 40°$. Column 1 shows establishment of a pair of counterrotating conical tip vortices which join together at the front corner of the wing tip and feed into the vertical strands of the leading and trailing edge starting vortices. Additional trailing edge starting vortices develop in column 2 and join the tip vortex, which now assumes a multihelical structure. Column 3 is characterized by strong turbulence and the pulling of trailing edge vortices into the leading edge tip vortex region. Turbulence finally blurs any detail. Clearly, the connectivity of the vortex system is initially preserved but finally cannot be tracked in column 3 due to turbulence. The initial vortex system has been sketched in Fig. 1 of [26].

Figure 4.13 shows the complex vortex system developing over and behind a rectangular wing with aspert ratio 2. The wing is exposed to a steady wind and pitches periodically around the quarter chord axis at angles of attack 0° to 30°. A curious

(Text continues on p. 83)

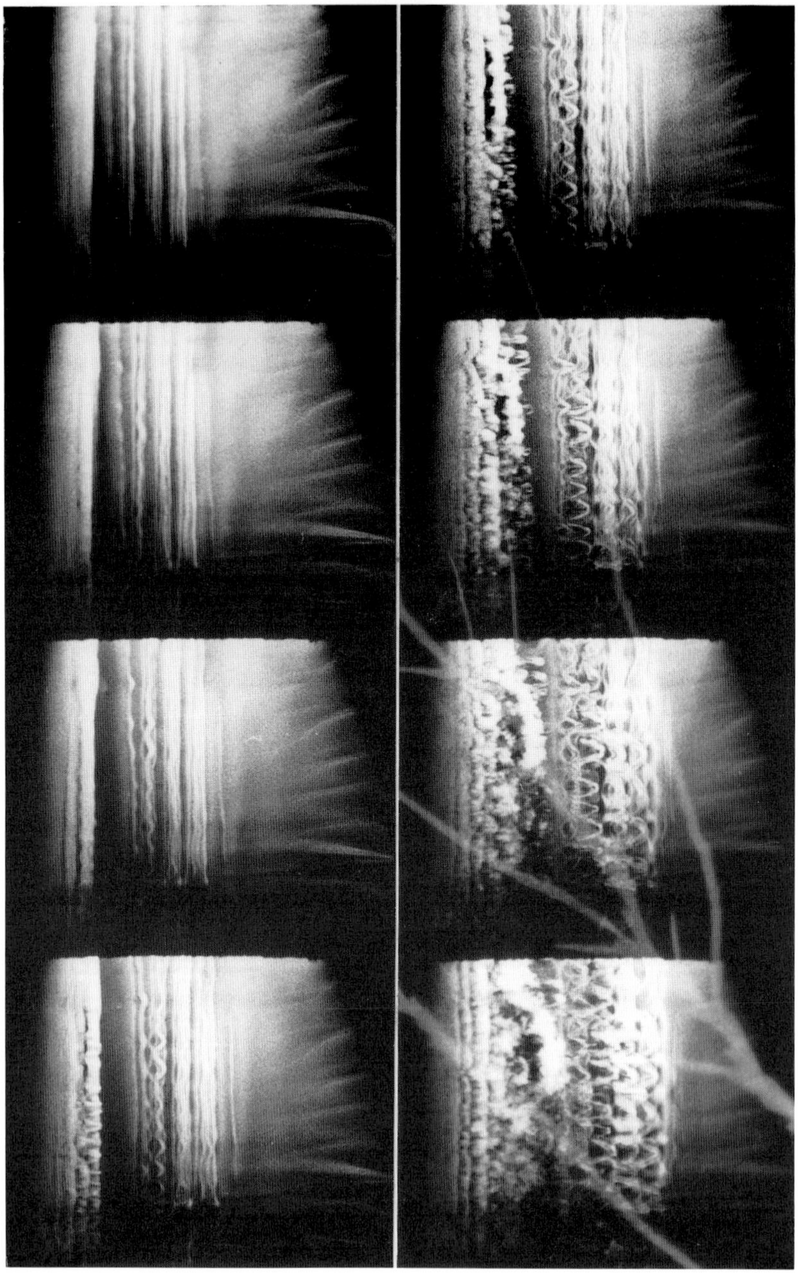

Fig. 4.11. Accelerated starting flow over an airfoil, global view, showing transition. $\alpha = 20°$, $a = 2.4 \text{ m/sec}^2$, $\Delta t = \frac{1}{64}$ sec, Re = 52,000.

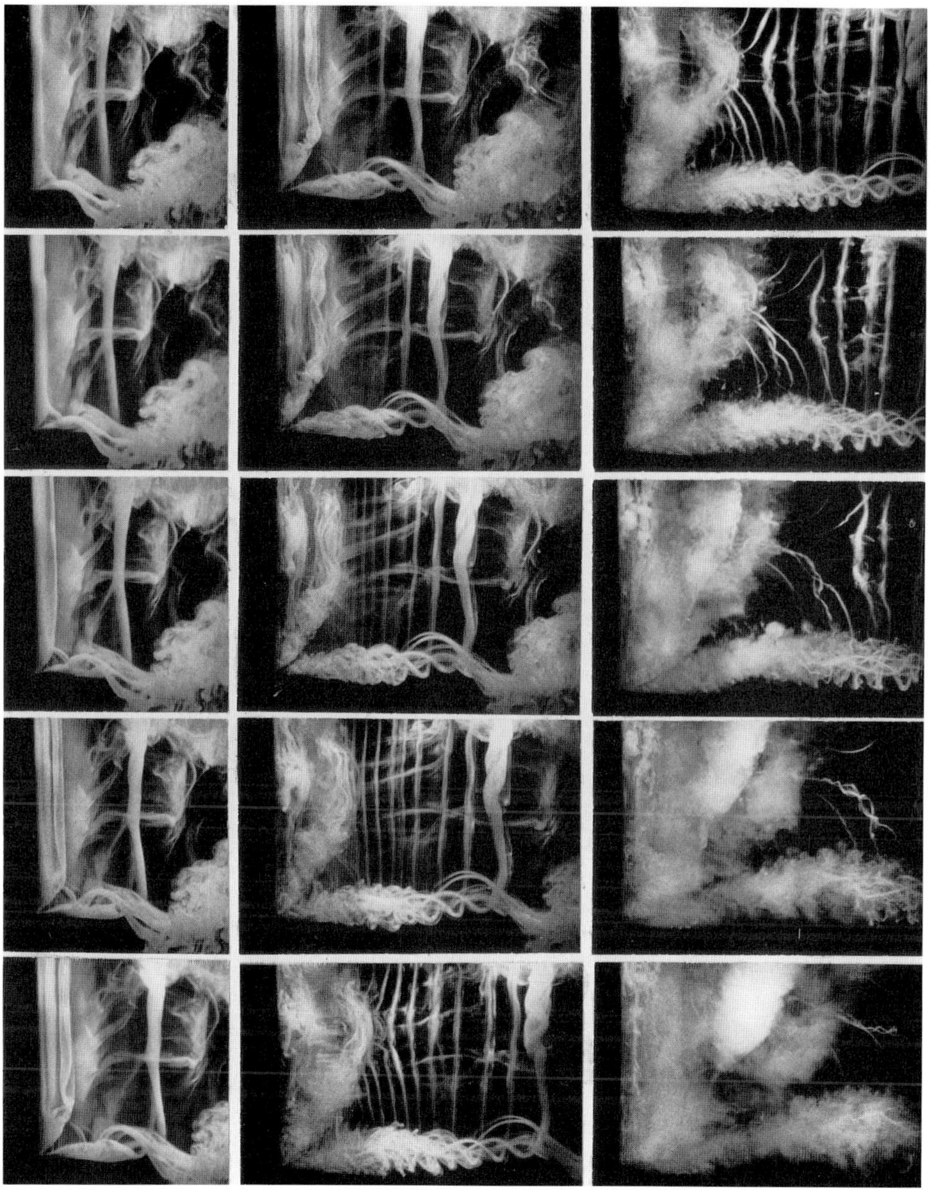

Fig. 4.12. Accelerated starting flow over a half-wing, global view of the wing tip region. $\alpha = 40°$, $c = 15.2$ cm, $a = 2.4$ m/sec^2, $\Delta t = \frac{1}{16}$ sec, Re = 5,200.

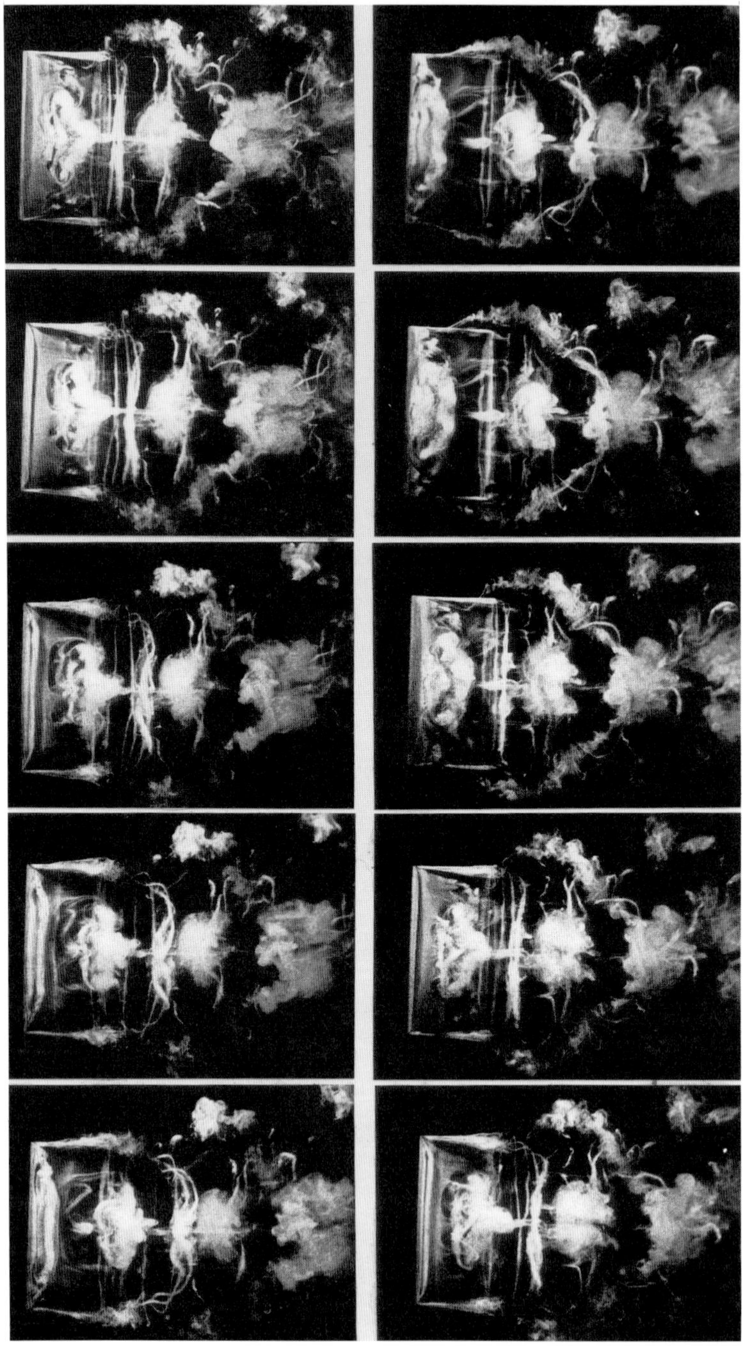

Fig. 4.13. Steady flow over a periodically pitching wing, global view. $U_0 = 61$ cm/sec, $c = 15.2$ cm, span I = 30.5 cm, $f = 2$ Hz, $k = 1.6$, pitching is between 0° to 30°, $\Delta t = \frac{1}{16}$ sec.

mix of two- and three-dimensional vortex developments can be seen in the mid span band over and behind the wing.

Figure 4.14 depicts vortex development over an equilateral delta wing at a 20° angle of attack and after impulsive start of the wing. Development of the leading edge tip vortices and their connectivity to the trailing edge starting vortex can be readily seen. Very similar developments were previously found in accelerated starting flow over delta wings [29], [48]. Also, a similarity of the leading edge tip vortices of Fig. 4.14 with those of Fig. 4.12 should be noted, despite the difference in tilt toward the flow of the front corners of the wings.

Figure 4.15 shows development of leading and trailing edge starting vortices for a delta wing immersed in accelerated starting flow, in closeup, at $\alpha = 60°$. It is an attempt to document the connectivity of the vortex strands closely. It can be seen that the first trailing edge vortex connects to the leading edge vortex. Soon additional trailing edge vortices develop and connect but the exact nature of these junctures are still unresolved. A secondary vortex close to the leading edge is also faintly indicated and is likely to connect to the leading edge vortex close to the upper airfoil corner. In essence, there is an insatiable need to resolve finer details which will not easily be filled in by computer simulation and which show the need for further improvement in flow visualization. It can generally be said that the closer one gets to the turbulent regime the more one loses visualization of connectivity.

Let us finally focus on unsteady configurations which prominently display reconnection processes.

Figure 4.16 shows vortex ring formation behind a delta wing in reverse, pitching from 0° to 60° around the $c/2$ axis [49]. The vortex loop of the trailing edge vortex necks down and in column 2, due to viscous-turbulent annihilation reconnection closes the vortex ring, leaving the remainder of the loop behind. Turbulence unfortunately blurs some of the detail. If the delta wing was mounted with a corner heading into the wing and pitched a series of intertwined trailing edge vortex loops resulted but no vortex ring emerged.

Figure 4.17 shows vortex ring creation for a rectangular wing in a quick pitch-up-down maneuver from 0° to 30° to 0°. In this case the two counterrotating trailing edge vortex loops, which were generated during the pitch-up-down maneuvers, reconnect into a vortex ring downstream of the airfoil. The remaining parts of the trailing edge vortex loops which originally connect to the leading edge corners annihilate each other. Also the leading edge separating vorticity seems to form a vortex ring but it is so turbulent that it gives the appearance of a blob. Vortex ring formation near the leading edge had been shown more clearly in accelerated starting flow in [48].

The first impressive attempts to model a reconnection process in infinite space are reported by Melander and Zabusky [50].

To obtain a more three-dimensional feel for the vortex ring motion of Fig. 4.17 we also took a movie sequence looking obliquely down and downstream into the wind tunnel from the top. The test section was lighted from the right side. This complementary view is shown in Fig. 4.18 where movement of the vortex ring toward the left side wall of the tunnel and simultaneously convection in the downstream direction can be seen. The shadow of the vortex ring on the left side wall can also be seen in frame 4, column 1 and frame 1, column 2.

(Text continues on p. 86)

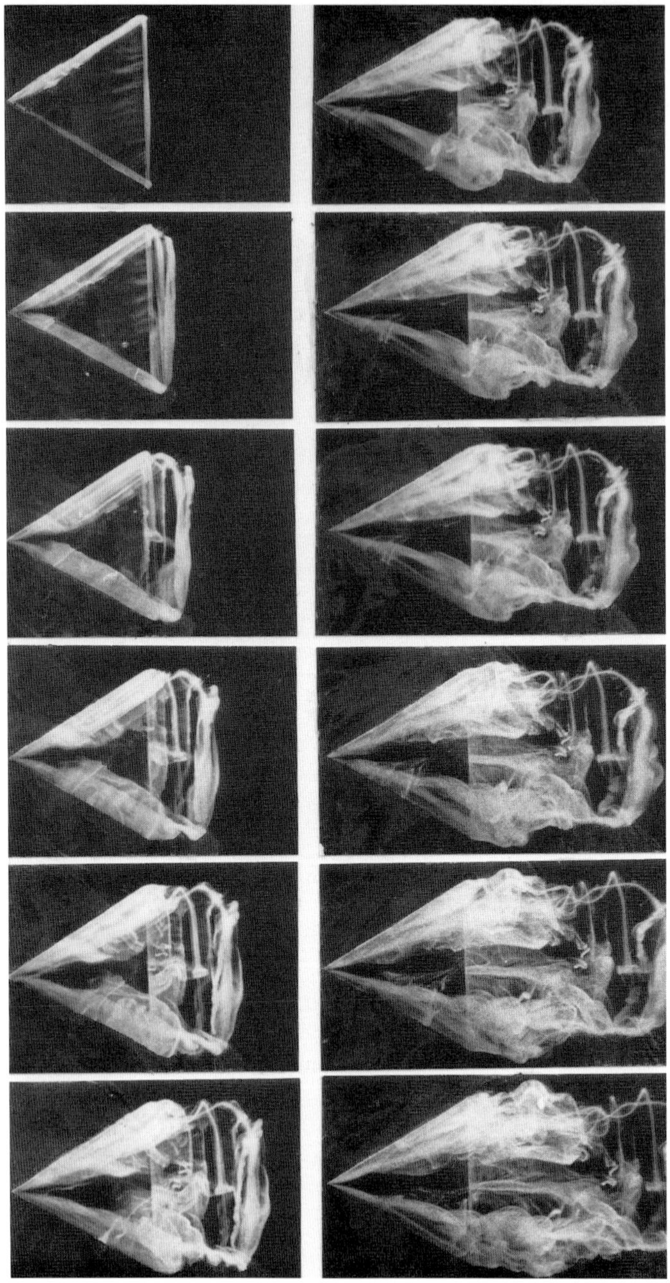

Fig. 4.14. Impulsive starting flow over an equilateral delta wing at $\alpha = 20°$, global view. $U_0 = 39$ cm/sec, side length $c = 15.2$ cm, $\Delta t = \frac{1}{16}$ sec, Re = 3,300.

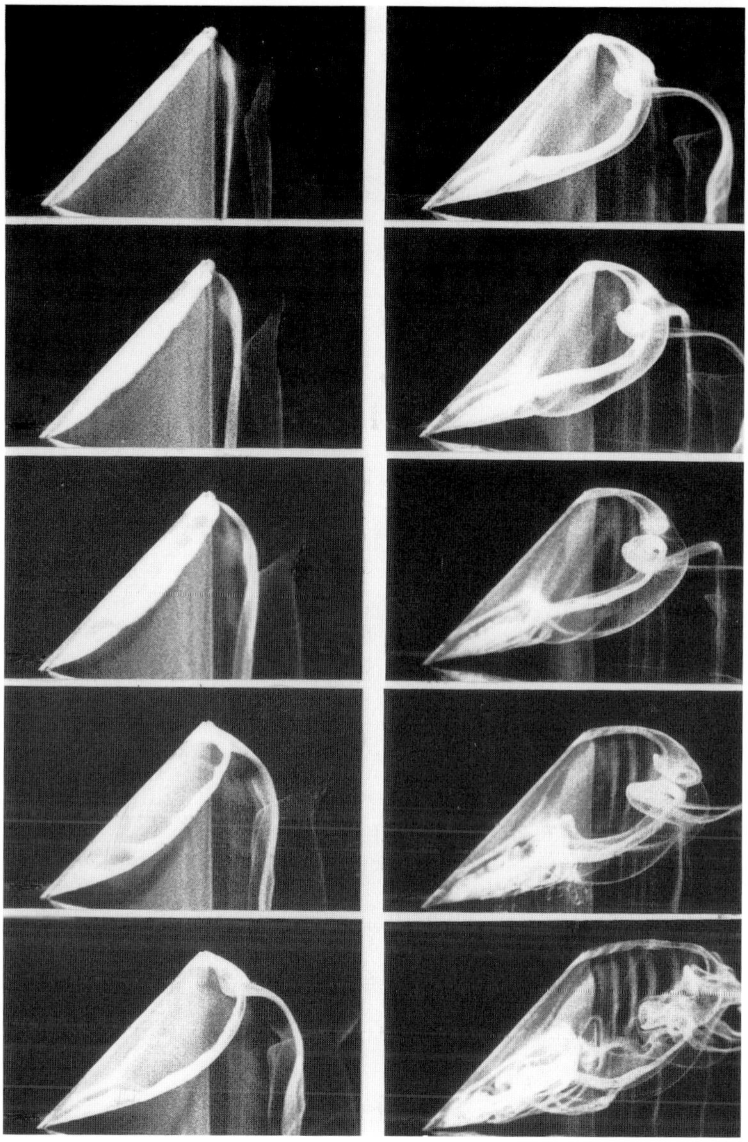

Fig. 4.15. Accelerated starting flow over an equilateral delta wing at $\alpha = 60°$, closeup global view. $c = 15.2$ cm, $a = 2.4$ m/sec^2, $\Delta t = \frac{1}{16}$ sec, Re = 5,200.

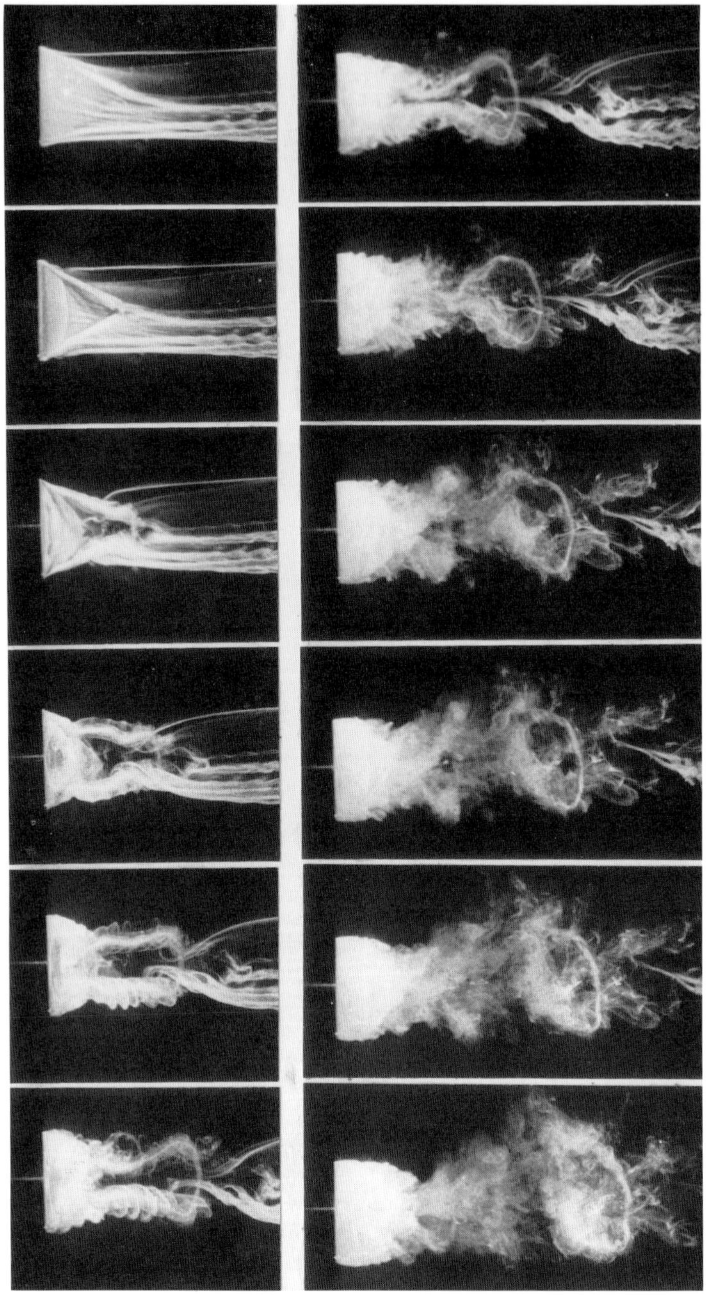

Fig. 4.16. Delta wing in reverse, pitching from 0° to 60° in 0.5 sec. Side length $c = 15.1$ cm, $U_0 = 61$ cm/sec, $\Delta t = \frac{1}{8}$ sec, Re = 5,100, view through the side wall.

As a variation on Figure 4.17 we let the vortex ring impinge on the transparent tunnel wall, as is shown in Fig. 4.19. Impingement leads to an increase in vortex ring diameter during its convection downstream, as the first column of Fig. 4.19 shows

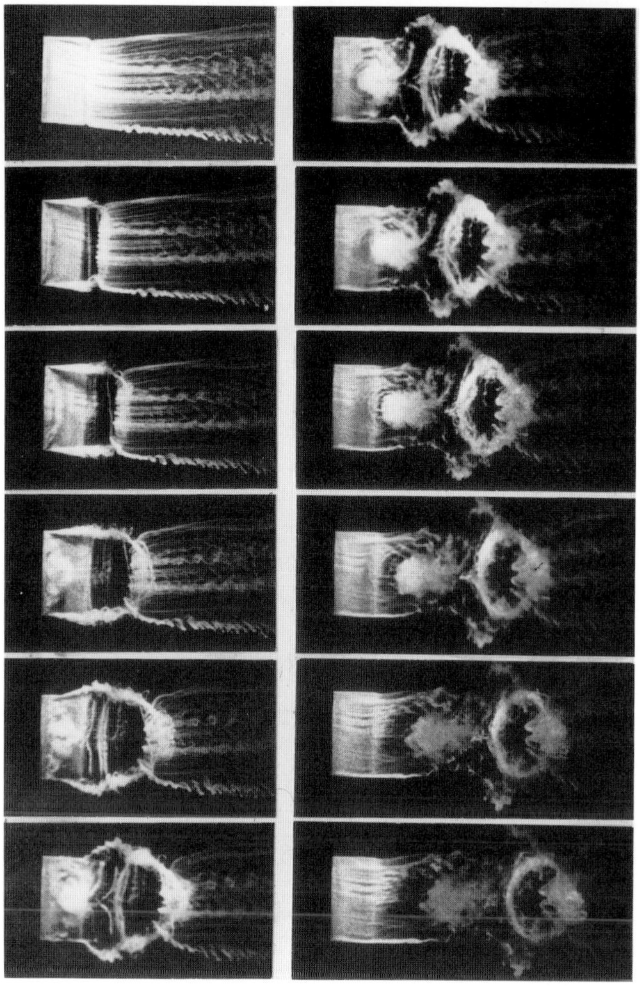

Fig. 4.17. Rectangular wing pitching from 0° to 30° and back to 0° in 0.5 sec. c = 15.1 cm, aspect ratio is 2, U_0 = 61 cm/sec, $\Delta t = \frac{1}{8}$ sec, Re = 5,100, view through the side wall.

and as can be expected from inviscid theory. Later on in column 2 the vortex ring penetrates the boundary layers of the wall and convection of the ring almost stops in frames 4–6 in column 2. At this point the smoke visualization gets hard to interpret since the ring now becomes part of the vortical environment of the wall boundary layer, which is not visualized. The impingement process of a vortex ring on a wall boundary layer represents an interesting modeling problem.

Figure 4.20 focuses on vortex loop formation behind a rectangular airfoil after pitch from 0° to 60° angle of attack in steady flow in side view. The loop necks down, similar to Fig. 4.16, although ring formation is less complete. A complementing oblique top view shown in Fig. 4.21 shows how three-dimensionally curved the loop is, as would undoubtedly also be found by three-dimensional computer modeling. It appears that complementary or stereoscopic visualization of vortex development is a valuable qualitative tool for comprehending three-dimensional vortex development.

(Text continues on p. 90)

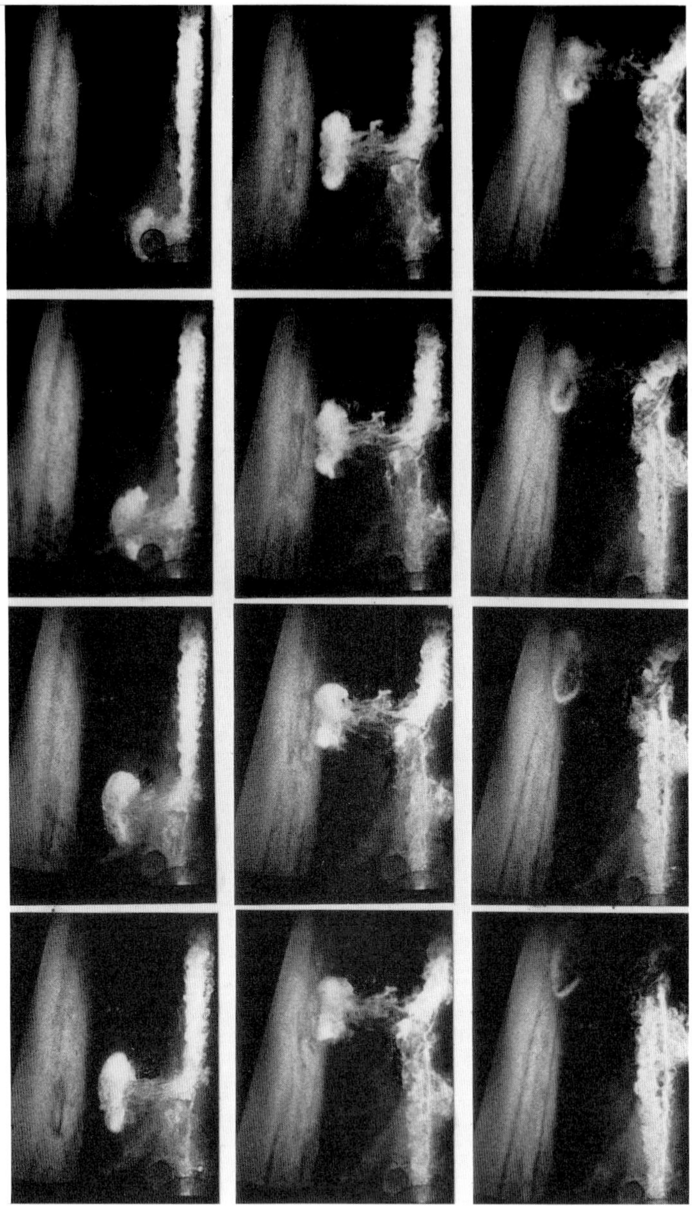

Fig. 4.18. Rectangular wing in pitch as in Fig. 4.17 but viewed through the top wall of the tunnel obliquely downward-downstream.

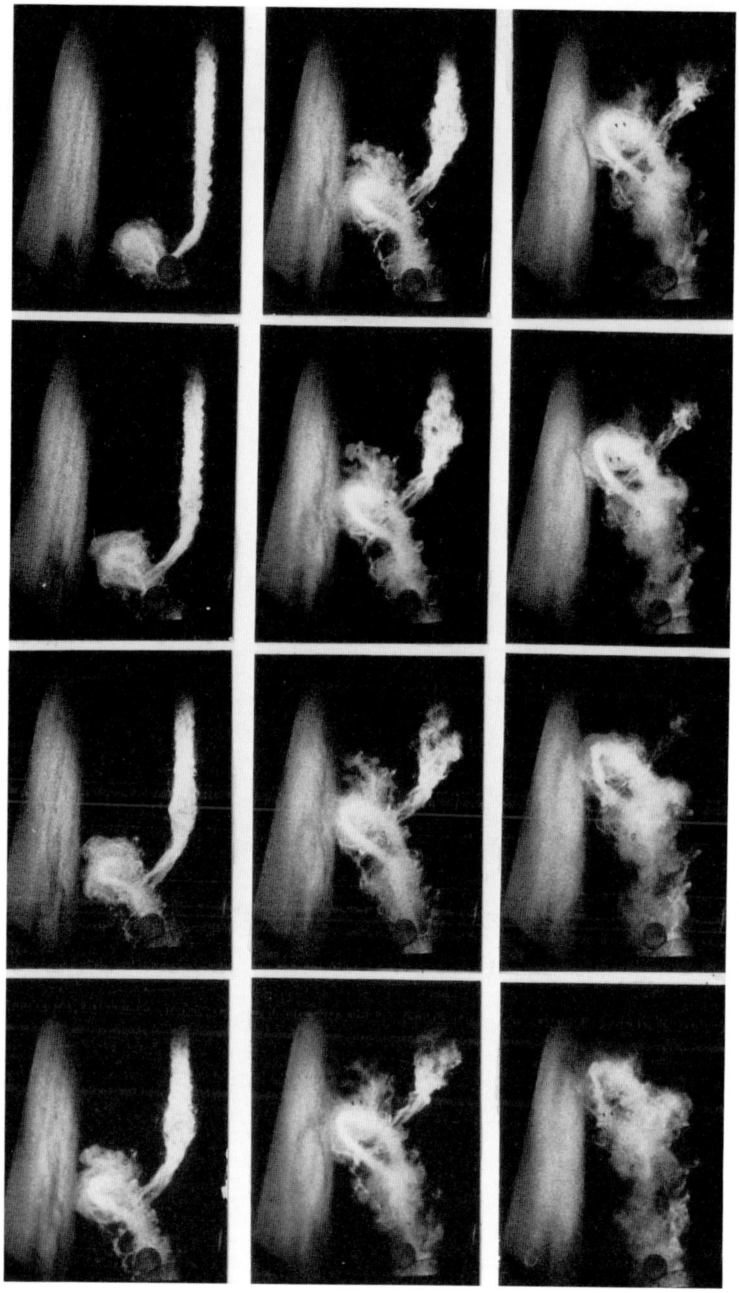

Fig. 4.21. Same process as in Fig. 4.20 but viewed through the top wall of the tunnel obliquely downward-downstream.

5. Conclusions

Two- and three-dimensional physical flow visualizations have been presented as a reference and challenge for computer simulations. Existing computer visualizations are most promising and pave the way for a better understanding of unsteady vortex flows in a not so distant future.

Acknowledgment

The help of Dr. Finaish with Fig. 4.14 is appreciated. An AFOSR contract F49620-84-C-0065 is acknowledged.

References

[1] T. SCHWENK, *Das Sensible Chaos*, Verlag Freies Geistesleben, Stuttgart, 1962; 5th ed., 1980.

[2] H. WERLE, *Hydrodynamic flow visualization*, Ann. Rev. Fluid Mech., 5 (1973), pp. 361–382.

[3] M. VAN DYKE, *An Album of Fluid Motion*, Parabolic Press, Stanford, CA 1982.

[4] T. J. MUELLER, *Flow visualization by direct injection*, in Fluid Mechanics Measurements, Hemisphere, New York, 1983, pp. 307–375.

[5] W. MERZKIRCH, *Flow Visualization*, Academic Press, New York, 1987.

[6] M. GAD-EL-HAK, *Review of flow visualization techniques for unsteady flows*, in Flow Visualization IV, Hemisphere, New York, 1987, pp. 3–14.

[7] G. S. SETTLES, *Modern developments in flow visualization*, AIAA J., 24 (1986), pp. 1313–1323.

[8] P. FREYMUTH, *Progress in visualization unsteady separation*, in Proc. Workshop II on unsteady separated flow, Colorado Springs, Frank J. Seiler Research Laboratory, 1988, pp. 197–210.

[9] N. J. ZABUSKY, *Computational synergetics*, Physics Today, 37 (1984), pp. 36–46.

[10] A. LEONARD, *Computational three-dimensional incompressible flows with vortex elements*, Ann. Rev. Fluid Mech., 17 (1985), pp. 523–559.

[11] N. J. ZABUSKY, *Visualizing mathematics: evolution of vortical flows*, Phys. D, 18 (1986), pp. 15–25.

[12] ———, *Grappling with complexity*, Physics Today, 40 (1987), pp. 25–27.

[13] M. V. MELANDER, E. A. OVERMAN, AND N. J. ZABUSKY, *Computational vortex dynamics in two and three dimensions*, Appl. Numer. Math., 3 (1987), pp. 59–80.

[14] M. V. MELANDER, N. J. ZABUSKY, AND J. C. MCWILLIAMS, *Symmetric vortex merger in two dimensions: causes and conditions*, J. Fluid Mech., 195 (1988), pp. 303–340.

[15] K. IZUMI AND K. KUWAHARA, *Unsteady flow field, lift and drag measurements of impulsively started elliptic cylinder and circular-arc airfoil*, paper AIAA-83-1711, American Institute of Aeronautics and Astronautics, NY, 1983.

[16] O. INOUE AND K. KUWAHARA, *A new approach to flow problems past a screen*, paper AIAA-85-1645, American Institute of Aeronautics and Astronautics, NY, 1985.

[17] Q.-Y. SHEEN, *Potential flow analysis of unsteady Joukowski airfoil in the presence of discrete vortices*, Ph.D. thesis, University of Colorado, Boulder, CO, 1986.

[18] H. J. LUGT AND H. J. HAUSSLING, *Laminar flow past an abruptly accelerated elliptic cylinder at 45° incidence*, J. Fluid Mech., 65 (1974), pp. 711–734.

[19] H. J. LUGT AND S. OHRING, *Rotating elliptic cylinders in a viscous fluid at rest or in a parallel stream*, J. Fluid Mech., 79 (1977), pp. 127–156.

[20] R. R. LEBEN, *Multigrid calculations of boundary-fitted orthogonal coordinates with applications to unsteady flows*, Ph.D. thesis, University of Colorado, Boulder, CO, 1986.

[21] K. GUSTAFSON AND R. R. LEBEN, *Robust multigrid computation and visualization of separation and evolution in aerodynamic flows*, paper AIAA-88-3605, First National Fluid Dynamics Congress, American Institute of Aeronautics and Astronautics, NY, 1988.

[22] K. OHMI, *Etude de la formation du sillage autour d'un profil en oscillation*, Ph.D. these, Université de Poitiers, Poitiers, France, July 1987.

[23] M. R. VISBAL, *Evaluation of an implicit Navier–Stokes solver for some unsteady separated flows*, paper AIAA-86-1053, American Institute of Aeronautics and Astronautics, NY, 1986.

[24] VISBAL AND J. S. SHANG, *Numerical investigation of the flow structure around a rapidly pitching airfoil*, paper AIAA-87-1424, American Institute of Aeronautics and Astronautics, NY, 1987.

[25] T. TAMURA, K. TSUBOI, AND K. KUWAHARA, *Numerical simulation of unsteady flow patterns around a vibrating cylinder*, paper AIAA-88-0128, American Institute of Aeronautics and Astronautics, NY, 1988.

[26] P. FREYMUTH, F. FINAISH, AND W. BANK, *The wing tip vortex system in a starting flow*, Zeitschrift für Flugwissenschaften, 10 (1986), pp. 116–118.

[27] ———, *Visualization of wing tip vortices in unsteady and steady wind*, paper AIAA-86-1096, American Institute of Aeronautics and Astronautics, NY, 1986; also in shorter form in AIAA J. Aircraft, 23 (1986), pp. 730–733.

[28] ———, *Parametric exploration of unsteady wing tip vortices*, in Flow Visualization IV, Hemisphere, New York, 1987, pp. 419–424.

[29] ———, *Further visualization of combined wing tip and starting vortex systems*, AIAA J., 25 (1987), pp. 1153–1159.

[30] M. GAD-EL-HAK, *Unsteady separation on lifting surfaces*, Appl. Mech. Rev., 40 (1987), pp. 441–453.

[31] P. FREYMUTH, W. BANK, AND M. PALMER, *Use of titanium tetrachloride for visualization of accelerating flow around airfoils*, Flow Visualization III, Hemisphere, New York, 1985, pp. 99–105.

[32] P. FREYMUTH, *Vortices*, in Handbook of Flow Visualization, Hemisphere, New York, 1989, pp. 459–479.

[33] ———, *Air flow visualization using titanium tetrachloride*, Advances in Fluid Mechanics Measurements, M. Gad-el-Hak, ed., Lecture Notes in Engineering, 45 (1989), pp. 1–31.

[34] C. C. LANDRETH, R. J. ADRIAN, AND C. S. YAO, *Double pulsed particle image velocimeter with directional resolution for complex flows*, Experiments in Fluids, 6 (1988), pp. 119–128.

[35] P. MONNET AND M. CONTANCEAU, *The use of visualization as a guide in the numerical determination of the flow around an abruptly accelerated elliptic cylinder or airfoil*, in Flow Visualization III, Hemisphere, New York, 1985, pp. 363–368.

[36] P. FREYMUTH, F. FINAISH, AND W. BANK, *Visualization of the vortex street behind a circular cylinder at low Reynolds numbers*, Phys. Fluids, 29 (1986), pp. 1321–1323.

[37] P. FREYMUTH, *Propulsive vortical signature of plunging and pitching airfoils*, AIAA J., 26 (1988), pp. 881–883. (See paper AIAA-88-0323 for more details.)

[38] ———, *On transition in a separated laminar boundary layer*, J. Fluid Mech., 25 (1966), pp. 683–704.

[39] ———, *The vortex patterns of dynamic separation: a parametric and comparative study*, Progr. Aerospace Sci., 22 (1985), pp. 161–208.

[40] P. FREYMUTH, W. BANK, AND F. FINAISH, *Surveying unsteady flows by means of movie sequences—a case study*, ASME Internat. Symposium on Physical and Numerical Flow Visualization, Albuquerque, NM, June 24–26, 1985; FED, 22 (1985), pp. 27–37.

[41] F. FINAISH, *Experimental study of two-dimensional vortex patterns for impulsively started bodies in comparison with other configurations*, Ph.D. Thesis, University of Colorado, Boulder, CO, 1987.

[42] F. FINAISH AND P. FREYMUTH, *Aerodynamic visualization for impulsively started airfoils*, La Recherche Aerospatiale, 4 (1988), pp. 55–62 (English edition).

[43] F. FINAISH, P. FREYMUTH, AND W. BANK, *Starting flow over spoilers, double steps and cavities*, J. Fluid Mech., 168 (1986), pp. 383–392.

[44] J. N. ADLER AND M. W. LUTTGES, *Three-dimensionality in unsteady flow about a wing*, paper AIAA-85-0132, American Institute for Aeronautics and Astronautics, NY, 1985.

[45] M. GAD-EL-HAK AND C. M. HO, *Unsteady vortical flow around three-dimensional lifting surfaces*, AIAA J., 24 (1986), pp. 713–721.

[46] F. M. PAYNE, T. T. NG, AND R. C. NELSON, *Visualization and flow survey of leading edge vortex structure on delta wing planforms*, paper AIAA-86-0330, 1986.

[47] P. FREYMUTH, F. FINAISH, AND W. BANK, *Three-dimensional vortex patterns in a starting flow*, J. Fluid Mech., 161 (1985), pp. 239–248.

[48] P. FREYMUTH, *Visualizing the combined system of wing tip and starting vortices*, TSI Flow Lines, Premier Issue, May 1986.

[49] ———, *Visualizing the connectivity of vortex systems for pitching wings*, J. Fluids Engineering, III (1988), pp. 217–220.

[50] M. V. MELANDER AND N. J. ZABUSKY, *Entanglement produces an apparent reconnection of 3D-vortex tubes*, Phys. Rev. Lett., submitted.

[51] T. MAXWORTHY, *The fluid dynamics of insect flight*, Ann. Rev. Fluid Mech., 13 (1981), pp. 329–350.

[52] M. LEE AND C. M. HO, *Vortex dynamics of delta wings*, Lecture Notes in Engineering, 46 (1989), pp. 365–427.

FOUR PRINCIPLES OF VORTEX MOTION

5

Karl E. Gustafson

Introduction

To aid in the understanding of the vortex motions of a viscous incompressible fluid flow, I would like to advance here four (new) principles of vortex motion. My intention [1] in formulating these four principles is to provide the beginnings of a framework for a theory of vortex dynamics. These four principles will be illustrated by examples from cavity flows and airfoil flows, principally the former. Extension and validation of these four principles, suitably modified to three dimensions, should follow.

So much has been written on vortex motion. If one takes a wide view, e.g., as in Lugt [1], all of fluid dynamics, hence all of its literature, is about vortex motion. Much has been learned and applied in countless important ways. Name almost any field of physical or biological science and you will find fluid dynamics, and hence vortex motion. The beautiful recent book by Lugt [1], and its predecessor by Van Dyke [2], are sufficient testimony to this point of view.

A vast literature of fluid dynamics has accumulated over the last several hundred years. Some of the recent treatises are those of Ladyzhenskaya [3], Batchelor [4], Serrin [5], Schlichting [6], Temam [7], Lamb [8], Joseph [9], and Peyret and Taylor [10]. There are numerous others I cannot mention. And there is no way in this chapter to discuss even the main discoveries, theorems, facts, and observations known of fluid dynamics today.

What I do wish to present here is the following *framework for a theory of vortex motion* [11], [12]. By its very name, vortex motion involves unsteady (as contrasted to steady) fluid dynamics. I will restrict my attention to the viscous incompressible Navier–Stokes equations

$$\frac{\partial u}{\partial t}=\frac{1}{\mathrm{Re}}\Delta u-(u\cdot\nabla)u-\nabla p, \qquad \text{(0.1)}$$

$$\operatorname{div} u=0$$

in a region Ω. I have assumed constant density $\rho = 1$, no body forces, i.e., $f = 0$. Usually I have in mind Dirichlet, e.g., no slip, boundary conditions, and a flow in which the initial condition is quite simple, e.g., an impulsive or accelerated start, the flow then tending to a much more complex final condition. There is no temperature dependence, for putting that in allows for all kinds of complicated dynamics which would mask the fundamental features I want to stress here.

I will look at the flow as occurring in four stages:

(1) Generation of vortex motion;
(2) Evolution of vortex motion;
(3) Dynamics of vortex motion;
(4) Limits of vortex motion.

Of course these stages overlap one another throughout the flow. For example, in any real flow in which the initial and final states are significantly different, vortex motion is being generated or regenerated throughout the evolving flow history. Dynamics is constantly taking place, as are spatial limiting features. Nonetheless there is some clarification obtained by thinking, first, in terms of the initial generation of the individual vortex motions, second, the basic large-scale development of vortex structures within the flow, third, those especially interesting intervals in which important pairwise vortex interactions take place, and fourth, the long-time tendency of the aggregate of vortex motions toward steadiness, periodicity, aperiodicity, or chaos.

As I discuss these four stages, or if you wish, from these four viewpoints or perspectives toward a flow development, I will advance four (new) principles of vortex motion. Although very definitely interrelated, I will bring them out one by one, one for each of the above four stages. These are generic principles, not necessarily applicable to all flows, but common to many. Stated briefly:

Principle 1. (Space Filling). The vortex motions will continue to be generated to fill up all available (e.g., reachable) region space.

Principle 2. (Self-Organizing). The vortex motions will self-organize according to local (e.g., semigroup) properties and to global (e.g., self-consistent) constraints.

Principle 3. (Parity Adjusting). The vortex motions will adjust the (e.g., information) content of the flow according to parity rules governing acceptable (e.g., bifurcation) state selection.

Principle 4. (Spectrally Refining). Within the above three principles and their limits lie the mechanisms which guide the flow through spectral refinement to its final (e.g., steady, periodic, aperiodic, turbulent) states.

The figures that follow will illustrate the above four flow stages for two flows: driven cavity flow and flow about an airfoil. I will refer to these flow histories, and indirectly to other known flow histories, to bring out the four principles formulated above. Each of these (new) principles provides a useful context for further discussion of a theory of vortex motion.

The references are broken into four divisions. After the general references and particular references are listed references for cavity flow and for airfoil flow. These are all meant to be helpful, indicative, but far from complete. Essentially I have tried to pick key recent publications, from which more extensive bibliographies may be traced.

(Text continues on p. 114)

Fig. 5.1(a). $t = 1$ to 12 seconds.

Fig. 5.1. Vortex generation, evolution, dynamics, and limits in lid driven cavity flow at Reynolds number Re = 10,000 and aspect ratio A = depth/width = 2. The lid moves uniformly to the left. 5.1(a)–(h) show the developing vortex dynamics from $t = 0$ through $t = 96$ (nondimensionalized) sec, at 12-second intervals. 5.1(i) and (j) show the final oscillation achieved for $336 < t \leqq 360$ on this 40×80 computational grid (Gustafson and Halasi [22]). A further study (Goodrich, Gustafson, and Halasi [23]) on a 48×96 grid showed that more time is needed to achieve a fully periodic asymptotic state. 5.1(k)–(n) show the stream function surface, normalized velocity plots, kinetic energy surface, and vorticity surface for that state at time $t = 1800.25$. The period was found to be $3.48 \leqq T \leqq 3.50$ on that grid. The lid moves uniformly to the right.

Fig. 5.1(b). t = 13 to 24 seconds.

Fig. 5.1(c). $t = 25$ to 36 seconds.

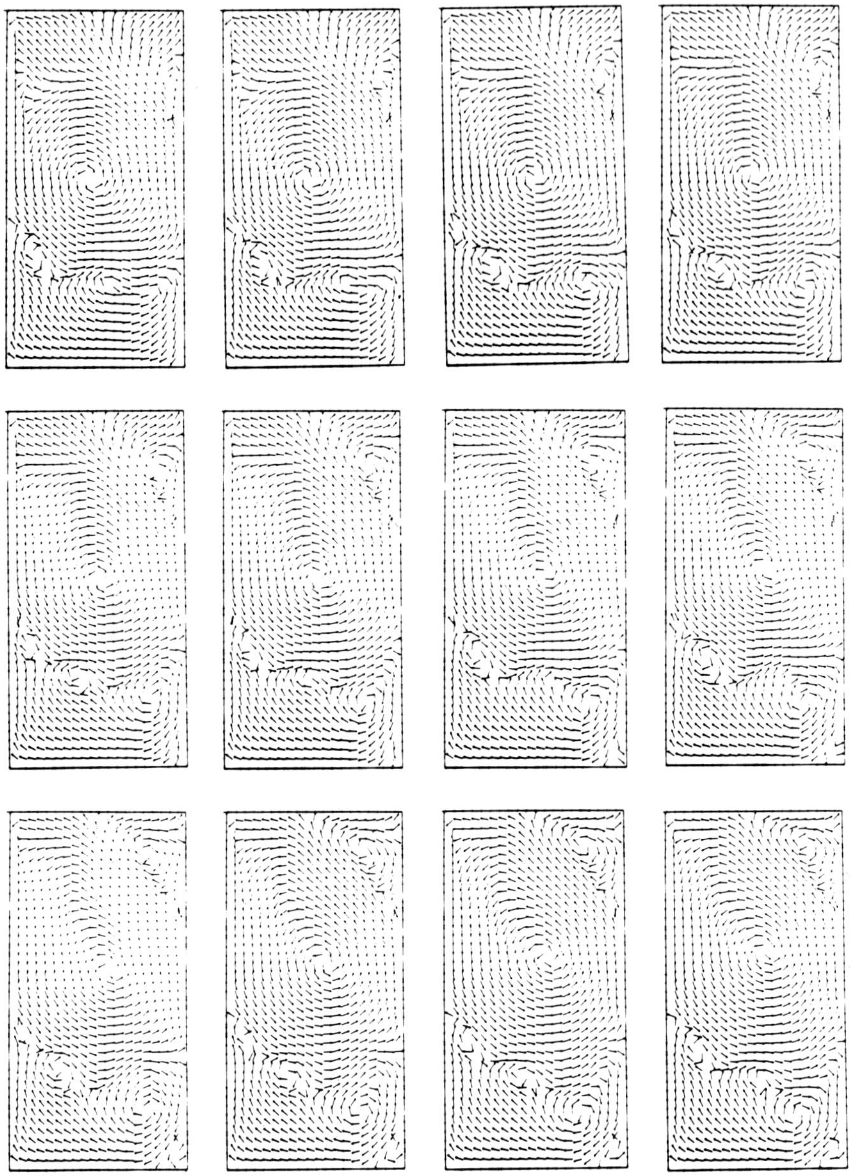

Fig. 5.1(d). t = 37 to 48 seconds.

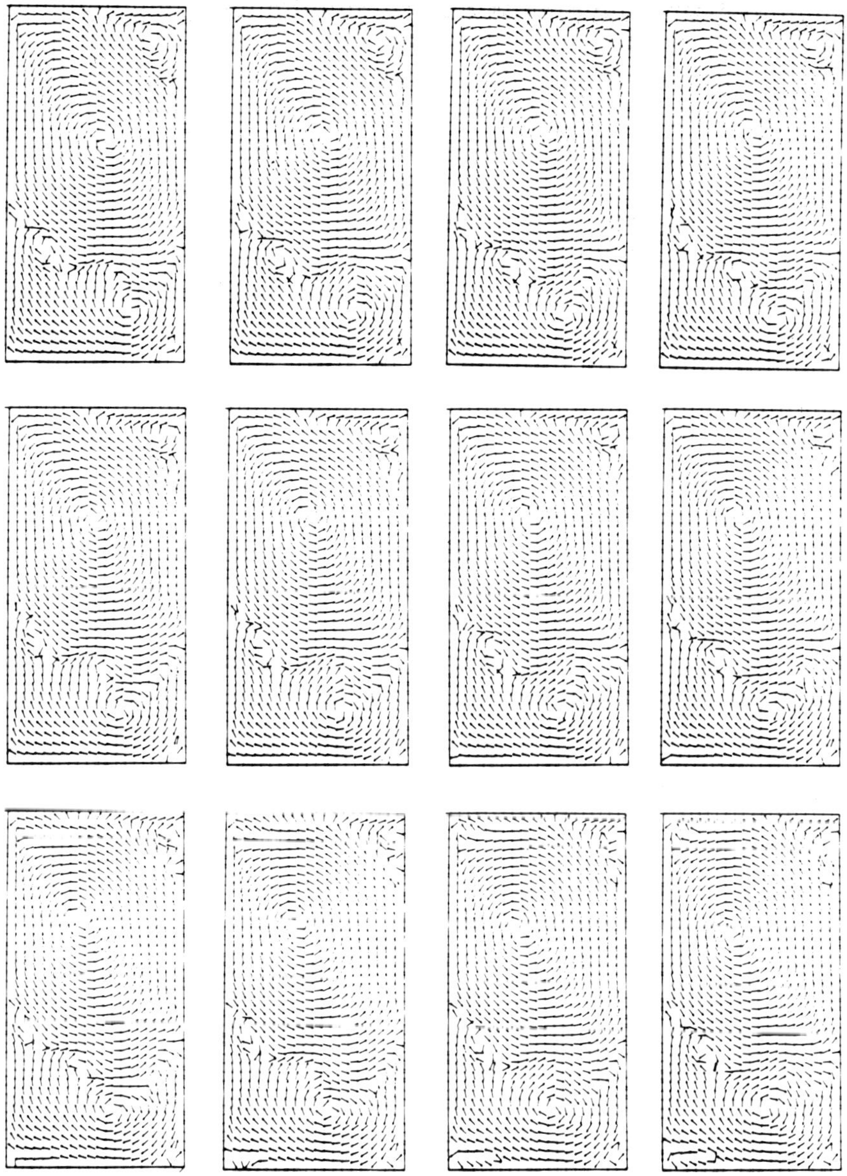

Fig. 5.1(e). t = 49 to 60 seconds.

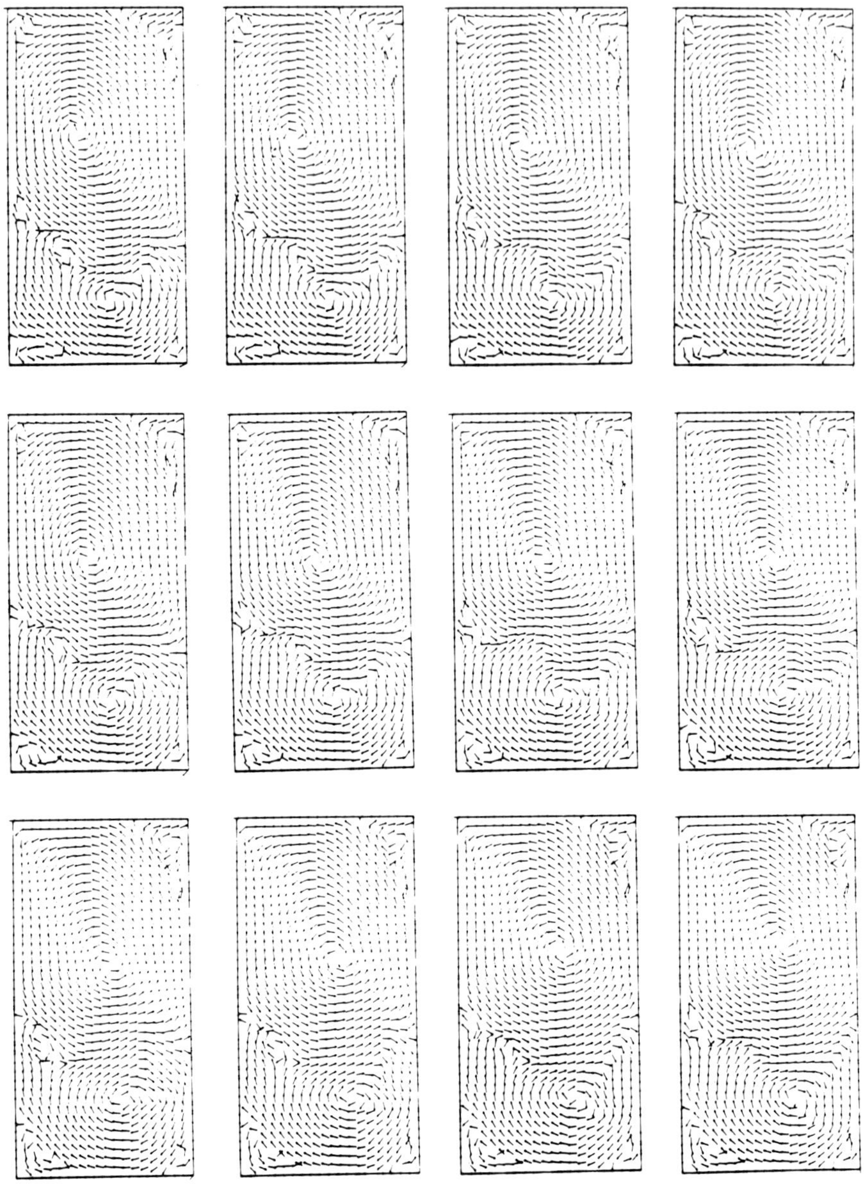

Fig. 5.1(f). $t = 61$ to 72 seconds.

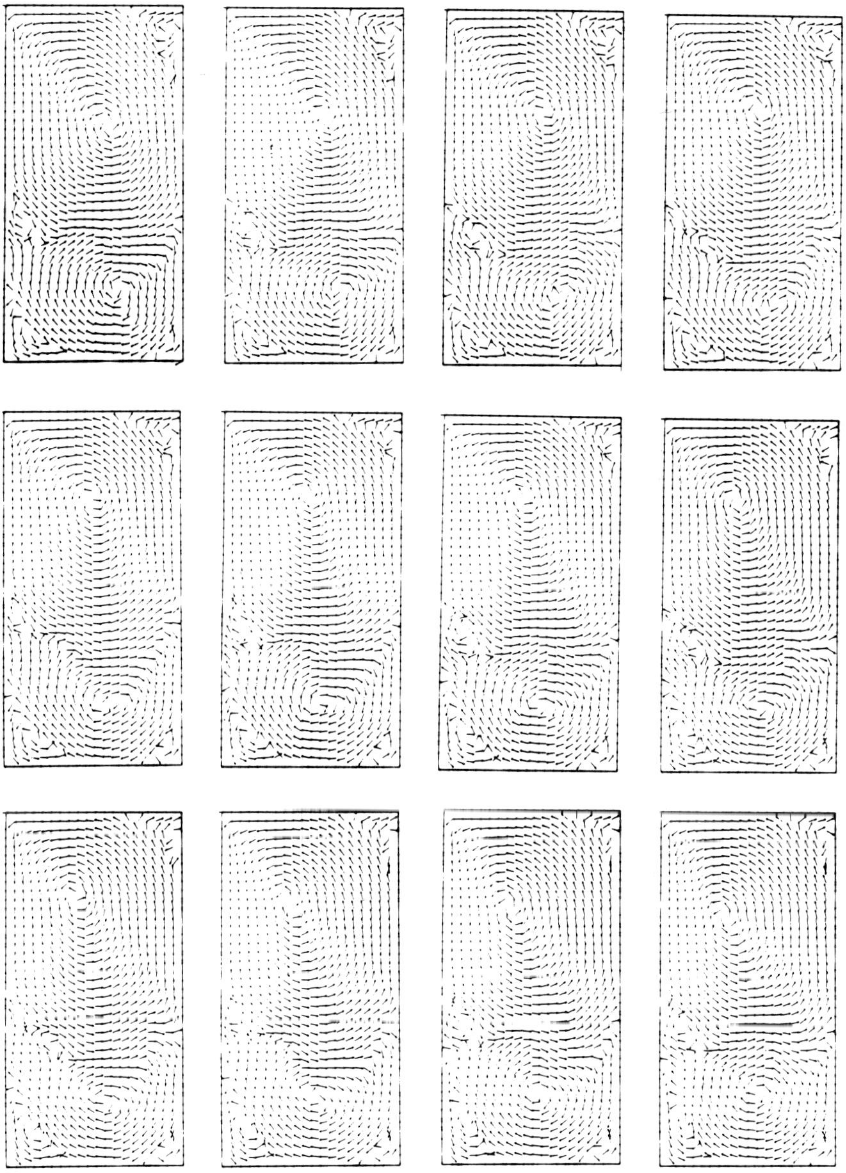

Fig. 5.1(g). t = 73 to 84 seconds.

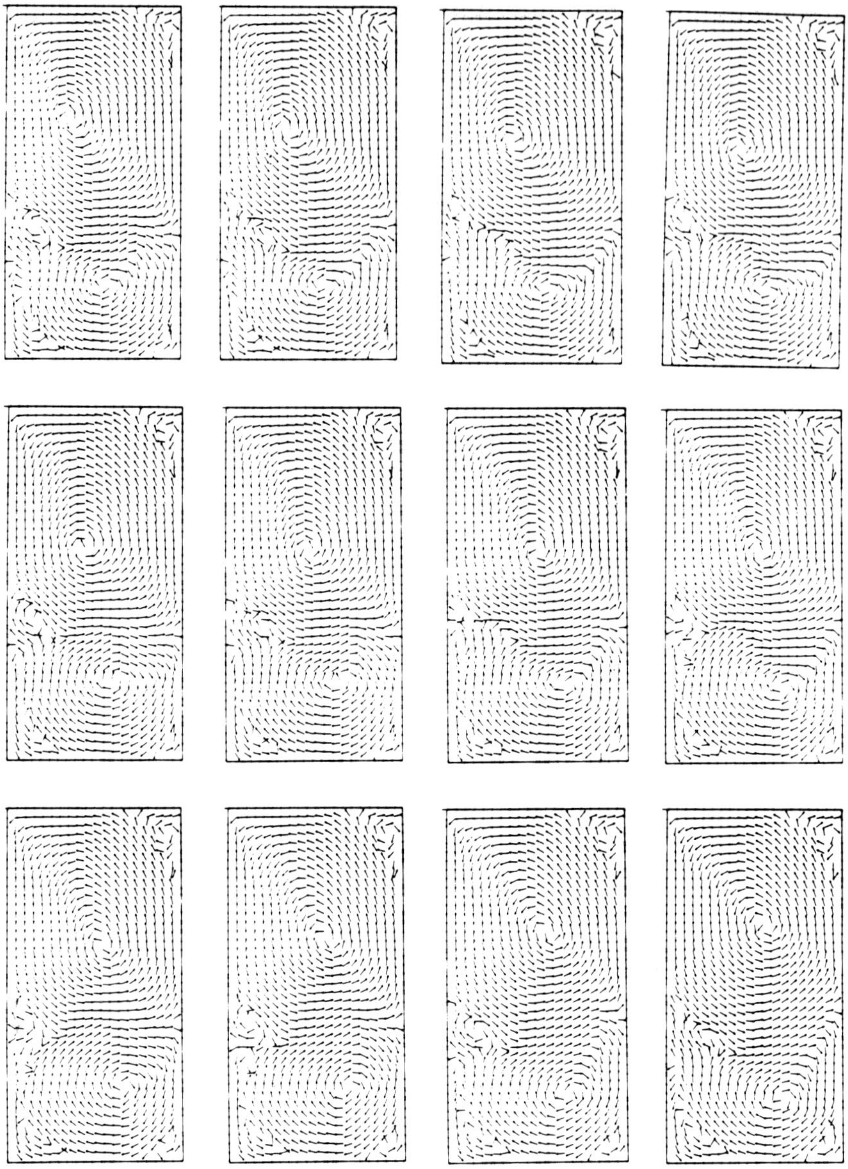

Fig. 5.1(h). t = 85 to 96 seconds.

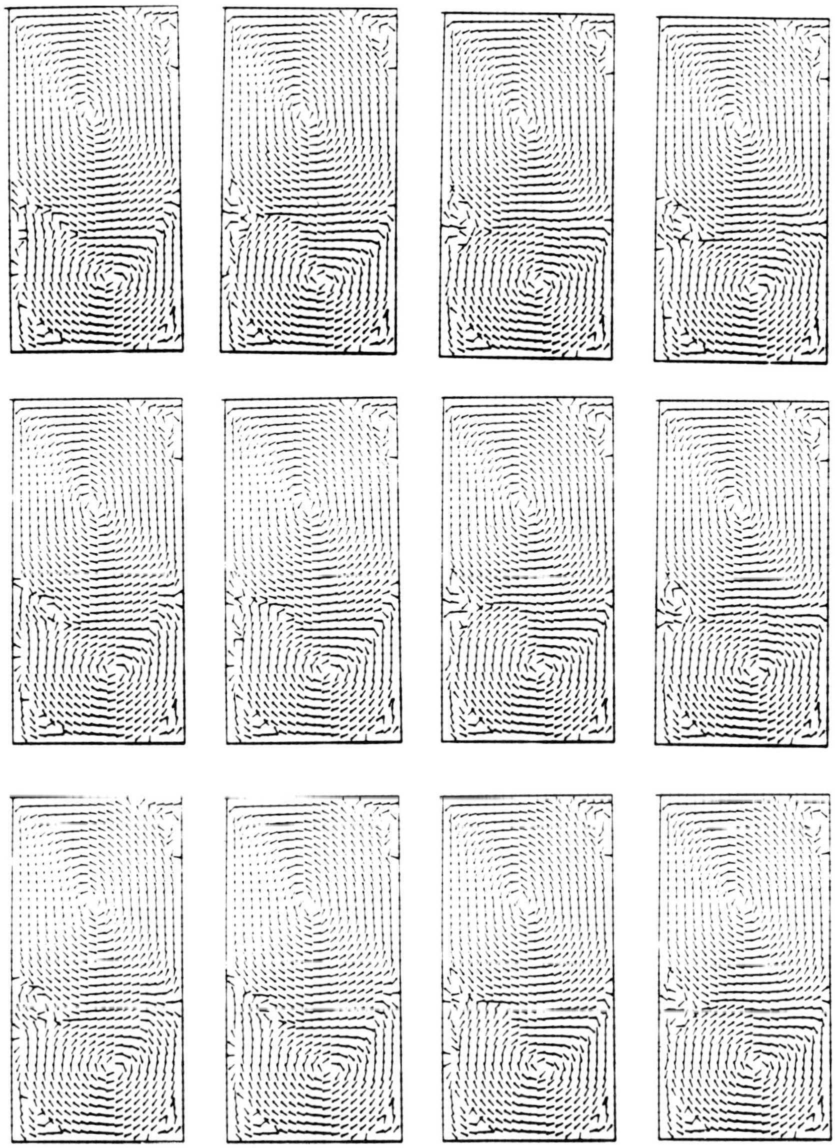

Fig. 5.1 (i). t = 337 to 348 seconds.

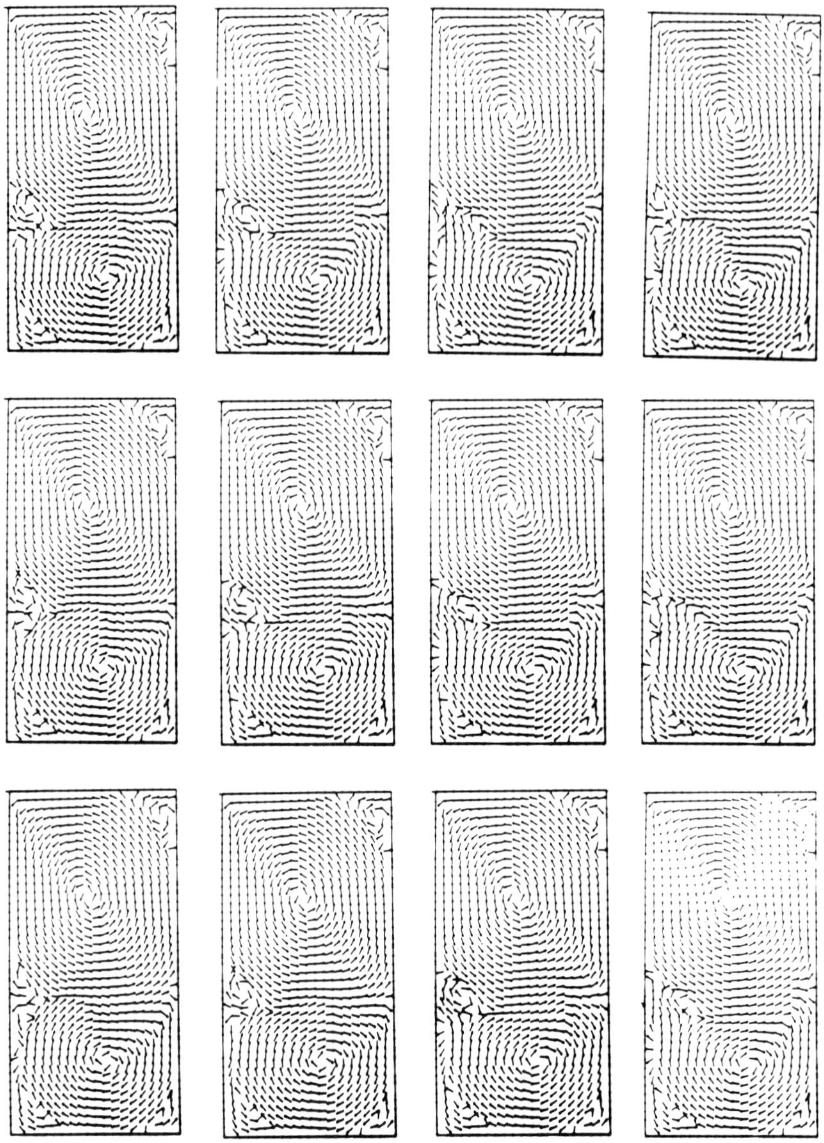

Fig. 5.1(j). t = 349 to 360 seconds.

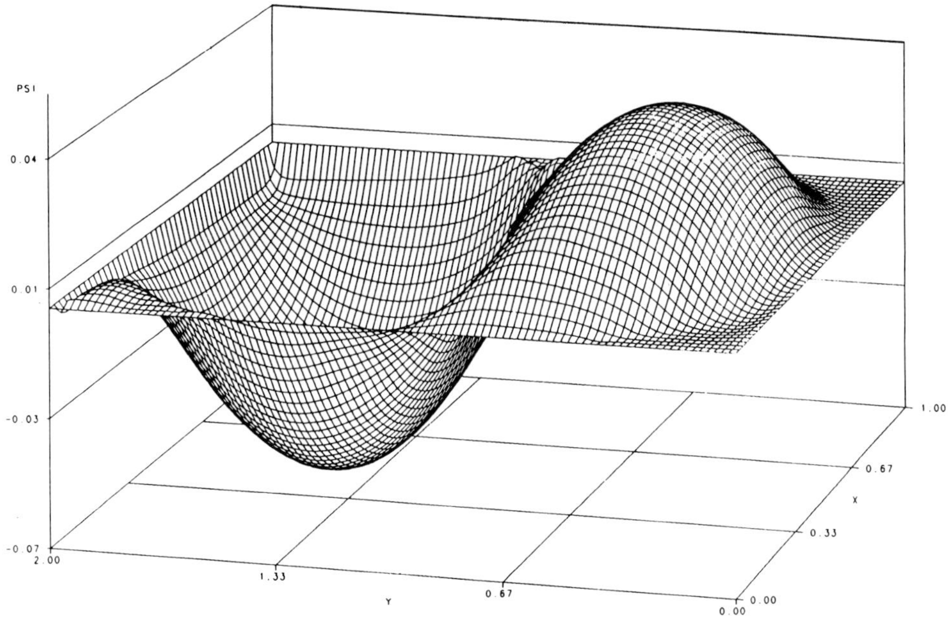

Fig. 5.1(k). Stream function surface.

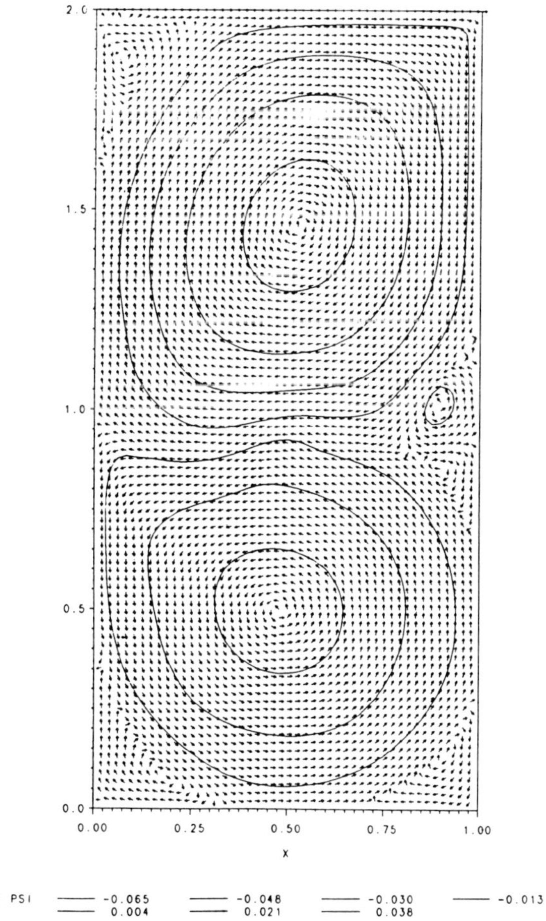

Fig. 5.1(l). Normalized velocity plots.

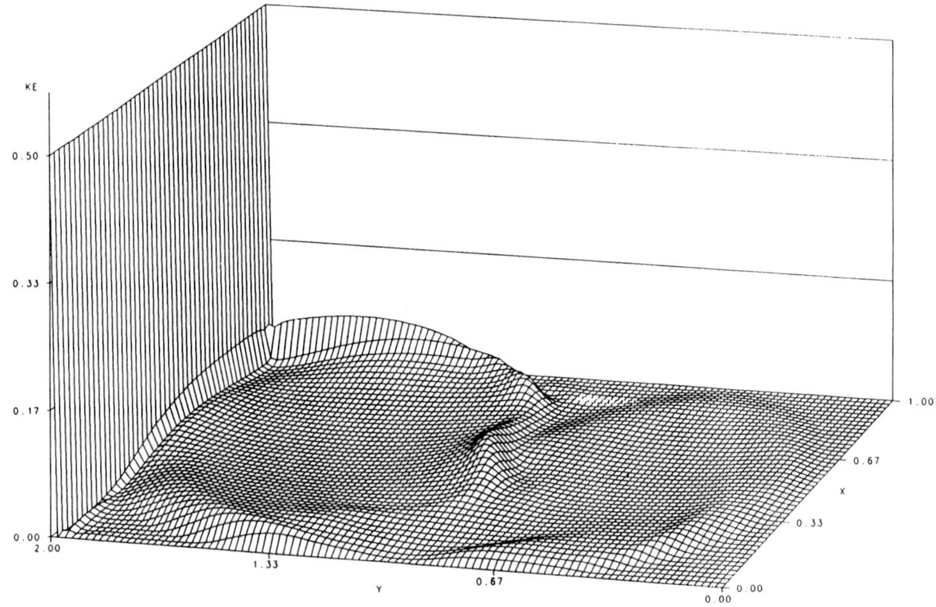

Fig. 5.1(m). Kinetic energy surface.

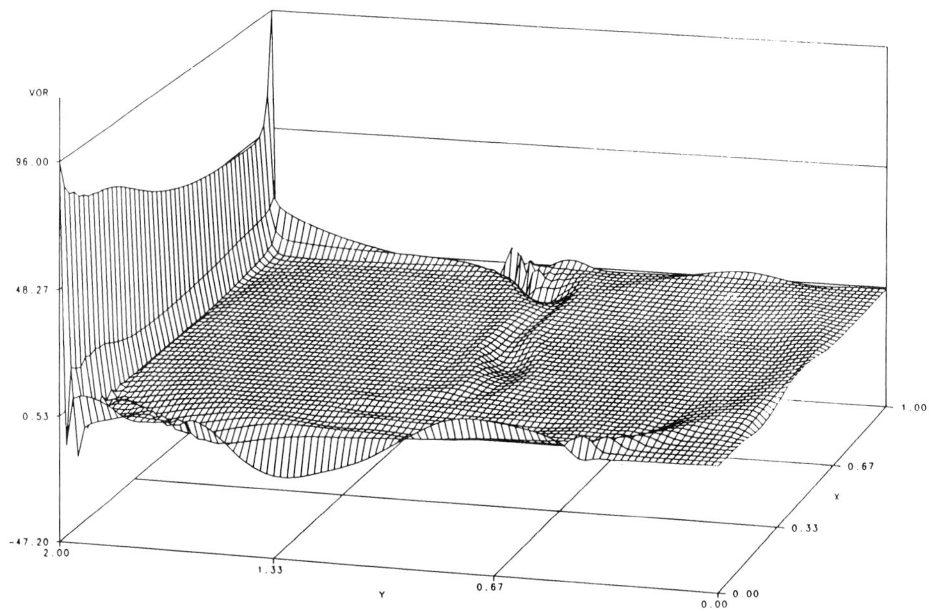

Fig. 5.1(n). Vorticity surface.

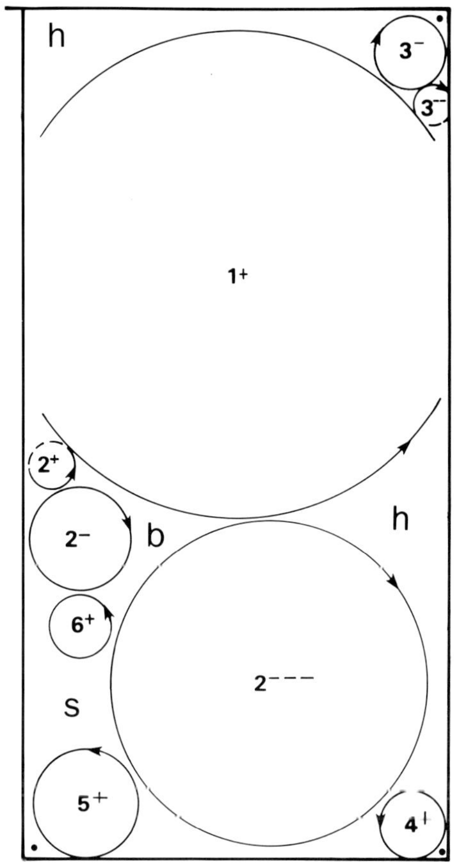

Fig. 5.2. Schematic of lid-driven cavity flow showing eventual vortex states.

Illustrating Principle 1 (Space Filling).
- (i) Separation Compensation: 2^-, 3^-, 4^+, 5^+, 2^+.
- (ii) Corner Sequences: 3^-, 4^+, 5^+, $\cdot$, $\cdot$, $\cdot$.
- (iii) Conic Tendency: 1^+, 2^-, 2^{---}, 3^-, 4^+, 5^+.
- (iv) Burst Providing: 2^-, 6^+, 2^+, b.

Illustrating Principle 2 (Self-Organizing).
- (i) Semigroup Properties: For example, the $2^{---\cdots---}$ sequence, each evolution therein.
- (ii) Self-consistent Properties: For example, the basic two modes 1^+ and 2^{---}, constrained in cavity.
- (iii) Self-similarity Properties: For example, the corner sequences $4^+\cdots$ and $5^+\cdots$.
- (iv) Stability Exchanging: For example, the construction of 2^{---} from 2^-, 2^{--}, and 2^{---}.

Illustrating Principle 3 (Parity Adjusting).
- (i) Initial Sources: 1^+, $2^{---\cdots---}$, 3^-, 3^{--}, $4^+\cdot$, $5^+\cdot$, $6^{+++\cdots+++}$.
- (ii) Bifurcation Dynamics: $1^+ \rightarrow 2^- \rightarrow 2^{--} \rightarrow 2^{---} \rightarrow 4^+, 5^+$, etc.
- (iii) Computational Parameters: 2^+, 3^{--}, b, h, s, $\cdots$.
- (iv) Parity Rules: 1^+, 2^{---}, 4^+; 1^+, 3^-; 1^+, 2^-, etc.

Illustrating Principle 4 (Spectrally Refining).
- (i) Energy Cascade: lid to 1^+; 2^-, 2^{--}, 2^{---} to 2^{---}.
- (ii) Enstrophy Cascade: 1^+ to 2^-; 2^{---} to 4^+ and 5^+; 1^+ to 3^-.
- (iii) Limiting Scales: $\cdot$, $\cdot$, $\cdot$, 2^+, 3^{--}.
- (iv) Dissipative Convergence: h, s, b.

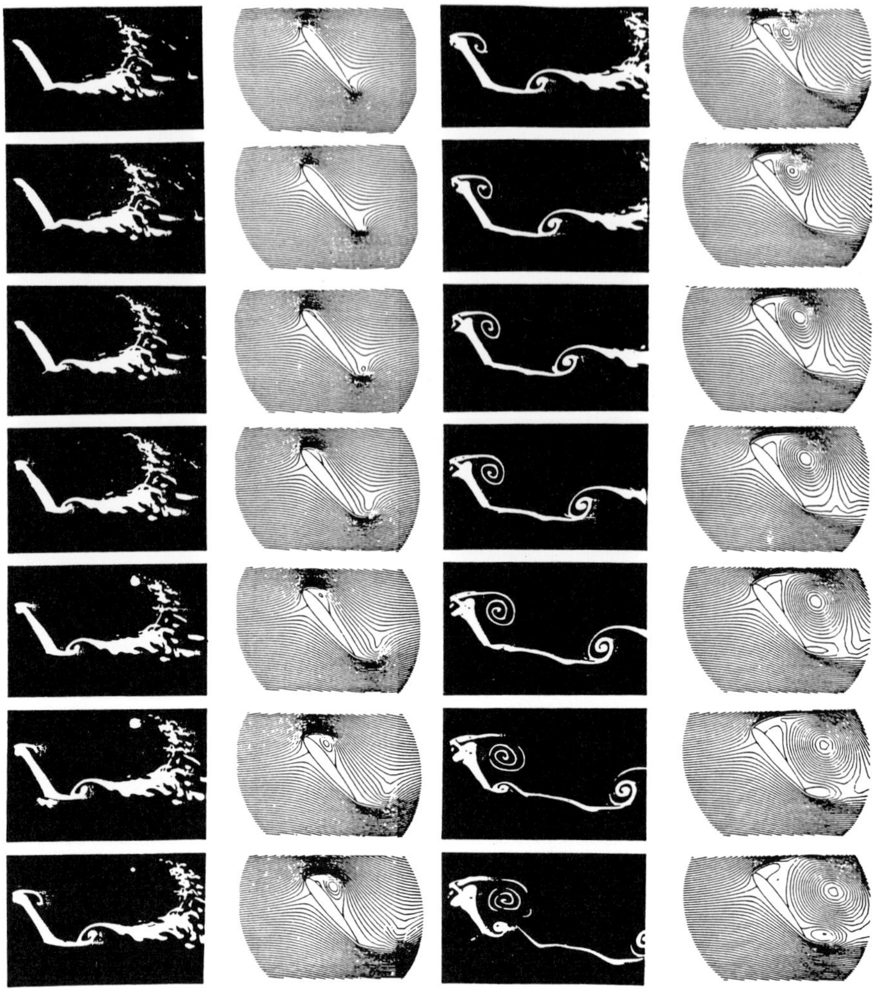

Fig. 5.3(a).

Fig. 5.3. Vortex generation, evolution, dynamics, and limits in airfoil flow: (a) Constantly accelerating flow from rest, $R_{acc} = 835$, $\alpha = 50°$; (b) Impulsively accelerated flow from rest, $R = 1{,}000$, $\alpha = 30°$; (c) Constantly accelerating flow from rest, $R_{acc} = 500$, $\alpha = 40°$.

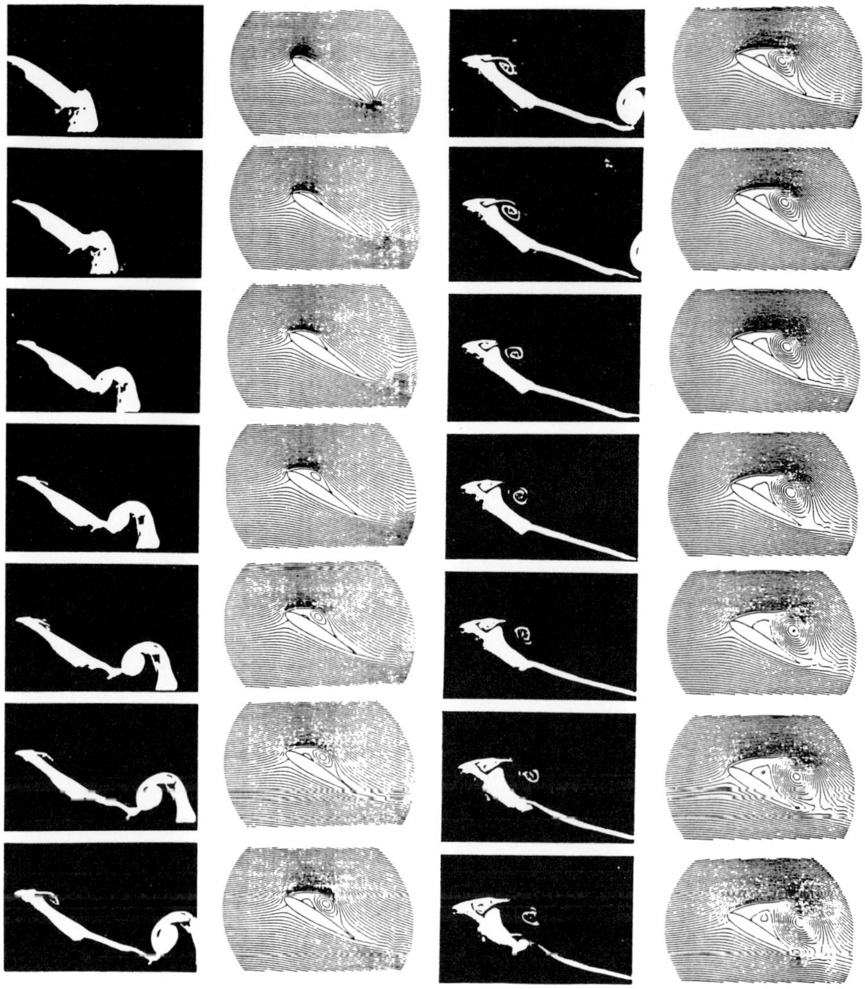

Fig. 5.3(b).

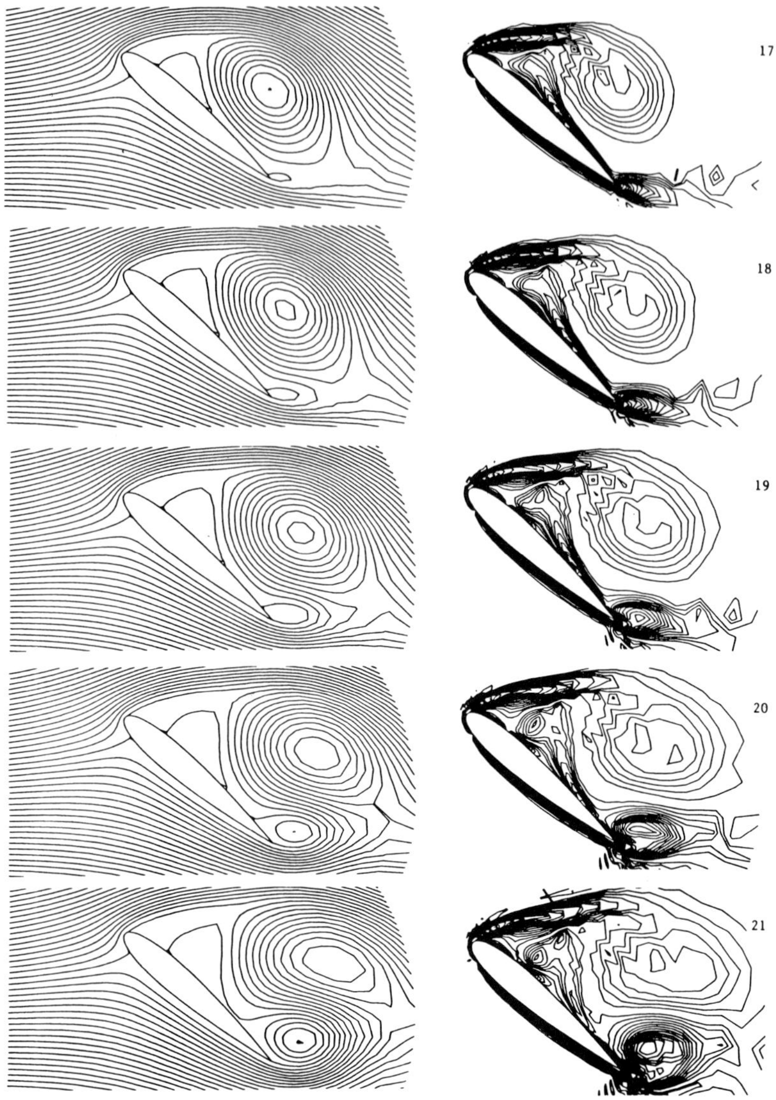

Fig. 5.3(c).

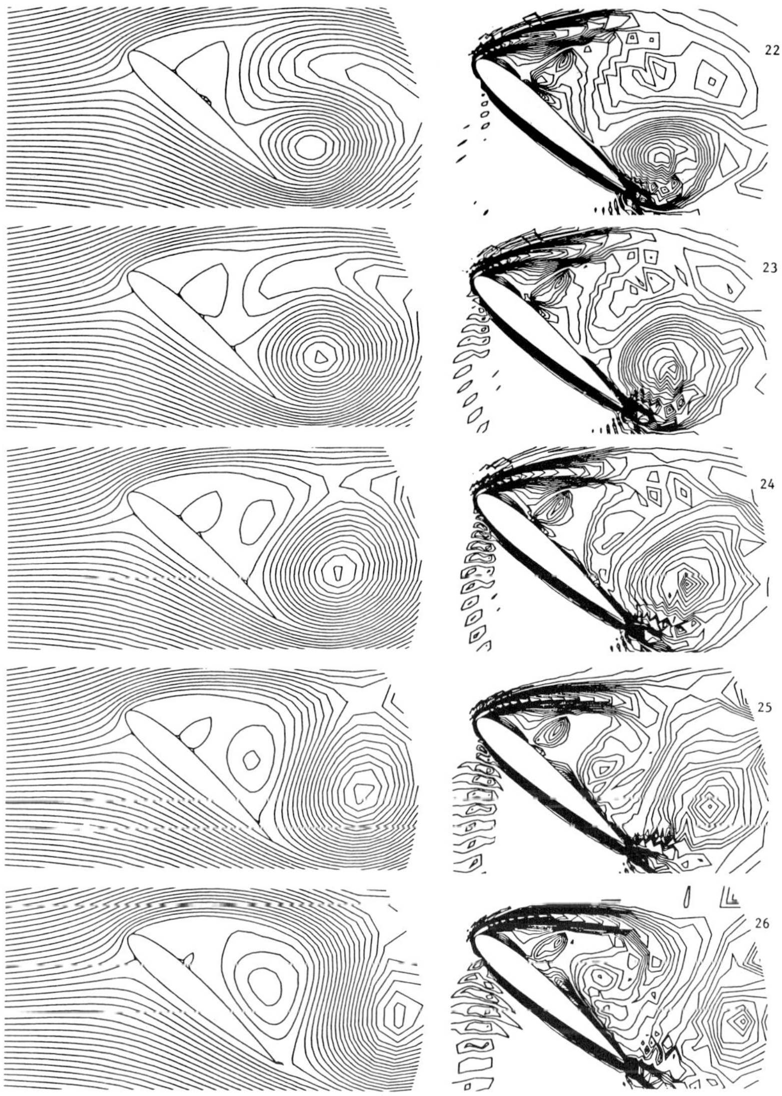

Fig. 5.3(c, continued).

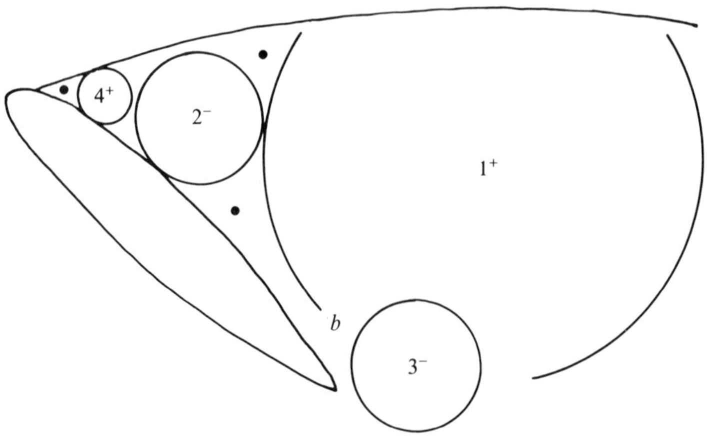

Fig. 5.4. Schematic of constantly accelerating airfoil flow showing vortex states.

Illustrating Principle 1 (Space Filling).
- (i) Separation Compensation: 1^+.
- (ii) Corner Sequences: 4^+.
- (iii) Conic Tendency: $1^+, 2^-, 3^-, 4^+$.
- (iv) Burst Providing: b.

Illustrating Principle 2 (Self-Organizing).
- (i) Semigroup Properties: For example, the evolutions of 1^+, then $4^+, \cdots$, from initial separation points.
- (ii) Self-consistent Properties: For example, the dominance of the 1^+ and 3^- modes.
- (iii) Self-similarity Properties: For example, $1^+, 4^+, \cdots$.
- (iv) Stability Exchanging: For example, the splitting of 1^+ into a stable and unstable part, under the action of 3^-.

Illustrating Principle 3 (Parity Adjusting).
- (i) Initial Sources: $1^+, 3^-$.
- (ii) Bifurcation Dynamics: 3^- splitting 1^+.
- (iii) Computational Parameters: $\cdot, \cdot, \cdot, b$.
- (iv) Parity Rules: $1^+, 2^-, 4^+$, etc.

Illustrating Principle 4 (Spectrally Refining).
- (i) Energy Cascade: 1^+ and 3^- enlargements.
- (ii) Enstrophy Cascade: 3^- splitting 1^+, 1^+ shredding 2^-.
- (iii) Limiting Scales: $\cdot, 4^+, b$.
- (iv) Dissipative Convergence: Downstream.

1. The Generation of Vortex Motion

Following Lugt [1], we may define a *vortex* (motion) as a rotating motion of a multitude of material particles around a common center, and *vorticity* as the angular velocity of matter at a point in continuum space. Note that a flow can possess vorticity without possessing vortices. Angular velocities can be aligned in laminar ways. In fact there is, roughly, vorticity everywhere the flow development reaches. This observation goes back at least as far as Descartes. Physically, all motions are quickly rotary due to mass conservation.

The generation of vortex motion, and the generation of vorticity, are intrinsically interlinked. Given viscosity in the fluid (and in this chapter we will always think in terms of the full Navier–Stokes equations rather than the special case of the Euler equations) the no-slip boundary condition and sharp enough corners generate vorticity.

Very early, Lord Rayleigh [13] recognized, as had Stokes and others before him, that when a viscous fluid is in contact with a fixed wall, certain features of the motion exist independent of the magnitude of the viscosity. From [13]:

> As Sir G. Stokes has shown, the steady motion is the same whatever be the degree of viscosity; and yet it is entirely different from the flow of an inviscid fluid in which no rotation can be generated. Considerations such as this raise doubts as to the interpretation of much that has been written on the subject of the motion of inviscid fluids in the neighborhood of solid obstacles.

In spite of the fact that one needs viscosity in order to generate rotation, most studies of vortex motion, even recent numerical studies, continue to focus on the Euler equations, in which the viscous term $\mathrm{Re}^{-1}\,\Delta u$ of equation (0.1) is omitted. The reason for this is of course the very useful theorems of Helmholtz and Kelvin for the transport of vorticity and circulation, respectively, in the absence of viscous diffusion. Roughly speaking, such investigations are consigned to the study of the propagation of vortex motion, not the generation of it.

Let us turn then to the generation of vortex motion. Immediately, we must distinguish the generation of vorticity and the generation of vortex motion. Note that vorticity is a well-defined mathematical concept: $\omega = \operatorname{curl} u$, where u is the velocity of the fluid. One difficulty in working with vortex motions is that the concept of a vortex apparently remains at the level of a notion. That is, it has no exact mathematical formulation.

Returning to the generation of vorticity, among the many treatises on fluid dynamics I like the discussion of Batchelor [4, pp. 278–279] of the source of vorticity in motions generated from rest. Batchelor views this as occurring in three stages. His discussion is so good that I want to paraphrase it here.

At $t = 0$ the fluid, under an "impulsive" start, instantaneously develops a tangential sheet of infinite vorticity at a no-slip boundary surface. This is the first stage. In the second stage, for small values of t, the vorticity just developed instantaneously on the boundary then diffuses into the fluid under the action of viscosity ν. This occurs in a thin layer of thickness $O(\nu t)^{1/2}$ in which the vorticity is finite. As the diffusion distance $(\nu t)^{1/2}$ grows, the vorticity is convected away from the boundary. This is the third stage, whereby the vorticity is injected into the flow.

By contrast, a vortex motion is not generated until it acquires an orientation and an organized structure. That is the viewpoint I will take here. To obtain such, it first requires some time to acquire its orientation, and then more time to stabilize into an organized structure. In other words, all three stages of Batchelor's portrayal of the generation of vorticity are needed in my stage (1), the generation of vortex motion.

Such vortex motions are generated from:

walls	separations	viscosities
corners	cascades	decelerations
provocations	shears	accelerations
interfaces	strains	instabilities

What I have written down is meant to be suggestive of the causes of vortex generation: geometrical (first column) causes, fluid (second column) causes, and external (third column) causes. I have not tried to render a complete list, and it is certainly simplified.

Vortex motion generation continues throughout the flow. Even if the flow approaches a final state, the generation goes on, for otherwise viscous dissipation would reduce the flow to rest.

Generation of vortex motion leads to the first (of four) general principles of a theory of vortex motion [12].

Principle 1 (The Space Filling Principle)

The vortex motions will continue to be generated to fill up all available (e.g., reachable) region space. By reachable space I mean a number of situations. But mainly, I mean: in the downstream flow-affected region. For example, in the cavity flow of Fig. 5.1, the containing vessel is closed and the vortex motions are able to fill (almost) all region space. In the second example, the airfoil flow of Fig. 5.3, the flow is an exterior flow clearly not filling all (Euclidean) 2-space. Small unreachable spaces in a flow region will be dealt with in § 4 on limits of vortex generation.

In connection with and as reinforcement for Principle 1, I also want to synthesize into four subprinciples four important observed fluid phenomena.

Four (Space Filling) Subprinciples

(i) Separation Compensation. The space opened up by an outward separating action is filled by an inward (vortex inducing) action.

(ii) Corner Sequences. Vortices try to fill a corner by (self-) generating a vortex sequence descending into it.

(iii) Conic Tendency. Vortices prefer a circular (conic, spherical, smooth curvature) shape, up to space limitations.

(iv) Burst Providing. Secondary small vortices in the wall spaces between larger vortex rolls are precursors to bursting.

We briefly describe these subprinciples.

Subprinciple (i) follows Prandtl–Tietjens (1929). An excellent description of how boundary layer separation induces a counterrotating vortex action is given by Schlichting [6, pp. 28–29]. It will be interesting to see how far the notion of separation compensation, as viewed here as a space filling feature, can be extended, for example, to high shear interfacial regions interior to a fluid flow.

Subprinciple (ii) follows Rayleigh (1911) and later Dean, Montagnon, Moffatt, Lugt, and Schwiderski [14] and others. Stokes flow under simplifying symmetry assumptions predicts an infinite sequence of self-similar vortices into a corner of sufficiently small angle. In a series of numerical multigrid (Stokes) experiments [21] we found up to 25 of these, the 25th with intensity smaller than 10^{-100}. A better understanding of the (nonlinear) theory of corner vortices is needed [24]. The point I am making here is more physical than analytical, the view being that of the space filling principle as a mechanism for the generation of corner vortex sequences.

Subprinciple (iii) follows Kirchoff (1869) who noted that constant vorticity elliptical patches preserved their shape in inviscid flow. What I have in mind about circular tendency is illustrated in Figs. 5.2 and 5.4: one can "round off" to approximate circles the final flow pattern, as constrained by the region's boundaries, and see the "hard to fill" remaining space. Somehow, vortices with high curvature portions do not survive the grinding down viscous lubrication actions, e.g., to make the point extreme, one does not see long-lived vortices possessing sharp corners or other inefficient irregularities. Conic cross sections prevail in three-dimensional vortex tubes as well.

Subprinciple (iv) stems from my interpretation (see § 4) of the recent analysis of Aubry et al. [15] of the bursting phenomenon. They conclude from a (highly) truncated Galerkin system of ordinary differential equations that wall rolls exist and eject bursts of turbulent energy filaments into the flow. In my interpretation here (see § 4), a mechanism for these bursts is a sequence of small secondary wall eddies (called $2^{---}\cdots^{---}$; § 2) induced by the major wall rolls (vortex 1^{+} and 2^{---} in my notation) and pulled at high acceleration into the adjacent flow.

Some additional detail and illustration of these four subprinciples will be found in the following sections. A larger set of references (example: subprinciple (iii) in further depth brings in analyses ranging from early configuration stability studies by Love (1893) to present day studies by Burbea (1982) and others) and a more complete exposition will be found in [12].

Further analysis and testing of these subprinciples is needed. Also I am presenting them here primarily from a two-dimensional perspective. Three-dimensional separation and corner theory will allow another degree of freedom and some escape from separation compensation by counterrotating vortex action alone, although the connectivity of vortices should still induce space filling properties. The conic tendency will be diluted in the third direction but some minimal surface energy preserving configurations should still be preferred. Bursting events will still pop out of ripple effects along walls and floors but that theory is in its infancy and so it is pointless to speculate on what will be found.

Much of Principle 1 and the general description of the generation of vortex motion can be illustrated in terms of the driven cavity flow history of Fig. 5.1. To complement that example (internal vortex dynamics) I will also briefly describe certain vortex events in the airfoil flow of Fig. 5.3 (external vortex dynamics). A more comprehensive description of these principles, including other flows, will be given in [12].

Let me conclude this section therefore with general descriptions of the cavity flow of Fig. 5.1 and the airfoil flow of Fig. 5.3. A more detailed history of the evolution of each individual vortex motion will be given in § 2, further information on their specific dynamical features in § 3, and limiting aspects of the vortex motions in § 4. For now I will thus confine myself to explaining the basic flows. The space filling interpretations of Principle 1 and its subprinciples are listed in the legends of the schematics, Figs. 5.2 and 5.4, which go along with the actual flows, Figs. 5.1 and 5.3, respectively.

Basic Description of Lid-driven Cavity Flow. The generation of vortex motion in the lid-driven cavity shown in Fig. 5.1 proceeds as follows (for more information see [22]). Here the Reynolds number Re = 10,000 is fairly high. At the lower Re = 2,000 the same flow converges to a final steady pattern, but this is not the case at Re = 10,000, as will be seen. Most cavity flow studies (e.g., see [25]–[28]) focus on the unit cavity, aspect ratio depth/width = 1. But we found the A = 2 cavity much more interesting, a much richer vortex dynamics being generated there. Also A = 2 sufficiently well represents effects of deeper cavities.

(a) The lid is impulsively started, moving to the left uniformly, $(u, v) = (-1, 0)$ on top throughout the continuing flow history, the fluid initially at rest, the side and bottom walls kept at a no-slip condition $(u, v) = (0, 0)$. The initial condition is weakly divergence-free and the flow remains so. A MAC (marker and cell) primitive variables method was used [22]. Other computational schemes [23] confirm the same vortex motions.

(b) A flow boundary layer forms just under the lid but no apparent separation occurs on the lid. An explanation for this can be given in terms of Hiemenz flow (see § 3).

(c) There is, however, an immediate rotation induced in the fluid just below the lid, already at two sec there is a single recirculation of roughly the length of the lid and just under it. This corresponds to the explanation by Batchelor, which I have described above, the sheet of infinite vorticity, the ultimate initial source of vorticity in motions generated from rest. This is in my opinion caused by the lid start alone, and not by the upper left corner, which does however immediately then come into play as a provocation point as follows.

(d) The underlid recirculation quickly impinges on the upper left corner and spawns a clearly defined vortex (at three sec) in the corner just slightly down the left wall. The upper left corner has acted as all of: provocation point, singular point, and stagnation point. But its main role is to direct this small vortex down and quickly into the flow. Experiments allowing leakage, others smoothing this upper corner effect, have always produced the same basic cavity flow features, so the exact nature of this singularity does not appear essential in the ensuing flow evolution.

(e) This lid-wall impact vortex (my terminology) caused by the underlid recirculation sheet being pulled swiftly against the perpendicular left wall of the cavity moves considerably down the left wall and then separates and heads into the interior of the cavity. At lower Re it heads inward at an earlier stage, i.e., from nearer the top of the left wall. This is clearly an inertial (advection) consideration. At all Re tested thus far, this first appearing vortex enlarges and eventually settles down to become the principle vortex 1^+ depicted in Fig. 5.2. Energy is constantly fed into it by the driving lid.

Let us agree on a sign convention: counterclockwise vortices will be positively oriented $(+)$.

(f) Immediately after vortex 1^+ separates from the left wall, a counterrotating vortex 2^- forms. In fact, an infinite sequence of 2^- vortices will shed, and it takes three of them (see § 2) to form the basic secondary circulation 2^{---}.

(g) Meanwhile, the main motion vortex 1^+ spins off another secondary system in the upper right corner, indicated on Fig. 5.2 by 3^- and 3^{--} (see § 2).

(h) After vortex 2^{---} establishes itself it spins off the tertiary vortices 4^+ and 5^+ in the lower two corners. It should be noted that the order and manner of formation of vortex motions is Reynolds number dependent, a consideration which should cause pause in Re-continuation methods used to produce steady flow bifurcation diagrams. Here, at $\text{Re} = 10{,}000$, the main secondary vortex 2^{---} is produced solely from accumulations of the shed 2^- system (see § 2). At lower Re (see [22]) the secondary vortex 2^{---} is created in fusion of a 2^- wall vortex combined with 4^- and 5^- secondary lower corner vortices. Then only after that 2^{---} main secondary motion is established, the counterrotating tertiary vortices 4^+ and 5^+ appear.

(i) The two large central vortex motions 1^+ and 2^{---} and the three corner motions 3^-, 4^+, and 5^+ now in place, the 2^- system continues to be spun off the principal vortex 1^+, the secondary circulation 2^{---} spinning off short-lived tertiary vortices 6^+, and apparently (see § 3) the 2^- system continues to attempt an induced very small counterrotation 2^+ in the space between 2^-, 1^+, and the wall, even in face of the onrushing powerful 1^+ flow.

(j) I have also indicated on Fig. 5.2 the spaces h, h, b, s, $\cdot$ not filled by vortex motions at the 40×80 resolution of this [22] experiment. These will be discussed in § 4.

Basic Description of Airfoil Flows. I will say very little about these flows, using them mainly to show similarities and distinctions of the theory for interior and exterior flows. The main point about the latter is that long-time studies reveal less because the circulation patterns are quickly transported downstream and out of the dynamics. On the other hand, the vortex motions immediately above the airfoil are important and show the same basic dynamics as in interior flows.

Here I have taken three airfoil flow histories from [31]. Our other work on such flow problems may be found in [32] and [33] (see also [21] and [34]). Other fundamental studies of similar airfoil flow dynamics are [36]–[40].

Figure 5.3 contains three airfoil flow histories. These are all low-speed flows and our first goal was to match Freymuth's physical visualization experiments (see [36]), shown on the left in Figs. 5.3(a) and 5.3(b), with our computational visualizations of the stream function profiles shown on the right in Figs. 5.3(a) and 5.3(b). This effort was quite successful, as can be seen, and higher resolution computational experiments are under way [33]. To better understand important airfoil dynamics such as vortex splitting and shredding, in Fig. 5.3(c) on the left we have the computed stream function profiles and on the right the corresponding computed constant vorticity profiles. These also matched the physical flow visualizations (not shown) very well. For flow detail visualization, sometimes the stream function profiles are better, other times the vorticity profiles show more.

Here for airfoil flows I will use a convention of (+) to indicate clockwise vortex motion. It turns out that the clockwise (+) motions are lift enhancing so the (+) is a convenient memory device for that important consequence. (One could make the cavity signs of the same orientation by driving the lid to the right.)

In the flow (a) a principal vortex 1^+ forms from a separation point just behind the leading edge of the airfoil. At roughly the same time a reasonably large (total circulation preserving) trailing edge vortex 1^- can be seen moving downstream in the physical visualizations. As the principal vortex 1^+ enlarges, a counterrotating secondary vortex 2^- can be seen forming just ahead of it on the airfoil surface. As the principal vortex 1^+ moves downstream it encounters a new trailing edge vortex 3^-. From these (coarse grid) pictures finer dynamics (e.g., the interaction between 1^+ and 3^-, even possible partial induction of 3^- by 1^+) are not clear.

Flow (b) differs from flow (a) in that an impulsive start is followed by no further acceleration. The principal vortex 1^+ is followed by a vortex 2^- as before, with initial and later trailing edge vortices 1^- and 3^- also as before. The physical visualization shows a new vortex 4^+ forming later on the forward top wing surface. We used only a 32×64 mesh on the computer experiments shown here but finer meshes are currently being employed [33].

Flow (c) shows more vortex motion detail. By using a lower Reynolds number the vortex dynamics details were not as quickly washed downstream. By simulating a constantly accelerating airfoil at high angle of attack α we found details of vortex splitting and shredding not seen before. This was the first numerical calculation of this flow field and geometry. The splitting of vortex 1^+ by vortex 3^- allows a portion of 1^+ to return toward 2^- to shred the latter, thereby reducing the loss of lift and preventing a stall of the airfoil.

In spite of their importance, the details of vortex dynamics over airfoils are not as well understood as one might think. One finds at regular intervals new accounts of airplane crashes due to ice forming on wings. Until a few years ago the public would have thought "weight" but now probably most know it as a question of "shape." I quote from the most recent account (in today's newspaper [35], in fact):

> Ice eyed as cause of B-1B crash.
>
> The ($280 million) B-1 bomber that crashed . . . Nov. 17 . . . was neither designed nor equipped with de-icers. . . . Ice on the wings could cause the airplane to crash, not from

excess weight, but because it affects the shape of the airfoil, resulting in loss of lift, these pilots said. A former military test pilot said ice accumulation is not necessarily a significant factor at high speed, but is a much bigger problem when an airplane slows down and increases its angle of attack for landing.

2. The Evolution of Vortex Motion

Before stating Principle 2 (self-organization) and its subprinciples, let us follow the evolution of each individual vortex motion of the driven cavity. With reference to the flow history in Fig. 5.1, times are all rounded to the approximate second. Vortex motions are ordered according to their initial occurrence time.

A word before we begin. Figure 5.1 runs out only to 96 sec and then skips to the "final" patterns from 337–360 sec. But nothing really dramatic happens in the flow evolution in between. There is a general smoothing of the flow (action of Δu) and an elimination of unwanted parities (see § 3). The main final question in the evolution of the vortex motion is that of whether a periodic final motion is ever achieved (see § 4).

The flow dynamics, to be discussed in § 3, and the flow evolution being discussed here, are very much coupled. But there is so much in the dynamics of a flow of vortex motions that I wanted to take two different perspectives.

Here in this section the view is that of an initial value problem. The solutions of the Navier–Stokes equations have the global semigroup properties $u(t) = S(t)u_0$, u_0 the initial state, of nonlinear evolution equations. The overall flow can also be thought of as a combination of local group actions such as translations, rotations, dilations, parametrized by the semigroup parameter t. But more particularly, one can think of each individual vortex motion as a $v(t) = S(t)v_0 = S(t - s_0)S(s_0)v_0$, where v_0 is a localized (cutoff) source state for the motion, and s_0 is the initial time interval required so that $S(s_0)v_0$ acquires its status as a fully organized vortex. In other words, each vortex motion to be described below can be thought of as a single (nonlinear) mode, ignoring as much as possible the interactions with the other mode (vortex) evolutions.

Meanwhile the full system seeks to move among self-consistent (in the sense of a self-consistent field theory, as in quantum physics) patterns. In this global sense the evolution is self-organizing, in the general sense as used, for example, by Prigogine and Nicolis [16].

Evolution of Cavity Vortex Motions

The Principal Vortex 1^+

Source:

The impulsive start underlid recirculation.

Evolution:

underlid recirculation eddy $t = 1$ sec

↓

upper left corner eddy $t = 2$ sec

↓

upper left wall vortex $t = 3$ sec

↓

separates into main cavity interior $t = 10$ sec

Destination:

Role as the principal vortex, to finally occupy over 50 percent of cavity, as primary upper recirculation.

The Secondary Vortex 2^-

Source:

The separation of vortex 1^+ from the left wall.

Evolution:

counterrotation induced by the separating principal vortex 1^+	$t = 10$ sec
↓	
middle left wall vortex	$t = 13$ sec
↓	
reduced (almost lost) to eddy	$t = 17$ sec
↓	
recovers vortex structure	$t = 18$ sec
↓	
separates into interior	$t = 21$ sec
↓	
makes contact with the cavity bottom	$t = 22$ sec
↓	
clearly leaving the left wall dynamics to enter lower cavity interior	$t = 24$ sec
↓	
makes contact with lower right corner	$t = 27$ sec
↓	
extends contact to lower right wall	$t = 28$ sec
↓	
continues toward right wall but weakening	$t = 30$ sec

Destination:

Gone, mostly dissipated by $t = 32$ sec, a small portion of its vestige perhaps fuses with 2^{--}.

The Secondary Vortex 2^{--}

Source:

The separation of vortex 1^+ from the left wall, combined with sufficient (separation) distance of 2^- from left wall.

Evolution:

counterrotation induced by shear from principal vortex 1^+, no slip wall, and separating vortex 2^-	$t = 21$ sec
↓	
separates into interior	$t = 27$ sec
↓	
moving quickly to right, strengthened by absorbing remains of 2^{--}	$t = 30$ sec
↓	
still a small vortex approaching right wall, but weakening	$t = 36$ sec

Destination:

Gone, partially by dissipation, but also fused at $t = 37$ sec with 2^{---}, strengthening the latter.

The Secondary Vortex 2^{---}

Source:

The shear induced counterrotation from vortex 1^+, sufficient distance from 2^{--}.

Evolution:

counterrotation from 1^+, no slip wall, and	$t = 27$ sec

sufficient distance from (separating) 2^{--}

↓

well-formed vortex, similar to 2^{--} at $t = 25$ sec — $t = 30$ sec

↓

ready to fuse with 2^{--} — $t = 36$ sec

↓

fusion 2^{---} and 2^{--} — $t = 37$ sec

↓

strengthened 2^{---} heading toward right wall — $t = 38$ sec

↓

to be further strengthened by approaching 2^{----} — $t = 42$ sec

↓

2^{---} has absorbed 2^{----}, 2^{---} now strong enough to induce 4^{+} in lower right corner — $t = 43$ sec

↓

2^{-----} about to be absorbed into 2^{---} — $t = 47$ sec

↓

2^{-----} being absorbed by 2^{---} — $t = 48$ sec

↓

2^{------} approaching the dominating 2^{---} to be absorbed — $t = 51$ sec

↓

$2^{-------}$ approaching 2^{---} to be absorbed — $t = 56$ sec

↓

$2^{--------}$ approaching 2^{---} to be absorbed. The vortex 2^{---} now strong enough to begin inducing 5^{+} in lower left corner — $t = 61$ sec

↓

$2^{---------}$ about to be absorbed by the enlarging 2^{---} vortex circulation system — $t = 65$ sec

↓

$2^{----------}$ about to be absorbed — $t = 70$ sec

Destination:

The secondary vortex 2^{---} has survived, its role to become the main lower cavity vortex, absorbing the $2^{---\cdots---}$ vortices as they arrive.

The Secondary Vortices $2^{---\cdots---}$

Source:

See above, counterrotation off principal vortex 1^{+} after sufficient departure distance of previous $2^{---\cdots---}$ vortex.

Evolution:

Shed from left wall at time intervals starting at about $\Delta t = 6$ sec apparently decreasing to approach a constant Δt. — $3 < \Delta t < 6$ sec

Destination:

To be absorbed into vortex 2^{---}.

The Secondary Vortex 3^{-}

Source:

The separation of vortex 1^{+} from the upper right wall.

Evolution:

counterrotation induced by principal vortex — $t = 33$ sec

1^+, space needing filling due to 1^+ in its deepest cavity position

↓

well defined wall eddy — $t = 36$ sec

↓

separates from upper right wall — $t = 40$ sec

↓

have moved into and filled upper right corner region — $t = 47$ sec

↓

pushed into smaller corner area due to rise of vortex 1^+ to its (essentially) highest position — $t = 56$ sec

↓

3^{--} spawned, will compete — $t = 57$ sec

↓

3^- weakening, appears to split — $t = 60$ sec

Destination:

Residual upper portion 3^- will remain, lower vestige portion to fuse with 3^{--}, their interaction dependent on up down oscillations of principal vortex 1^+.

The Secondary Vortex 3^{--}

Source:

Separation of 1^+ from upper right wall, sufficient distance from vortex 3^- due to space opened up by principal vortex 1^+ starting a minor descent.

Evolution:

counterrotation off vortex 1^+ — $t = 54$ sec

↓

fully formed vortex — $t = 57$ sec

↓

has fused with lower vestige of 3^- — $t = 61$ sec

↓

appears fused with upper residual of 3^- — $t = 65$ sec

↓

being absorbed by residual of 3^- — $t = 74$ sec

↓

fully absorbed by 3^- — $t = 77$ sec

Destination:

Gone, lost to 3^-, due to rise of eye of 1^+ which decreased corner space.

The Secondary Eddies $3^{---\cdots}$

Source:

Unlike the vortices $2^{---\cdots---}$, which are apparently infinite in number, there appear to be only a finite number of $3^{---\cdots}$ eddies, their size decreasing below some critical diameter as vortex 1^+ eye position settles into one location. However, the situation in this corner is far from clear, and only long-time very fine grid simulations could verify the finite number, or infinite number, of these tertiary eddies.

Destination:

Eventually they occupy a small region below 3^-, in a barely organized structure, at least in the current grid resolution.

The Tertiary Vortex 4^+

Source:

The separation of dominating secondary vortices 2^{---} from lower right wall.

Evolution:

counterrotation induced by 2^{---} $t = 43$ sec

↓

full vortex structure established $t = 46$ sec

↓

almost lost as lower recirculation pattern has shifted to left of cavity $t = 56$ sec

↓

has recovered $t = 61$ sec

Destination:

Will remain, slightly enlarged, a backwater in the total evolution.

The Tertiary Vortex 5^+

Source:

The separation of 2^{---} vortex recirculation system from cavity bottom.

Evolution:

counterrotation induced by 2^{---} $t = 56$ sec

↓

established as lower left corner vortex $t = 59$ sec

↓

has organized full vortex structure and occupied its basic flow region $t = 70$ sec

Destination:

Will remain, in contrast to right corner vortex 4^+, as a vital dynamic vortex in the ensuing evolution, as it upwells along lower left wall in interaction with the $2^{---}\cdots^{---}$ vortex sequence.

The Quartiary Vortices $6^{+++}\cdots^{+++}$

Source:

A complex interaction between the dominating vortex 2^{---}, the tertiary vortex 5^+, and the shedding vortex sequence $2^{---}\cdots^{---}$

Evolution:

requires sufficient upwelling from lower left corner vortex 5^+ $t = 92$ sec

↓

still present in well-defined vortex form $t = 94$ sec

↓

quickly dissipated into the high shear interface between principal recirculation 1^+ and 2^{---}. Possibly partially absorbed by 1^+, partially absorbed by 5^+ $t = 95$ sec

↓

reappears and disappears sporadically until total cavity final motion is established.

Destination:

Establishes an infinite tandem shedding relationship with the opposite-signed sequence $2^{---}\cdots^{---}$.

The Tertiary Eddy 2^+

Source:

Counterrotations off the sequence $2^{---}\cdots^{---}$.

Evolution:

"phase collision (two arrows) above new vortex 2^- $t = 14$ sec

↓

attempted eddy $t = 16$ sec

Destination:

Reappears intermittently throughout the flow history, too weak to survive against vortex 1^+.

The latter eddy brings us to the resolution limit of the 40×80 grid used to produce the vortex motion evolution shown in Fig. 5.1. Finer grids are being used in the runs [23] currently under way. Resolution and mesh aspects of vortex motion are elaborated on in § 4.

Given the above vortex mode histories, and in the spirit of the discussion at the beginning of this section, let me state the second principle of vortex motion.

Principle 2 (The Self-Organizing Principle)

The vortex motions will self-organize according to local (e.g., semigroup) properties and to global (e.g., self-consistent) constraints. The semigroup property refers to the point of view, as mentioned earlier, of a vortex motion $u(t) = S(t)u_0 = S(t-s)S(s)u_0$ viewed as an individual semigroup evolution of a "vortex" $S(s)u_0$, so denoted once it has enough local vortex structure to be thought of as an individual vortex. There are some heuristics here, e.g., could we always get such a vortex $S(s)u_0$ from an isolated initial state u_0, and if so, in how much time s? In the next section we will see that the separation-to-vortex structure time s is always about three sec in this flow. The self-consistent constraints refer to the balancing off of the vortex modes one against the other toward a stable equilibrium within the whole region. For example, the primary global constraint is the boundary walls of the vessel, a secondary global constraint is the arrangement of the vessel's geometry, a third global constraint is the nature of the driving force, a fourth global constraint is that of (all of) the other vortex motions (e.g., see the discussion of the origin of couples, in the next section). The last (fourth) subprinciple is the most important to understand once the flow and its vessel are given. It is, in point vortex approximation, a "variable n-body problem." I like to think of it, without approximation, as a "territory allocation problem."

An arrangement of important subprinciples is the following.

Four (Self-Organizing) Subprinciples

(i) Semigroup Properties. The whole flow is a nonlinear evolution semigroup. The individual vortex motions evolve as local semigroups, approximately, from the point of view of local mode analysis.

(ii) Self-consistent Properties. Overall flow patterns are governed by the need to achieve self-consistent nonlinear vortex mode equilibria under wall and other vessel constraints and as a balance (territory control) of each vortex mode against the influence of all others.

(iii) Self-similarity Properties. The flow's self-organization from large to small, and small to large, will follow group theoretic and physical self-similarity principles.

(iv) Stability Exchanging. The changing number and configuration of vortex motions during the flow evolution will trace a path in time in accordance with a principle of exchange of stabilities.

I will not go into any further specifics on subprinciples (i) and (ii), which have already been discussed above, except to mention the following.

Semigroup properties (i) in parameters other than t are sometimes (tacitly) assumed in computational experiments on cavity, airfoil, step, Taylor, and other flows. A better statement and understanding of these could be important. As an example, a computational solution u should be thought of as having a parameter dependence: $u = u(u(0), \mathrm{Re}, A, t, \delta x, \delta y, \Delta x, \Delta y)$. Reynolds number continuation methods (advance from final state $u(\infty)$ at Re to initial state $u(0) \equiv u(\infty, \mathrm{Re})$ at incremented $\mathrm{Re} + \Delta\,\mathrm{Re}$) tacitly assume that the solution has the semigroup property in Re. So do aspect ratio continuation methods. The validity of such continuation methods will depend on how robustly they can pass over flow instabilities to achieve the next state (see the next section).

A good example of the self-consistency subprinciple (ii) is the Green function approach studied by Lin [17] for the case of incompressible irrotational flow.

I do not know all the self-similar properties (iii) but there are two points to be made here. First, concerning initial conditions, there has been recent success with Lie group methods to find particular exact solutions (particular flow modes). These are usually scale invariant and can be expected to reproduce locally within the flow when conditions are right. Thus smaller eddies will look like the larger eddies off which they are spun. A very definite self-similarity is found in the corner eddy sequence. Second, concerning boundary conditions, the vessel's boundary shape will influence in an overall way the symmetries and self-similarities present in the vortex motion configurations. So will the values of the boundary variables (e.g., velocity, pressure, vorticity) themselves. Generally, getting the boundary conditions into the self-similarities is not well understood.

As the flow develops the classical instabilities, new vortices are generated to accommodate the principle of exchange of stabilities (iv), a principle widely enough known in the fluid dynamics literature but, it seems, not very well understood mathematically. This happens, for example, when Taylor–Couette flow becomes the Taylor-rolls. As time-dependent unsteady flow simulations become more common, this principle will become more important for their understanding, in both two- and three-dimensional flows.

Evolution of Airfoil Vortex Motions

Due to the coarseness of the mesh that we have used to date, and the transientness of the vortex motion histories, there is not a great deal to say beyond the histories (a) and (b) shown in Fig. 5.3 and the description of the general features of these flows already given in § 1.

However, let us trace the individual vortex motions in the more detailed flow (c) of Fig. 5.3, in which we may observe the important feature of vortex splitting to provide lift enhancement. Avoidance of stall, as mentioned at the end of the previous section, is an important aerodynamics problem, very likely involving critical phenomenon in vortex motions.

The Principle Vortex 1^+

Source:

Separation of incoming main flow U near top front of airfoil (in this sense, 1^+ is already a secondary vortex motion, off U).

Evolution:

beginning to move downstream	frame 17
↓	
split by 3^- evident in vorticity plot	frame 20
↓	
nearer wing portion 1^+ residual encounters vortex 2^-	frame 24
↓	

shredding of 2^- completed, 1^+ residual moving downstream — frame 26

Destination:

Downstream.

The Secondary Vortex 2^-

Source:

Counterrotation spin off of Vortex 1^+.

Evolution:

following 1^+ downstream — frame 17

↓

entrained by the 1^+ and 3^- interaction (see streamline) — frame 21

↓

shredding by 1^+ begins in earnest — frame 24

Destination:

Shred.

The Secondary Vortex 3^-

Source:

The trailing edge singularity.

Evolution:

enhancement off lower surface of airfoil — frame 18

↓

splits 1^+ in interaction — frame 20

↓

achieves maximal extent — frame 24

Destination:

Downstream.

The self-organizing interpretations of Principle 2 and its subprinciples are listed in the legends of the schematics, Figs. 5.2 and 5.4, corresponding to the individual vortex evolutions seen within the global evolutions in Figs. 5.1 and 5.3, respectively.

3. The Dynamics of Vortex Motion

As mentioned in the previous section, the evolution of a fluid flow, and its dynamics, are very interrelated concepts. In Principle 2, I emphasized the *organizational* aspects of a flow evolution. The actual dynamics of the flow were relegated to a secondary status, even though they underlie the individual vortex mode evolutions, and even though they are fundamental in achieving self-consistent intermediate and final states. The emphasis in Principle 2 was on global geometric, scalar, algebraic, organizational, overall vortex mode balance, as a decentralized but total system.

In this section, I want to emphasize the role of *vortex interactions.* Here we are, historically at least, on firmer ground. Vortex interactions have been the basis of intense and widespread study for over 100 years now. An early impetus was Lord Kelvin's (1867) theory of vortex atoms, which lost out to the quantum mechanical model of matter. Let me use the latter two theories, however, to make a vital point: *the dynamics of vortex motion is an n-body problem*. In view of the difficulty (general impossibility, for $n \geqq 3$) of analytically solving n-body problems, we can expect the general dynamics of vortex motion to be a difficult subject.

Going further, it is a *variable n-object* problem, with nonuniform objects whose individual configurations and number change in time. This would at first sight appear to make the problem even more intractable. However, "in every cloud there is a silver lining."

What do I mean by this? A number of thoughts are represented; let me limit them to four.

(1) All of the fluid, meteorology, mechanics studies, in which a body, object, or blob is replaced, approximated, by a point mass. Many myriad scientific studies have done this, going all the way back to Newton's idea of how to view a body at a distance just in terms of its "center." All of the vortex motion studies in terms of point vortices. By way of generalization, all of the recent studies replacing point vortices by vortex patches of constant vorticity (Batchelor–Prandtl theory), contour dynamics (Christianson, Zabusky, Overman, others). The random vortex (blob) methods (Chorin et al.). The more recent studies (Lax, Majda, Constantin, Titi, others) following the evolution of the boundaries of vortex patch equations as they relate to possible blowup of solutions to the Euler equations in two and three dimensions (see, for example, [18] and others). All such studies believe the silver lining will carry the cloud.

(2) The extent of the relation of such studies as precursors to those such as (this) the present one, in which I want to keep in mind every step of the way, even if it cannot be handled, the presence of the viscous term $\nu\Delta u$, no matter how small. The fact that all vortex dynamics stems originally from (delta-function-like, turbulent, or bursting (see § 4)) inputs generated by the full (viscous) Navier–Stokes fluid in boundary layers near the region walls, as described in § 1. The fact (see [19] and references therein) that in combustion heat balance models many blowup conclusions drawn from (singular) approximating equations do not carry over to the full (more physical) equations. On the other hand, the fact that observed subprinciples of vortex motion such as the conic tendency subprinciple of the space-filling Principle 1 are basically Euler equation or at least inertially dependent. The fact that the self-consistent subprinciples inherent in the self-organizing Principle 2 are also significantly inertially dependent even while their driving and vortex creation forces are viscously dependent as is their degradation to acceptable final patterns. The fact that only the computer has really opened up the full Navier–Stokes equations, with its essentially viscous effects, to us. That this tool may permit us to get inside the silver lining, into the cloud itself, to reveal basic phenomena.

(3) Returning to the subject of vortex (interaction) dynamics, the admissions in the literature of some of the best recent studies [20] of coherent vortex motions in turbulent flow, that we still do not understand the fundamental mechanisms of even just pairwise vortex interactions, not even the merger of like sign vortices, whether in Euler approximation or with Navier–Stokes or even higher order viscosities present. The corollary conclusion that even though vortex motion dynamics has been studied long and hard, its theory is in its infancy.

(4) The (Webster) dictionary "similism" for the above quote: A consoling or hopeful prospect.

By this I do not mean that we should console ourselves in the thought that the "silver lining" studies will be all that we will ever get. I would rather believe the hopeful prospect of understanding the full "cloud" dynamics so as to better appreciate the silver lining as an explainable boundary feature of the cloud.

Leaving these philosophical thoughts, I will next take the position that the study of the dynamics of vortex motions is such a vast and extensive subject that any formulation of a basic Principle 3 may offend and exclude large numbers of practitioners. Nonetheless I will formulate a Principle 3 and four subprinciples that I believe to be important in the context of the other principles, then explain why, then comment on a few of the observed dynamics in §§ 1, 2, and 4, and then move on to § 4. In other words, I will not get into particulars of vortex interactions themselves, leaving that to a more extensive study [12].

Principle 3 (The Parity Adjusting Principle)

The vortex motions will adjust the (e.g., information) content of the flow according to parity rules governing acceptable (e.g., bifurcation) state selection.

Four (Parity Adjusting) Subprinciples

(i) Initial Sources. All vortex dynamics trace back to their (viscous) origins and depend essentially on their (viscous) interactions with vessel boundaries and among themselves.

(ii) Bifurcation Dynamics. All physically meaningful vortex motions should be charted through their qualitative changes as they progress in time toward meaningful final (e.g., steady, periodic) states.

(iii) Computational Parameters. To be fully understood, the dynamics of the vortex motions of flow in a given configuration must be subjected to full (e.g., Re, A, Δx, Δy, δx, δy, u_0) parametric study.

(iv) Parity Rules. Generation, evolution, dynamics, and limits of vortex motions are governed (and simplified) by "parity rules" having both physical and mathematical grounding.

Let me begin by giving an example of the first subprinciple (i). As already described in § 1, the descriptions (e.g., Batchelor, Prandtl, Schlichting) of the sources of vorticity and vortex motion rely on boundary layer phenomena, which are viscous. We saw in the lid-driven cavity experiments that all vortex motion begins with a translational motion of the instantaneously created underlid (viscous) recirculation toward the corner where the (still basically translational) energy is transferred to rotational motion (to preserve continuity of the fluid, i.e., to avoid tearing it apart at the corner vertex). This transfer of energy from translational to rotational motion was of sufficient amount to lead to a fully organized small vortex motion which then continued to absorb energy and became the principal vortex 1^+. The other vortex motions were generated from that vortex motion.

Stated another way, this is a regularity property, in the terminology of the theory of partial differential equations [34]. Due to the viscous term $\nu \Delta u$ the solutions are immediately regularized at time $t = O^+$. The advection term has little transport effect in the first instants as the at rest fluid just under the lid dominates the mean velocity. So the flow equations act parabolically with the elliptic term Δu dominating and, like the heat equation, the solutions instantly smooth initial discontinuities.

Similar remarks could be made concerning the overriding importance of the (viscous) generation of the vortex sequence $2^{---} \cdots {}^{---}$ which leads to all of the lower cavity recirculation features, as described in § 2.

The connections between bifurcation theory and fluid dynamics are long standing (see, e.g., Joseph [9]). Often the approach is to try to develop a *bifurcation diagram* (e.g., see [34, § 1.8] for an introduction) for the *stationary* fluid equations. In such a diagram one plots some norm $\|u\|$ of the solution, against some key parameter (e.g., Re) governing basic flow characteristics. In this way one can track first the failure of uniqueness as the parameter reaches a critical value (meaning, the existence of two solutions to the stationary equations, one usually unstable, the other the physical one). Hopefully, one can go further and identify a Hopf bifurcation indicating the onset of a periodic solution in the flow problem. (In spite of claims to the contrary, it is my impression that no one has yet *proven theoretically* a Hopf bifurcation for the full Navier–Stokes equations.)

By *bifurcation dynamics* in subprinciple (ii) I intend a turn from the bifurcation diagram approach of analyzing only the stationary flow equations (in which the acceleration u_t does not even appear), an approach of little detail (the norm $\|u\|$ is scant information), to a more comprehensive notion: a theory which will account for all the important *qualitative changes* in solutions $u(t)$ as a *function of time*. For example, for a single solution from a given initial condition $u(0)$, a very barebones program would be to plot the maximum acceleration $\|u_t(t)\|$, which charts the progress of the solution $u(t)$ toward either a final steady, or periodic, solution. We have done this recently [22], [23]. Vortex events are correlated in interesting ways to relative maximums and minimums of $\|u_t(t)\|$. For example, the $2^{---}\cdots^{---}$ shedding correlates perfectly (see discussion below). An expanded program of bifurcation dynamics would keep track of the number of vortices at time t. A full parametric study, which is the way to fully understand a flow, would carry out this program for a span of all significant physical parameters such as Reynolds number Re, aspect ratio A. This being done for a single initial condition $u(0)$ is a departure from the usual bifurcation theory, which based only on the stationary flow equation amounts to asking which initial conditions actually lead to the (multiple) solutions found in the (steady solutions only) bifurcation diagram.

It is the power of computational methods (iii) which will permit such parametric studies. Large premiums will have to be placed, and are already being placed, on the development of fast, robust numerical algorithms. Once in place, the full parametric studies follow. An example of the importance of this new view to science and engineering is the gradual replacement of wind tunnels, where parametric airplane design studies are expensive, by computational models, where one merely changes an input parameter.

Turning then to the *parity rules* (iv), a full theory of these rules does not yet exist, although I am certain [12] many of their more coarse features can now be synthesized from the numerous recent computational studies of vortex dynamics. For example, a first simple parity rule is that newly spun off secondary vortex motions have opposite sign. A second simple parity rule is that a necessary condition for vortex merger is that the two interacting vortices have like sign. A more complicated parity rule would give a necessary condition (e.g., the existence of at least four vortex motions) for the production of a persistent vortex couple (opposite signs). There must be a "minimum parity" rule (theorem) which tells us that an internal flow will, even if at first increasing the information content of the flow (the number of vortex motions) as a means of preserving (in rotational form) energy, will nonetheless in the end seek a configuration of "minimal parity state" (minimum number of vortex motions). Such rules clearly impinge on stability considerations (more vortices are less stable; proximity rules for merger), dynamical considerations (the tendency of a flow to seek the unstable manifold), energy considerations (the attempt by the flow to avoid heat death)—as such, they will be very important rules in the theory of vortex motion.[2]

To leave the (very unfinished) business of such parity rules, which will be developed more fully in [12], one way to view them is as a higher order (and expanded) "super bifurcation diagram," which at the same time excludes many branches (and bifurcation points of wrong qualitative type) of a (stationary) bifurcation diagram as (physically) unacceptable final states due to their coming into violation of the set of parity rules. Such *parity diagrams* would measure the orientation in a solution and would in effect carry out a delineation of stable secondary bifurcations. They would perhaps implement a principle of exchange of stabilities (Principle 2). Such a parity rule theory would be *information theoretic* and could create a new chapter in bifurcation theory.

Some Comments on Dynamics in the Observed Flows. Why are there no further secondary motions generated when the two main vortices 1^+ and 2^{---} thrust against the right cavity wall? A theoretical answer for this may be found in Hiemenz flow (Schlichting [6, p. 96], Batchelor [4, p. 287]) at a stagnation point. The relatively high combined velocities of 1^+ and 2^{---} impact the right wall in (approximately) the fashion of Hiemenz flow. The available analytic solution to the latter will approximate the physical flow. The Hiemenz flow has no secondary recirculations.

We may go further and give the same explanation for why no further vortices (after the initiation of vortex 1^+) are generated in the upper left corner of the cavity. Imagine a flow 1^- "virtual" to the principal vortex 1^+, the virtual vortex 1^- a mirror image of 1^+, but above the lid. Again we have a Hiemenz flow, qualitatively, the analytic solution enduring into the physics.

The dynamics in the upper right corner is more complicated. Although an infinite sequence 3^{---} tries to form, as the eye of vortex 1^+ settles into a final position, the generation of this sequence ceases. Because long-time flow studies [22], [23] indicate a final solution which is periodic (see § 4), there continues to be sufficient vorticity generated at the separation point of 1^+ just below 3^- to maintain 3^- against the counterpull of the lid just above. The subvortex 3^{--} does not seem to maintain enough organization to qualify as a vortex, however, in the later stages of the flow. This "combined suction-impingement" corner flow deserves further study.

The vortex attempts, 2^+, continually spun off from the $2^{---}\cdots^{---}$ sequence, also appears to fail to maintain sufficient vortex organization due to its facing directly into the oncoming strong vortex 1^+. Nonetheless this $2^{+++}\cdots^{+++}$ semicoherent structure reappears intermittently throughout the flow history. Such a feature really asks for a higher resolution study.

Other features can be discerned. Consider the dynamical question of the *time of eddy to vortex formation.* We note:

	Eddy to vortex		Time (in seconds) for formation
Vortex 1^+	$t = 0$	$t = 3$	3
Vortex 2^-	$t = 10$	$t = 13$	3
Vortex 2^{--}	$t = 21$	$t = 24$	3
Vortex 2^{---}	$t = 27$	$t = 30$	3
Vortex 3^-	$t = 33$	$t = 36$	3
Vortex 3^{--}	$t = 54$	$t = 57$	3
Vortex 4^+	$t = 43$	$t = 46$	3
Vortex 5^+	$t = 56$	$t = 59$	3

Other dynamical features of cavity flow will be treated in § 4 and in [12].

Turning to the airfoil flows of Figs. 5.3, the effect of vortex splitting (fission) and its importance to stall avoidance have already been mentioned in §§ 1 and 2. Let me add (see [31]) that a separate computation of the lift force coefficients shows that, after the initial maximum lift which occurs in frame 17, a subsequent loss of lift occurs until frame 20. This can be seen in the flow portraits to correspond to the entrainment of a significant amount of fluid from the lower surface of the airfoil, evidenced in frame 18 by the recirculation about 1^+ of the first streamline from below the airfoil. The loss of lift continues until the splitting of 1^+ by the trailing edge vortex 3^-. Lift increases moderately during the splitting process and then increases dramatically as the shredding of 2^- by the returned coherent portion of 1^+ takes place.

Other dynamical features of airfoil flow will be treated in [12] and [33].

There needs to be a developing vocabulary of vortex dynamics as this theory moves ahead. Freymuth [36] gave such a vocabulary for physically observed dynamic separation patterns. Let me close this section with an incomplete beginning of such a vocabulary for vortex motions.

A Partial Vocabulary of Vortex Actions in Qualitative Vortex Dynamics

Two vortices of same sign combine into a single vortex of the same sign: *fusion*, *merger*, *coalescence*, *condensation*, *absorption*, *pairing*, *amalgamation*. Basic mechanisms: proximity, and no opposite-signed vortex between.

A vortex separates into two smaller vortices, all of the same sign: *fission*, *splitting*, *tearing*, *breakup*. Basic mechanisms: an approaching strong opposite-signed vortex, induced strain.

A vortex forms and leaves a given boundary, wall, surface: *shedding*, *vortex separation*. Basic mechanisms: a developing flow instability, acceleration, deceleration.

A vortex is reduced to disorganized form by an opposite-signed vortex: *shredding*, *grinding*. Basic mechanisms: a spatial entrapment by walls, opposite-signed vortices.

Two interacting vortices of opposite sign: *couples*, *V-state*, *dipole*, *modon*. Basic mechanisms: vorticity concentrated in isolated structures.

Two interacting vortices of same sign: *pair*. Basic mechanisms: repeated shedding situation, or distinct nonproximity spawning.

4. The Limits of Vortex Motion

This chapter of fluid dynamics has yet to be written. But I will try to describe here what I believe to be some of its general features. Further exposition will be found in [12].

There are many limiting processes in viscous vortex motions. I list four:

Time:	long-time behavior, $t \to \infty$,
Space:	small-scale behavior, $l \to 0$,
Parameter:	high-speed behavior, $\mathrm{Re} \to \infty$,
Discretization:	truncation error behavior, $\Delta x \to 0$.

Of course, all of these may be order reversed: $t \to 0$, e.g., effect of initial pressure impulse; $l \to \infty$, e.g., aspect ratio $\to \infty$, boundary conditions at ∞, or just downflow; $\mathrm{Re} \to 0$, e.g., Stokes flow; $\Delta x \to \infty$, far-field cutoffs.

In fact these four limiting processes are linked. Consider, for example, the currently most widespread subject of investigation, the possible evolution of a flow dynamics toward a turbulent state. A currently widely held view is the following:

$$t \to \infty$$

Possible long-time limits of flow:

(1) Steady: low Re, $u_t(t) \to 0$, $u(t, x) \to u_s(x)$.
(2) Periodic: medium Re, $u_t(t) \not\to 0$, $u(t, x) \to u_p(t, x, T)$.
(3) Aperiodic: higher Re, $u(t) \to u_a(t, x, \lambda_1, \cdots, \lambda_n)$.
(4) Turbulent: limiting Re, $u(t) \to u_{st}(t, x)$ stochastic behavior.

This view is that of the flow as a dynamical system, a theory reasonably well worked out for smaller systems of ordinary differential equations, not so well understood for partial differential equations.

Note that time and parameter limits are linked. One must be careful in uncoupling them. For example, the Hopf bifurcation theory attempts to study the transition from (1), steady time limits, to (2), periodic time limits, by means of varying Reynolds numbers within the stationary equations. The Euler equation approximations attempt to study time limits of the Navier–Stokes equations in the Reynolds number limit.

Continuing, I next consider certain spatial limits:

$$x \to 0$$

Possible small-scale structures of flow:

(1) Eigenmode macroscales: λ_L; underlying linear mode constraints; vortices.
(2) Taylor miniscales: λ_T; nonlinear coherent structures; eddies.
(3) Kolmogorov microscales: λ_K; deterministic organization limits; chaos.
(4) Dissipation molecularscales: λ_D; incoherent, disorganized flow; turbulence.

This breakdown may need improvement (e.g., the roles of chaos versus turbulence) but there is little doubt that vortex motions choose certain scales. A modelexample is the four scales (from meteorology): hurricane, tornado, dustdevil, and wisp.

Perhaps the best known example of a parameter limit

$$\text{Re} \to \infty$$

is the theoretical approximation of full Navier-Stokes flow by Euler flow, with the accompanying difficulties of matching boundary conditions.

A good example of discretization limit

$$\Delta x \to 0$$

is our finer mesh results [23] for cavity flow at Re = 10,000, under which the corner vortices 5^+ and 4^+ join to form a third principal recirculation region, as discussed briefly below.

Before turning to an attempted comprehensive formulation of the limiting processes in vortex motions (Principle 4, below), let me briefly indicate some aspects of the limiting considerations above that can be seen in the driven cavity flow. These are mentioned only by way of illustration and not in any sense of complete conclusion, but some important points can be brought out.

At all Re up to 2,000, the depth two cavity flows converged (numerically) to a steady final solution. In all of these flows (see specifically [22] Re = 400, 800, and 2,000), the left wall 2^- and lower corner vortices 4^- and 5^- merged to form a major lower cavity recirculation 245^-. Thereafter further corner eddies 4^+ and 5^+ appeared, and no sequence 2^{--} follows. On the other hand, the flow of Fig. 5.1 at Re = 10,000 formed the major lower cavity recirculation 2^{---} completely from wall vortices 2^-,

2^{--}, and 2^{---}. The corner eddies 4^+ and 5^+ then appeared, apparently with no hint of predecessors 4^- and 5^-. In other words, the final vortex states at Re = 2,000 and Re = 10,000 appear in different order. This means that one should be quite careful in drawing conclusions from continuation studies, either stationary or unsteady, to be sure that continued solutions can acquire the quite different vortex states found at higher Re.

This "interchange of order" of t and Re is also an aspect of the semigroup (tacitly assumed) laws among parameters mentioned after Principle 2 above. Similar such semigroup laws may or may not hold between combinations of time, space, parameter, discretization, and solutions. Counterexamples are given in [22].

Also [22], we tested the unsteady flow at Re = 2,000 on both a coarse and fine mesh. Both flows converged to the "same" steady solution. Repeating this procedure for Re = 10,000 has become a time consuming task. The original study [22] of the Re = 10,000 flow went to 360 sec. Because we used an explicit scheme in primitive variables, this study invoked 360,000 Poisson solves on a 40×80 grid. A final periodic solution was in evidence but not yet attained. Our later studies [23] of this flow have shown by a more efficient implicit method that indeed (on a 48×96 grid) a periodic solution is attained. But the period of about 4.6 sec estimated from [22] on a 40×80 grid at 360 sec becomes a converged period of about 3.5 sec on a 48×96 grid at about 1500 sec [23]. Furthermore, there are indications that at Re = 10,000, we are well beyond the Hopf bifurcation value and are attempting convergence to a higher critical state. In particular, an Re = 10,000 run [23] on a 96×192 grid has not yet converged to our satisfaction at this writing. In this run, the corner vortices 5^+ and 4^+ have joined to form a third recirculation region at the bottom of the cavity.

Because of this we lowered Re to 5,000 and have shown beyond a doubt [23] that we attain a periodic solution, the value of the period converging coarse (48×96, $t \approx 2.47$) to fine (96×192, $T \approx 2.31$). There is also some analytic theory to support qualitative continuity in the mesh parameter and, for the sake of definiteness, I conjecture that the analytic solution at Re = 5,000 is periodic with a period not far below these computational values. The finer mesh final solution resolves the boundary layer at the bottom of the cavity into a third "figure eight" principal recirculation. It is interesting to ask, if the analytic solution is indeed periodic, "how many" of the corner vortex sequences it has, finite or infinite, and if they are not locally stationary, how do they infinitesimally reverberate or otherwise interact with each other. These are questions in the limits of vortex motion. They usher in a new chapter in numerical analysis, methods for the investigation and determination of the qualitative properties of unsteady flows treated as infinite-dimensional dissipative dynamical systems.

As mentioned earlier, for steady Stoke's flow in the unit cavity we managed to numerically resolve 25 of the corner vortex sequences, the last ones achieving intensities less than 10^{-100}. This is below the Kolmogorov microscale η.

Another spatial limit interpretation, in the sense of the capability of a fast changing solution to maintain its regularity on small subdomains, is related to bursting phenomena. I alluded to this in discussing subprinciple (iv) in § 1. In the recent analysis of Aubry et al. [15], it is concluded from a truncated analysis of the partial differential equations that wall rolls exist and eject (turbulent) solution filaments into the flow. It appears to me that the wall shedding sequence $2^{---} \cdots {}^{---}$ of the cavity flow, caught between the counterrotating rolls 1^+ and 2^{---}, aided and abetted by the quickly shredded counterrotating bursts 6^+, and hindered only ineffectually by the disorganized

2^+ structures, are just such a mechanism. Were there no right wall, the effects of the $2^{---}\cdots^{---}$ bursts would filament into the flow there.

Looking at Fig. 5.1, at $t = 337$ we see the next local upsurge (in the region indicated by s in Fig. 5.2) beginning. Acceleration plots [22] show this relatively disorganized state to be one of minimum acceleration (the previous burst occurred at $t = 336$). At $t = 338$ the small eddy 6^+ has transferred the (translational) upsurge energy into an organized vortex motion. This is an enstrophy transfer from 2^{---} to 6^+. At the same time the new 2^- counterrotation is forming off the principal vortex 1^+. By $t = 339$ the 6^+ and 2^- have collided, forcing incoherent motion into the shear interface region extending laterally across the cavity between 1^+ and 2^{---}. Note, carefully comparing $t = 339$ and $t = 340$, the propagated disturbance in the arrows in the far (right-hand one third) of that interface. The acceleration plots show a maximum acceleration at $t = 340$. What is left of 2^- is absorbed by 2^{---} at $t = 341$ and the bursting cycle starts over again. Maximum accelerations are found again at $t = 345, 350, 354, 359$, approximately; compare the bursting events in Fig. 5.1.

At the very smallest (viscous) scales λ_D, dissipation is the predominant mechanism. When one reflects on the fact that all dissipation has been represented in the single term $\nu\Delta u$, one can conclude that at these scales the Navier–Stokes equations are near the limits of their validity. On the other hand, these equations are proving to be remarkably robust at small scales.

Combining the time, space, parameter, and discretization limiting processes leads me to my fourth and final principle of vortex motion.

Principle 4 (The Spectrally Refining Principle)

Within the above three principles and their limits lie the mechanisms which guide the flow through spectral refinement to its final (e.g., steady, periodic, aperiodic, turbulent) state.

Four (Spectrally Refining) Subprinciples

(i) Energy Cascade. Energy is transferred to larger scales, in an attempt to conserve the angular momentum in larger vortices.

Space-filling: the enlargements eventually extend to (reachable) region boundaries.

Self-organizing: the total flow patterns can be viewed "top-down" as organized hierarchically ordered from the larger vortex motions to the smaller ones.

Parity-adjusting: parity counting rules begin with the larger vortex motions, and total parity (total number of sign variations) is eventually minimized by the merger of smaller into larger vortices.

(ii) Enstrophy Cascade. Enstrophy is transferred to smaller scales, large eddies generating small ones (through viscous action) to preserve fluid continuity.

Space-filling: the smaller eddies eventually fill all (reachable) small corners, nooks, and crannies.

Self-organizing: the smaller secondary and tertiary eddies are driven locally by the larger adjacent ones.

Parity-adjusting: somehow, the signs of the smaller eddies should carry smaller weights than the larger ones.

(iii) Limiting Scales. The nonlinearity of the flow differential equation establishes advection-diffusion scale limits beyond which solutions lose regularity.

Space-filling: these scale limits will explain why certain (e.g., corners) region space is unreachable.

Self-organizing: the nonlinearity is essential in explaining the control of the higher frequencies (smaller eddies) by the lower frequencies (larger vortices).

Parity-adjusting: the advection-diffusion scale limits will involve sign limits in the allowable oscillations of the eigenmodes of linear stability equations.

(iv) Dissipative Convergence. The (viscous) linearity of the flow differential equations will eventually produce convergence over long enough time.

Space-filling: eventually the solutions will diffuse to *all* (reachable) region space, as is required of a solution of a differential equation.

Self-organizing: the numerical, or physical visualization meshes must go to zero to "accomplish" a complete convergence, i.e., this self-organizing action is one carried ultimately to the limiting sense. The final state possesses the most order possible.

Parity-adjusting: the velocity directions (in the sense of flow arrow orientations) will align themselves like those of most proximate vortex motions, thereby adding (like-signed) distortion boundary effects to those dominating nearby vortex motions.

The notions of energy cascades are well known in the fluid dynamics literature. Sometimes the cascade is meant to refer to the transfer to lower scales and then into dissipative loss due to viscosity. But here in the energy cascade Subprinciple (i) I am emphasizing the transfer of energy to larger scales (sometimes called the "reverse" energy cascade). I imagine that the momentum equation prefers this as a means of conserving energy. As a flow encounters an obstacle, wall, or other features of the fluid motion, friction is minimized by assuming a rolling motion. (Every gymnast, wrestler knows this.) Mathematically, this means to transfer as much energy as possible to the inertial term in the equation.

The transfer to larger scales is well illustrated in the cavity flow of Fig. 5.1: examples, the building (and maintaining) of vortex 2^{---} by the $2^{---} \cdots {}^{---}$ sequence, the mergers of 3^- and 3^{--}

In the enstrophy cascade subprinciple (ii) lie still unanswered fundamental questions related to the closure problems in turbulence theories. In terms of the discussion of the energy cascade above, the energy transferred to smaller eddies in the enstrophy cascade is the part, when it reaches small enough eddies at viscosity scale, that cannot be conserved.

The transfer to smaller scales is well illustrated by the cavity flow: examples, 3^+, 4^+, 5^+, and the smaller (unresolved) corner sequences in the corners below 4^+ and 5^+.

It is noteworthy that early cavity studies believed that as Re was increased, the major vortices like 1^+ would drive the corner region counterrotations down to very small size. This turned out to not be the case. An explanation for this is that even though the corner eddies maintain a significantly large spatial extent even as Re is increased, they contain only a rather small portion of the total energy.

I formulated the limiting scales subprinciple (iii) to emphasize a thought I had only this year, but one that probably has occurred to others thinking about these things: it is absolutely essential that the equations be *nonlinear.* Linear equations do not permit the kind of energy transfer between modes seen in fluids. More precisely, stated mathematically: the linearized (Stokes) flow equations have only $u = 0$ as an attractor, and it is unstable. Restated, the Kolmogorov theory, and the more recent (Temam, Foias, and others) inertial manifold theory, are of necessity nonlinear.[3]

For example, the "corner space filling" infinite eddy sequence predicted by Moffatt [14] depends on a linear argument that probably will not stand up to the (nonlinear) scales theory. I conjecture that there will only be a finite number of corner eddies allowed by the nonlinear equations. It is possible that not all original region space can be reached.

The solution regularity breakdown when scales are exceeded may occur as seen in combustion problems when a nonlinear eigenvalue parameter becomes too large. As in our results reported in [19], such instabilities can be tracked in terms of the number of zeros of associated linear eigenmodes. Parity rules may determine the limits. From a physical viewpoint, these scale limits will create a qualitative change in the solutions of the Navier–Stokes equations such that the continuity ($u \in C^1$) and continuity ($\text{div}\, u = 0$) can no longer be satisfied.

The ideas in the dissipative convergence subprinciple (iv) are all speculative. The diffusive linearity $\nu \Delta u$ is capable of propagating the solution (in analogy with the heat equation) at infinite speed into all as yet unfilled vessel space. Some of that vessel space (as was finally the case for the famous Dirichlet problem) may turn out to be unreachable.

5. Summary

A framework for a theory of viscous vortex dynamics has been advanced.[4] The view is fourfold and both *paradigmatic* and *archetypal*. The a posteriori analysis of the vortex motions of a flow is cast at four levels: generation, evolution, dynamics, and limits. The flow proceeds simultaneously within these hierarchies, ordered by time. Four corresponding underlying action principles are formulated on an a priori basis: space-filling, self-organizing, parity-adjusting, and dissipative-converging.

Endnotes

[1] I have (admittedly) written this chapter in a speculative or at least unorthodox tone. Indeed such a tone was encouraged in the original format of the challenging publication series (Frontiers) within which this book was invited about four years ago. Among the editorial goals were that the series would provide "a provocative intellectual forum," with chapters including "nonuniform viewpoints on state-of-the-art," emphasizing "controversial aspects to stimulate and provoke," with each chapter "most likely a highly opinionated account detailing the author's viewpoint." Taking advantage of that (unusual) charge, I completed the writing of this chapter about two years ago and presented the *Four Principles* at the IMACS World Congress in Paris in 1988.

In the two years of publication delay caused by the usual things, the editorial format became a bit more conservative. But I have decided to leave my attempt at a broad theoretical framework for the understanding of vortex motions, and by extension most of fluid dynamics, as written. The reader may take it as a personal view, an attempt to synthesize a theory which up to now does not exist.

A second added endnote[2] cites some known flow topological rules, related to but not the same as my parity rules, of which I was unaware at the first writing. I have also added a third endnote[3] including a little on the very recent inertial manifold theory of long-time behavior. In particular, I show it to be, in its stationary aspects, a Lyapunov–Schmidt alternative. Finally, an endnote[4] containing further closing perspectives has been added at the end of the chapter.

[2] During a recent visit (1989) to the NASA Lewis Research Center I learned from Dr. H. Kao there that certain topological rules have been employed to delineate plausible patterns in flow visualization studies. These rules apparently go back at least to R. Legendre, *Research*

Aerospatiale (1956). See also Lighthill's Chapter 2 in *Laminar Boundary Layers*, L. Rosenhead, ed. (1963). A more full account will be given in [12].

These topological rules are in the same spirit as the ideas for parity rules that I have put forth above. The topological rules are however purely kinematic and not dynamic and do not take into account pressure and viscous forces. As such they could be regarded as steady *constraint rules* which could implement basic connectivity restrictions under which the dynamic parity rules must evolve. In that way they could be a useful complement to the parity rules, especially for the latter in three dimensions where connectivity becomes a larger consideration.

[3] The inertial manifold theory is a recent, ambitious attempt to develop an analytic theory of a *finite-dimensional*, *inertial manifold* $\mathcal{M}$ that is positively invariant for the flow, contains all attractors, and that attracts all the orbits at an exponential rate. If such a manifold can be found for the time-dependent Navier–Stokes equations, it will control all of the long-time behavior of solutions of the flow equations. Moreover, the flow equations restricted to the finite-dimensional $\mathcal{M}$ will become ordinary differential equations amenable to ordinary differential equation solvers.

This theory is usually contexted in the semidiscrete (spectral) Galerkin computational frame. Excellent recent references are the book by R. Temam, *Infinite Dimensional Dynamical Systems in Mechanics and Physics*, Springer-Verlag, Berlin, New York, 1988 and the AMS/IMS/SIAM Research Conference Proceedings, The Connection between Infinite Dimensional and Finite Dimensional Dynamical Systems, *Contemporary Mathematics* 99 (1989), B. Nicolaenko, C. Foias, and R. Temam, eds., 1989.

It is my understanding that the physical notion of such manifolds came out of meteorology/oceanography where such large-scale controlling currents are observed. The ideas are related to those of Kolmogorov and Kraichnan concerning the exponential decay of all small eddies toward a small limiting value. Numerically it would mean that asymptotically obtained long-time states (e.g., as we are pursuing in [23]) on a finite enough grid are all that matter, i.e., smaller grids would not have any physical significance in the long-time behavior.

The procedure for seeking inertial manifolds (see Foias, Manley, and Temam, *Modelling of the interaction of small and large eddies in two-dimensional turbulent flows,* Math. Mod. and Num. Anal. (M^2AN), 22 (1988), pp. 93–114.) seems to me to be comparable at least in concept to the Lyapunov–Schmidt alternatives method for equations of the form $Lu + N(u) = 0$, L linear, N nonlinear. See, for example, Gustafson and Sather, *Large nonlinearities and closed linear operators,* Arch. Rational Mech. Anal., 52 (1973), pp. 10–19. In both of these theories, one projects the equation onto a finite-dimensional subspace P_mH of eigenfunctions of the linear part, with the rest of the equation projected into $(I - P_m)H$.

In the Lyapunov–Schmidt theory (see GS, just cited) one then tries (by estimates) to control the nonlinear part of the equation in $(I - P_m)H$ in such a way that a solution $u = p + q$, p in P_mH, q in $(I - P_m)H$, p dependent on q, is shown to exist, at least for small q. The finite-dimensional equation in P_mH is called the *bifurcation equation* and the key in any implementation of the Lyapunov–Schmidt technique is in the demonstrated dependence of p on q. For example, the projected infinite-dimensional q equation has the form $(I - P_m)[L(p + q) + N(p + q)] = 0$. In the case where L is self-adjoint and P_m is an eigenprojector, this becomes $L(q) = -(I - P_m)N(p + q)$, and in the case where L is invertible, $q = -L^{-1}(I - P_m)N(p + q)$.

In the inertial manifold approach (see FMT just cited, especially eqs. (1.11), (2.2)) one splits the space as above, the projected infinite-dimensional (time-dependent) q equation taking the form $dq/dt + L(q) + (I - P_m)N(p + q) = 0$. In FMT, L is νA, self-adjoint, and N is the Navier–Stokes inertial term B. For longer times q is shown small, and then it is assumed that the evolution has become quasistatic to the extent that dq/dt can be ignored, resulting in the approximate equation $q = -L^{-1}(I - P_m)N(p + q)$. This is a Lyapunov–Schmidt alternative problem, as I claimed above.

In FMT a further approximation is then proposed (on the basis of estimates), namely, that q is so small that the nonlinear term may be approximated by $(I - P_m)N(p)$. This results in the considerably simpler alterantive equation $q = -L^{-1}(I - P_m)N(p)$. This step may be described as approximating an implicit equation by an explicit equation. The graph of the resulting explicit dependence $q = \Phi_0(p)$ defines in H a smooth manifold $\mathcal{M}_0$ of dimension m which is called the *approximate inertial manifold.* This manifold is thus seen to be a considerably simplified Lyapunov–Schmidt alternative.

On the other hand, the actual sought inertial manifold $\mathcal{M}$ (see NFT cited above) is to be the graph of a function $q = \Phi(p)$ with additional properties: bounded support in P_mH, containing the global attractor $\mathcal{A}$, and possessing a Lipschitz property with respect to some power of the linear operator L (so that the exponential attraction will follow in the long-time flow). This feature is not in the usual Lyapunov–Schmidt alternative theory and would appear to delineate the two theories. The inertial manifold theory is intrinsically time-dependent, whereas the Lyapunov–Schmidt theory is basically stationary.

[4] My perspective in the chapter is clearly two-dimensional. Hence all of the admonitions in the literature about distinguishing the two- and thee-dimensional views apply. One is reminded, in particular, of Charles Truesdell's, *The Kinematics of Vorticity* (1954), poetic emphasis on the three-dimensionality of the vorticity vector, as quoted for example by Lugt [1, p. 45]. However, very few long-time histories of the dynamics of the vorticity vector in three-dimensional flows have yet been visualized. Vorticity is a higher derivative and most of what we have really seen or simulated about it has been in two-dimensional sections and at lower Reynolds number.

Second, my perspective is that of the whole Navier–Stokes equations, but in the solenoidal subspace. By the first noted, I mean never thinking in terms of the Euler equations, in which one has thrown away diffusion. I must mention that it bothered me from my very first physics courses how physicists always got their results so cleanly. My mind always bogged down with imagined side effects, what was happening at the boundary of whatever they were considering, and so on. Later on I realized that my intuition had been right, that so-called dissipative and constant forces are always present, indeed essential. On the other hand, as second noted above, the perspective of the chapter is limited to incompressible (solutions always in the solenoidal subspace) flows, for want of personal experience and intuition with compressible flows. Hence temperature dependence is also kept out. The rationale (excuse) is to "keep it to the basics," but the real reason, as with the physicists, is to make the context more manageable.

Third, the range of validity of the principles is unknown when one ventures away from Dirichlet, especially no-slip, boundary conditions. Most of my experience, both theoretical and in computation, has been for those boundary conditions. The Navier–Stokes equations, by themselves, have a multitude of solutions: what is selected out in applications depends critically on the boundary conditions (and how the latter are modeled). This point is made very nicely in Patrick Roache's fine early book *Computational Fluid Dynamics*, 1976. On the other hand, Dirichlet boundary conditions are at least more realistic than the "space-periodic" boundary conditions (under which the flow eigenmodes and the Fourier eigenmodes are made to coincide) currently used in many theoretical studies. Viscosity, if retained in the partial differential equations in the flow domain, needs also to act at the boundary of the flow.

Fourth, I believe that the importance of vortex motion and its mathematical codification, which I have attempted here, will greatly increase. This view is consistent with that of Peter Lax in his recent survey, *The flowering of applied mathematics in America*, SIAM Rev., 31 (1989), pp. 533–541, wherein fluid dynamics is projected as a central future discipline in applied mathematics. My four principles may be viewed as a synthesis and axiomatic framework to guide its testing and mathematical justification.

References

General References

[1] H. LUGT, *Vortex Flow in Nature and Technology*, John Wiley, New York, 1983.

[2] M. VAN DYKE, *An Album of Fluid Dynamics*, Parabolic Press, Stanford, CA, 1982.

[3] O. LADYZHENSKAYA, *The Mathematical Theory of Viscous Incompressible Flow*, Gordon and Breach, New York, 1963.

[4] G. BATCHELOR, *An Introduction to Fluid Dynamics*, Cambridge University Press, Cambridge, 1967.

[5] J. SERRIN, *Mathematical Principles of Classical Fluid Mechanics*, Handbuch der Physik VIII/1, Springer-Verlag, New York, 1959, pp. 125–263.

[6] H. SCHLICHTING, *Boundary Layer Theory*, McGraw-Hill, New York, 1979.

[7] R. TEMAM, *Navier–Stokes Equations*, North-Holland, Amsterdam, 1983.

[8] H. LAMB, *Hydrodynamics*, Dover, New York, 1945.

[9] D. JOSEPH, *Stability of Fluid Motions*, I, II, Springer-Verlag, New York, 1976.

[10] R. PEYRET AND T. TAYLOR, *Computational Methods for Fluid Flows*, Springer-Verlag, New York, 1983.

Particular References

[11] K. GUSTAFSON, *A Theory for Computational Vortex Dynamics*, IMACS Ann. Comput. Appl. Math., 1 (1989), pp. 265–269; Opening Lecture, Session on Computational Vortex Dynamics, 12th IMACS World Congress on Scientific Computation, Paris, July 1988.

[12] ———, in preparation.

[13] LORD RAYLEIGH, *On the question of the stability of the flow of fluids*, Philos. Mag., 34 (1892), pp. 59–70.

[14] K. MOFFATT, *Viscous and resistive eddies near a sharp corner*, J. Fluid Mech., 18 (1964), 1–18. W. DEAN AND P. MONTAGNON, *On the steady motion of viscous liquid in a corner*, Proc. Cambridge Philos. Soc., 45 (1949), pp. 389–394. H. LUGT AND E. SCHWIDERSKI, *Flows around dihedral angles*, Proc. Roy. Soc. London Ser. A, 285 (1965), pp. 382–422.

[15] N. AUBREY, P. HOLMES, J. LUMLEY, AND E. STONE, *The dynamics of coherent structures in the wall region of a turbulent boundary layer*, J. Fluid Mech. 192 (1988), pp. 115–173. See also Ersoy and Walker, *Viscous flow induced by counter-rotating vortices*, Phys. Fluids, 28 (1985), pp. 2687–2698, who found a similar phenomenon somewhat earlier.

[16] I. PRIGOGINE AND G. NICOLIS, *Self-Organization in Nonequilibrium Systems*, John Wiley, New York, 1977.

[17] C. C. LIN, *On the Motion of Vortices in Two Dimensions*, University of Toronto Studies, Applied Mathematics Series, No. 5, University of Toronto Press, Toronto, Ontario, Canada, 1943, pp. 1–39.

[18] P. CONSTANTIN AND E. TITI, *On the evolution of nearly circular vortex patches*, Comm. Math. Phys., to appear.

[19] K. GUSTAFSON, E. ASH, B. EATON, K. HALASI, AND R. LEBEN, *Nonblowups, Periodicities, Vortex Shreddings in Combustion and Hydrodynamic Flows: A Conference Report*, Sixth Army Conference on Applied Mathematics and Computing, ARO Report 89-1, Trans. Army Research Office, 1989, pp. 483–500. E. ASH, B. EATON, AND K. GUSTAFSON, ZAMP 41 (1990), pp. 558–578.

[20] J. C. MCWILLIAMS, *The emergence of isolated coherent vortices in turbulent flow*, J. Fluid Mech., 146 (1984), pp. 21–43. A. K. M. F. HUSSAIN, *Coherent structures and turbulence*, J. Fluid Mech., 173 (1986), pp. 303–356. A. BABIANO, C. BASDEVANT, B. LEGRAS, AND R. SADORNAY, *Vorticity and passive-scalar dynamics in two-dimensional turbulence*, J. Fluid Mech., 183 (1987), pp. 379–397.

References for Cavity Flow

[21] K. GUSTAFSON AND R. LEBEN, *Multigrid calculation of subvortices*, Appl. Math. Comput., 19 (1986), pp. 89–102. *Vortex subdomains*, in Proc. 1st Internat. Symposium on Domain Decomposition of Partial Differential Equations, R. GLOWINSKI, G. GOLUB, G. MEURANT, AND J. PERIAUX, eds., Society for Industrial and Applied Mathematics, Philadelphia, PA, 1988, pp. 370–380. *Multigrid localization and multigrid grid generation for the computation of vortex structures and dynamics of flows in cavities and about airfoils*, in Multigrid Methods, S. McCormick, ed., Marcel Dekker, New York, 1988, pp. 229–249.

[22] K. GUSTAFSON AND K. HALASI, *Vortex dynamics of cavity flows*, J. Comput. Phys., 64 (1986), pp. 279–319. *Cavity flow dynamics at higher Reynolds numbers and higher aspect ratio*, J. Comput. Phys., 70 (1987), pp. 271–283. *Parameters of Hopf bifurcation in driven cavity flow*, in Proc. IMACS 12th World Congress on Scientific Computation, Vol. 2, 1988, pp. 278–281. *The flow semigroup approximation problem*, to appear.

[23] J. GOODRICH, K. GUSTAFSON, AND K. HALASI, *Hopf bifurcation in the driven cavity*, J. Comput. Phys. (Sept., 1990). *Time asymptotic fluid flows*, Comput. Phys. Comm. (1991), to appear.

[24] K. GUSTAFSON, K. HALASI, AND R. LEBEN, *Controversies concerning finite/infinite sequences of fluid corner vortices*, in Proc. 1987 Joint AMS/IMS/SIAM Summer Research Conference on the Connection between Infinite Dimensional and Finite Dimensional Dynamical Systems, B. Nicolaenko, C. Foias, and R. Temam, eds., Contemp. Math., 99 (1989), pp. 351–357.

[25] O. BURGRAFF, *Analytical and numerical studies of the structure of steady separated flows*, J. Fluid Mech., 24 (1966), pp. 113–151. F. PAN AND A. ACRIVOS, *Steady flows in rectangular cavities*, J. Fluid Mech., 28 (1967), pp. 643–655.

[26] U. GHIA, K. GHIA, AND C. SHIN, *High* Re *solutions for incompressible flow using the Navier–Stokes equations and a multigrid method*, J. Comput. Phys., 48 (1982), pp. 387–411.

[27] R. SCHREIBER AND H. KELLER, *Driven cavity flows by efficient numerical techniques*, J. Comput. Phys., 49 (1983), pp. 310–333. R. GLOWINSKI, H. KELLER, AND L. REINHART, *Continuation-conjugate gradient methods for the least squares solution of nonlinear boundary value problems*, SIAM J. Sci. Statist. Comput., 6 (1985), pp. 793–832.

[28] P. GRESHO, S. CHAN, R. LEE, AND C. UPSON, *A modified finite element method for solving the time dependent incompressible Navier–Stokes equations. Part* 2: *applications*, Internat. J. Numer. Methods Fluid, 4 (1984), pp. 619–640.

[29] D. SERRE, *Equations de Navier–Stokes stationaires avec données peu reguliere*, Ann. Scuola Norm. Sup. Pisa Cl. Sci. (4), (1984), pp. 543–559.

[30] J. KOSEFF AND R. STREET, *The lid-driven cavity flow: a synthesis of qualitative and quantitative observations*, J. Fluids Engrg., 106 (1984), pp. 390–398. A. PRASAD AND J. KOSEFF, *Reynolds number and end-wall effects on a lid-driven cavity flow*, Phys. Fluids, (1989), pp.

References for Airfoil Flow

[31] K. GUSTAFSON AND R. LEBEN, *Robust multigrid computation and visualization of separation and vortex evolution in aerodynamic flows*, in Proc. 1st National Fluid Dynamics Congress, AIAA/ASME/SIAM/APS, Part 1, Cincinnati, OH, July 1988, pp. 174–182.

[32] ———, *Viscous flow around airfoils*, IMACS Ann. Comput. Appl. Math., 1, (1989), pp. 139–144.

[33] ———, in preparation.

[34] K. GUSTAFSON, *Partial Differential Equations*, 2nd ed., John Wiley, New York, 1987.

[35] Boulder Daily Camera, Boulder, Colorado, Nov. 25, 1988, p. 10A, "Ice eyed as cause of B-1B crash," reported via news nets via a Washington report via the Chicago Tribune.

[36] P. FREYMUTH, *The vortex patterns of dynamic separation: a parametric and comparative study*, Progr. Aerospace Sci., 22 (1985), pp. 161–208.

[37] U. B. METHA AND Z. LAVAN, *Starting vortex, separation bubbles, and stall: a numerical study of laminar unsteady flow around an airfoil*, J. Fluid Mech., 67 (1975), pp. 227–256.

[38] H. J. LUGT AND H. J. HAUSSLING, *The acceleration of thin cylindrical bodies in a viscous fluid*, J. Appl. Mech., 45 (1978), pp. 1–6.

[39] Y. LECOINTE AND J. PIQUET, *Unsteady viscous flow around moving circular cylinders and airfoils*, AIAA paper 85-1490, in AIAA-CP 85, Proc. 7th CFD Conference, Cincinnati, OH, American Institute of Aeronautics and Astronautics, NY, 1986, pp. 38–49. K. GHIA, G. OSWALD, AND U. GHIA, *Analysis of two-dimensional incompressible flow past airfoils using unsteady Navier–Stokes equations*, in Numerical and Physical Aspects of Aerodynamic Flows, III, T. Cebeci, ed., Springer-Verlag, Berlin, New York, 1986, pp. 318–340.

[40] W. J. MCCROSKEY, *Unsteady airfoils*, Ann. Rev. Fluid Mech., 14 (1982), pp. 285–311. M. GAD-EL-HAK, *Unsteady separation on lifting surfaces*, Appl. Mech. Rev., 40 (1987), pp. 441–453.

VISUALIZATION AND COMPUTATION OF HOVERING MODE VORTEX DYNAMICS

6

Peter Freymuth, Karl E. Gustafson, and Robert Leben

Introduction

An airfoil is operated in combined plunging and pitching motion to generate a thrusting jet in a still air environment. The device serves as a simple, generic, and two-dimensional model for hovering flight of small birds and insects. When properly tuned the device produces very large average thrust coefficients. As shown here, the visualized vortical signature of the tuned jet is a vortex street with reverse sense of rotation as the vortices of a Karman street.

Because this problem is two-dimensional, it is, in principle, amenable to detailed computation. However, a number of new considerations arise in simulating the combined translating-pitching motion. We give here first results of the computed vortex dynamics, which agree quite well with the physical visualizations.

1. Visualization

We found previously [1], [2] that an airfoil which is exposed to a steady wind and which executes pure plunging or pure pitching motions can generate thrust similar to a flapping bird wind or a pitching fish tail. The vortical signature of this thrust generation is a vortex street with vortex rotation reverse to a Karman street, i.e., a two-dimensional jet. To obtain this thrust signature in a still air environment, i.e., during hovering flight, it has been stated [1] that a combination of plunge and pitch motions of the airfoil would be needed. In this paper we follow up on this concept by designing and investigating a thin airfoil which executes the appropriate periodic plunge-pitch motions in a still air environment.

Our apparatus is intended as a basic, two-dimensional, generic model of the unsteady hovering flight of small birds and insects. In contrast, the mainstream literature [3]–[9] is very animal specific (hummingbirds, dragonflies, chalcid wasps, etc.), and therefore is very complex and hard to interpret from an aerodynamic point of view. It is hoped that these simple and complex approaches are synergistic.

Comments on Possible Thrusting Modes. The dominant thrusting motion of birds during sustained forward travel appears to be wing flapping, while for fishes it is pitching of the tail. Consequently, airfoils in pure plunge or in pure pitch motions were investigated in a wind tunnel [1], [2] to reveal the propulsive vortical signature for these modes of travel within a two-dimensional approximation. During sustained forward flight the thrust force has only to overcome the drag of the bird, which is an order of magnitude smaller than the weight balancing lift force. Therefore we expect the flow field around and behind the wing to be dominated by the bound and tip vortices of static lift generation, as facilitated by the mean angle of attack of the wing. Thrusting should only cause a perturbation of the attached flow field adjacent to the wing.

In contrast, during hovering in still air thrusting action by the wings must support the entire weight of the animal. No permanently bound vortex exists in this case. The wings must create their own relative wind, which makes hovering especially difficult for larger animals. It appears that hovering is mostly achieved by appropriate combined flapping and pitching motions. In order to model hovering in two-dimensional approximation, a small airfoil model was built and investigated which allowed combined plunging and pitching motions in still air. Because of the large pitch amplitudes and because of the strong unsteady wing loading by thrusting in the hover mode leading edge separation is the most important process to watch for.

Hovering is mainly achieved by insects and the question arises whether during sustained forward flight they use the forward flight mode of birds or whether they tilt their vertical hover thrust forward. In other words, during forward thrusting, do insects more resemble an airplane or a forward flying helicopter? The question seems not to have been generally answered and will not be investigated in this paper, neither will other modes of flight, such as starting, landing, upward flight, be considered. Many insects have pairs of wings on each side. Possible interactions of wings will not be investigated either, but all these omissions can be interesting future topics.

Dimensional Considerations and Modes of Hovering. Consider a thin, flat plate airfoil with chord length c exposed to still air and executing a translating (plunging) motion h in horizontal direction:

$$h = h_a \sin 2\pi f t, \tag{1.1}$$

where h_a is the amplitude of translation, f the frequency of oscillation, and t time. Consider the airfoil to simultaneously execute a pitching motion around the half-chord axis:

$$\alpha = \alpha_0 + \alpha_a \sin(2\pi f t + \phi), \tag{1.2}$$

where α is the angle of attack with respect to the horizontal line, α_0 is the average or offset angle of attack, α_a is the pitch amplitude, and ϕ is the phase difference between pitching and plunging. Our experimental apparatus is designed to allow, within some limits, combined plunging-pitching motions of the airfoil.

Appropriate dimensionless parameters of the system are: α_0, α_a, ϕ, the dimensionless plunge amplitude h_a/c, and a Reynolds number $R_f = 2\pi f h_a c/\nu$ based on maximum plunge speed and on c, where ν is the kinematic viscosity.

The simplest modes of hovering identified and investigated in this paper are: mode 1 or "water treading mode" characterized by $\alpha_0 = 0$ and $\phi = \pi/2$ (90°), as sketched in Fig. 6.1. In this mode leading and trailing edges switch their role during one cycle of oscillation. The resulting hover-jet as sketched is thrown upward, with thrust on

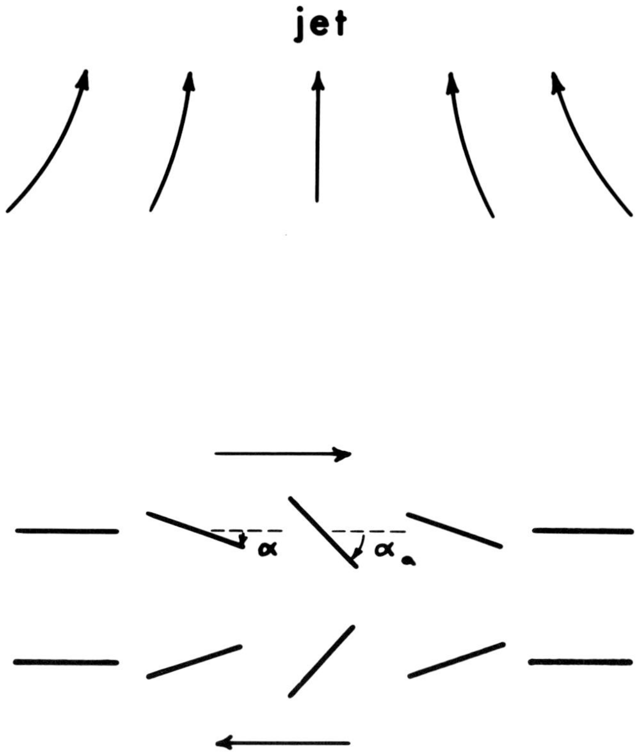

Fig. 6.1. Sketch of combined translating-pitching motion of the airfoil during one cycle for mode 1 hovering.

the airfoil pointing downward. Since the apparatus is mounted on the ground and obstructs the view in downward direction, we intentionally let the jet develop in upward direction, in contrast to insect hovering. The other mode investigated is mode 2 or "degenerate figure eight mode" characterized by $\alpha_0 = \pi/2$ (90°) and $\phi = -\pi/2$ (−90°). As sketched in Fig. 6.2, in this mode leading and trailing edges do not switch their role during one cycle.

It should be mentioned that switching sign of ϕ only switches the jet from "up" to "down."

Thrust Coefficient of the Hover-jet. Detailed experimental results on the hover-jet will be shown later. Our concept of such a jet is best illustrated at this point by a flow visualization example in Fig. 6.3. The hovering airfoil near the bottom of the figure generates a vortex street which is marked by smoke. The left column of vortices is staggered compared to the right column and these vortices induce on each other an upward motion. This upward moving somewhat turbulent hover-jet generates thrust on the airfoil in downward direction.

To characterize the time-averaged thrust T on the airfoil, let us define a thrust coefficient C_T:

$$C_T = \frac{T}{(\rho/2)\overline{V_t^2}C \cdot l}, \tag{1.3}$$

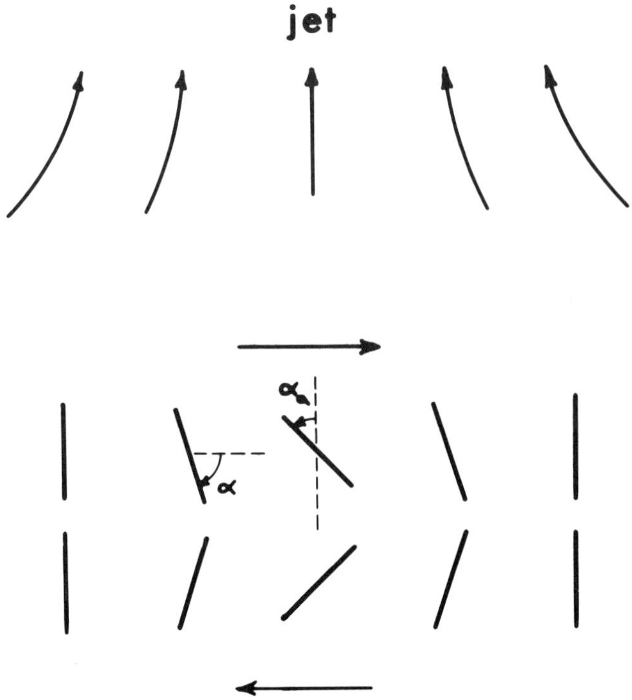

Fig. 6.2. Sketch of combined translating-pitching motion of the airfoil during one cycle for mode 2 hovering.

where ρ is the air density, $\overline{V_t^2}$ is the mean square speed of the horizontal airfoil motion, and $l \gg c$ is the span of the airfoil. From the kinematics of the translating motion

$$\overline{V_t^2} = 0.5(2\pi f h_a)^2 \tag{1.4}$$

and from the momentum theorem

$$T = \rho l \int_{-\infty}^{\infty} \overline{V^2}\, dx, \tag{1.5}$$

where $\overline{V^2}$ is the mean square velocity in the jet at a sufficient distance above the airfoil where the jet has acquired ambient pressure, x is the horizontal coordinate across the jet. The thrust coefficient of the jet therefore is

$$C_T = \frac{\int_{-\infty}^{\infty} \overline{V^2}\, dx}{(\pi f h_a)^2 c}. \tag{1.6}$$

Thrusting would, of course, be upward during actual hovering and the jet directed downward. The thrust coefficient C_T may therefore also be considered a lift coefficient of the hovering airfoil.

The purpose of our paper will be the experimental determination of C_T over a limited parameter range and the visualization of the vortical signatures of the hover-jet.

Experimental Apparatus and Procedures. The experiment allows a small airfoil to execute combined plunging and pitching motions and is sketched in Fig. 6.4. The

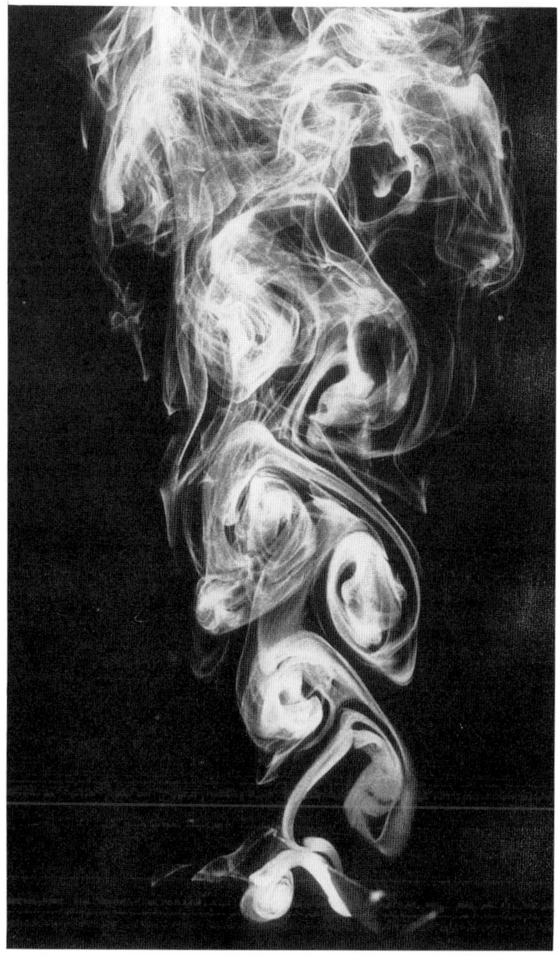

Fig. 6.3. Single photograph of a hover-jet for mode 1 hovering of the airfoil. $h_a = 2.54$ cm, $\alpha_a = 66°$, $f = 1$ Hz, $h_a/c = 1$, $R_f = 230$.

planar airfoil has a thickness of 1.6 mm with rounded edges, a chord $c = 2.54$ cm, and a span $l = 30.5$ cm. The airfoil is mounted on a slide frame which is driven in nearly sinusoidal horizontal translation by means of a dc-motor which is mounted on the fixed main frame of the apparatus. The motor rotates the drive wheel which connects to the slide frame by means of a drive rod as shown. In order to generate the simultaneous pitching motion for the airfoil a "slave wheel" is mounted on the slide frame and connects via another drive rod to the fixed main frame. This arrangement forces the slave wheel to execute the same rotation as the drive wheel. The drive rods are elastically anchored to avoid jamming of the drive mechanism. The axis of the slave wheel also carries the pitch wheel which connects via the pitch rod to the pitch lever of the airfoil.

The pitch wheel can be rotated against the slave wheel and the amount of preset rotation determines the phase angle (in our experiments $\phi = \pm 90°$). The airfoil chord can be rotated against the lever and this way α_0 can be adjusted (0° in mode 1, 90° in mode 2). The pitch amplitude α_a can be adjusted to values 25°, 33°, 42°, 52°, and 66° by inserting the drive rod axis into holes of different radial distance on the pitch wheel. The plunge amplitude h_a can be adjusted to values 1.27 cm, 2.54 cm, 3.81 cm, 5.08 cm, and 6.35 cm by linking both drive rods to holes at these radial distances in

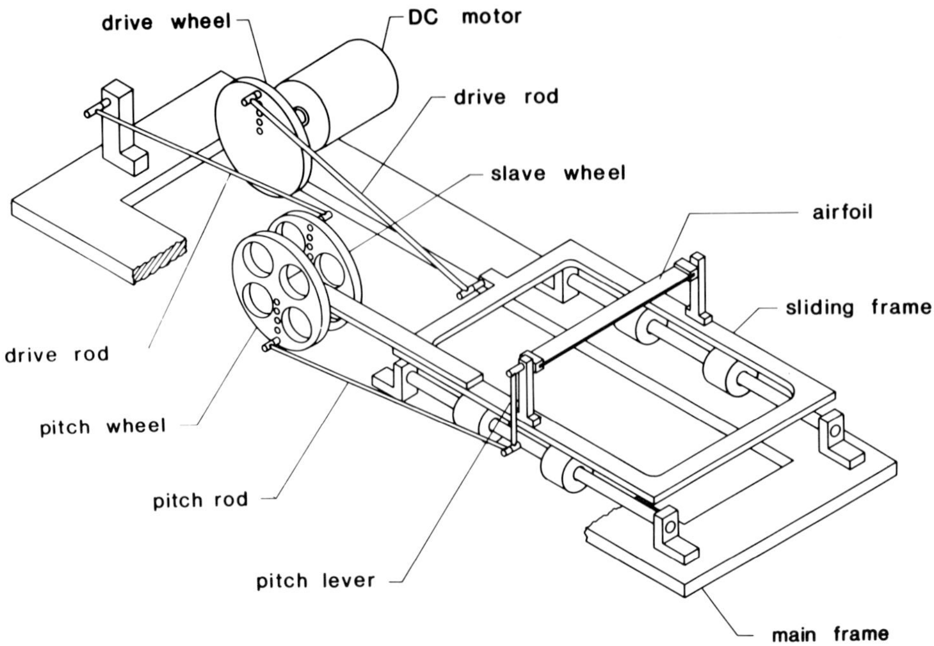

Fig. 6.4. Sketch of experimental apparatus.

the drive and slave wheels. The rotation rate f of the dc-motor can be regulated between 1 and 6 Hz and was measured by means of a reed switch connected to an electronic counter.

A photograph of the apparatus is shown in Fig. 6.5 for additional orientation. This photo also shows the Pitot tube above the apparatus which was used in connection with a sensitive pressure transducer to determine the dynamic pressure profile across the jet. This allows, according to (1.6), the determination of C_T. The pressure transducer was connected to a digital voltmeter with a time constant of 20 sec for averaging purposes.

Flow visualization is by means of the titanium tetrachloride ($TiCl_4$) method of vortex tagging [10]. A liquid film of $TiCl_4$ is deposited on the airfoil as a center strip using a brass pipette. The white fumes which develop tag the vorticity generated in the boundary layer of the airfoil and make separation and subsequent vortex developments visible, when photographed in side view. The airfoil was floodlit from above as well as below. Movies were taken with a Bolex 16 mm movie camera at a rate of 64 frames/sec.

Experimental Results. Velocities in the hover-jet were of order 1 m/sec at best, taxing somewhat the use of the Pitot tube for velocity profile determination. Most measurements of C_T were conducted at the highest Reynolds number our apparatus allowed, i.e., $R_f = 1{,}700$. Dependence of airfoil forces on Reynolds number is usually weak [11]. Velocity profiles were measured four chord lengths above the airfoil and a typical example is shown in Fig. 6.6. Squaring these profiles and graphic integration yields thrust coefficients C_T according to (1.6). In this way C_T was determined for the limited parameter range of h_a/c and α_a available to us both for mode 1 and mode 2 operation of the airfoil. Figure 6.7 summarizes the results obtained in mode 1 operation and Fig. 6.8 those in mode 2 operation. The most outstanding features of these results

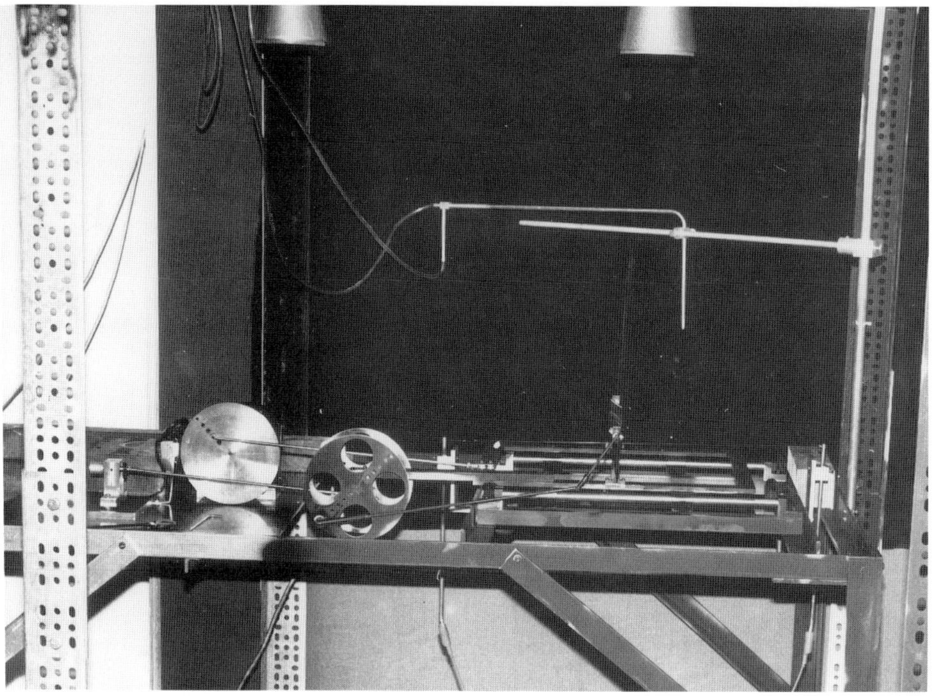

Fig. 6.5. Photograph of experimental apparatus.

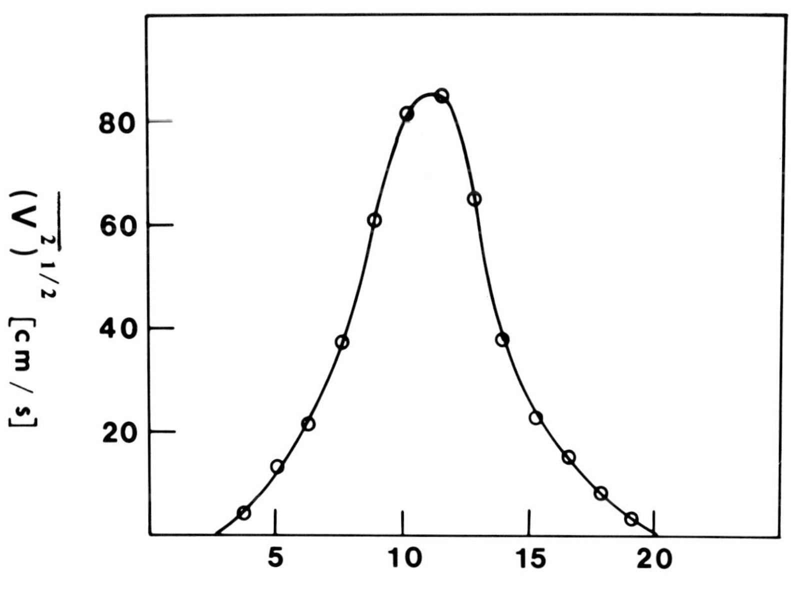

Fig. 6.6. Example of a velocity profile of the hover-jet measured with Pitot tube 8 chord lengths above the airfoil. h_a = 3.81 cm, α_a = 66°, f = 5 Hz, mode 1 hovering.

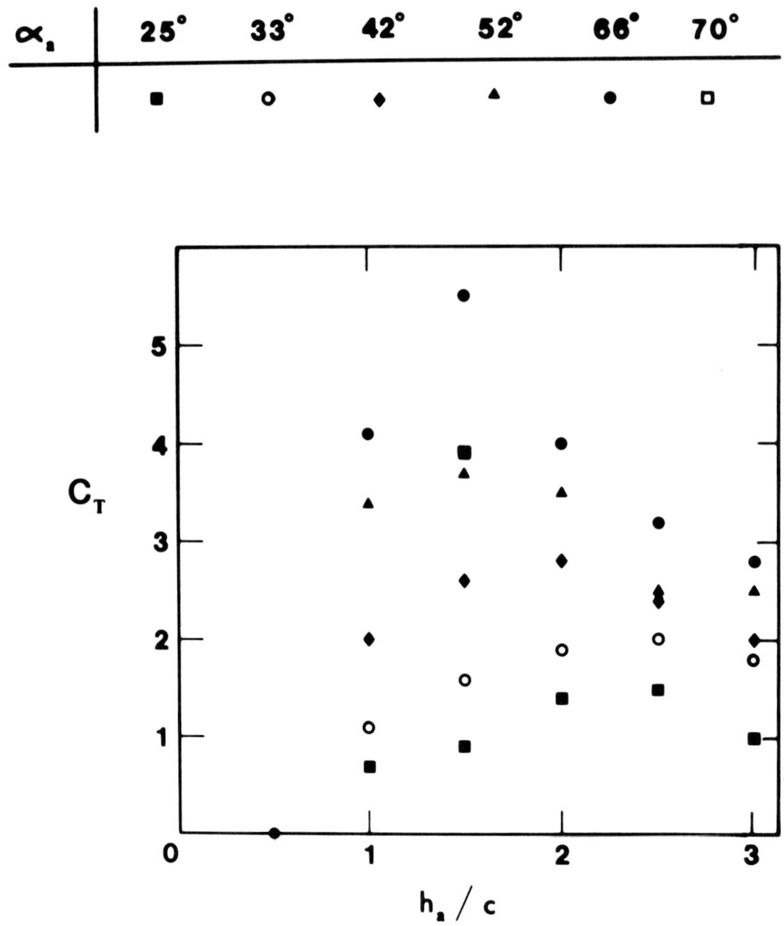

Fig. 6.7. Thrust coefficient C_T versus h_a/c and α_a for mode 1 hovering.

are the extraordinarily high thrust coefficients which can be reached, up to at least 5.5 in mode 1 and seven in mode 2. If correct, thrust coefficients of six reported for dragonflies by Norberg [5] would not seem out of line. Independent confirmation of high thrust coefficients found by us seems however to be an urgent need.

It has been thought that large thrust coefficients found for some insects could only be the result of special interactions of a pair of wings, like proper phasing in dragonfly wing beats [8], [9] or the clap and fling mechanism for the chalcid wasp [3], [4], [6]. It came as a surprise that high thrust coefficients could be reached by our generic hovering experiments with a single airfoil. According to Figs. 6.7 and 6.8 all that is required for large thrust coefficients are fairly small plunge amplitudes of order of a chord length and angles of attack of order of 60° reached at some point in the cycle. In other words, "tuning" of the hovering airfoil for high thrust coefficients is quite simple.

Figures 6.9 and 6.10 show single photographs of well-tuned hover-jets in mode 1 and mode 2 operations. Both cases show almost identical vortex features, i.e., alternate vortices forming a simple vortex street indicative of a jet. In contrast, Fig. 6.11 shows a hover-jet "out of tune," i.e., far from maximum thrust coefficient. In this case a

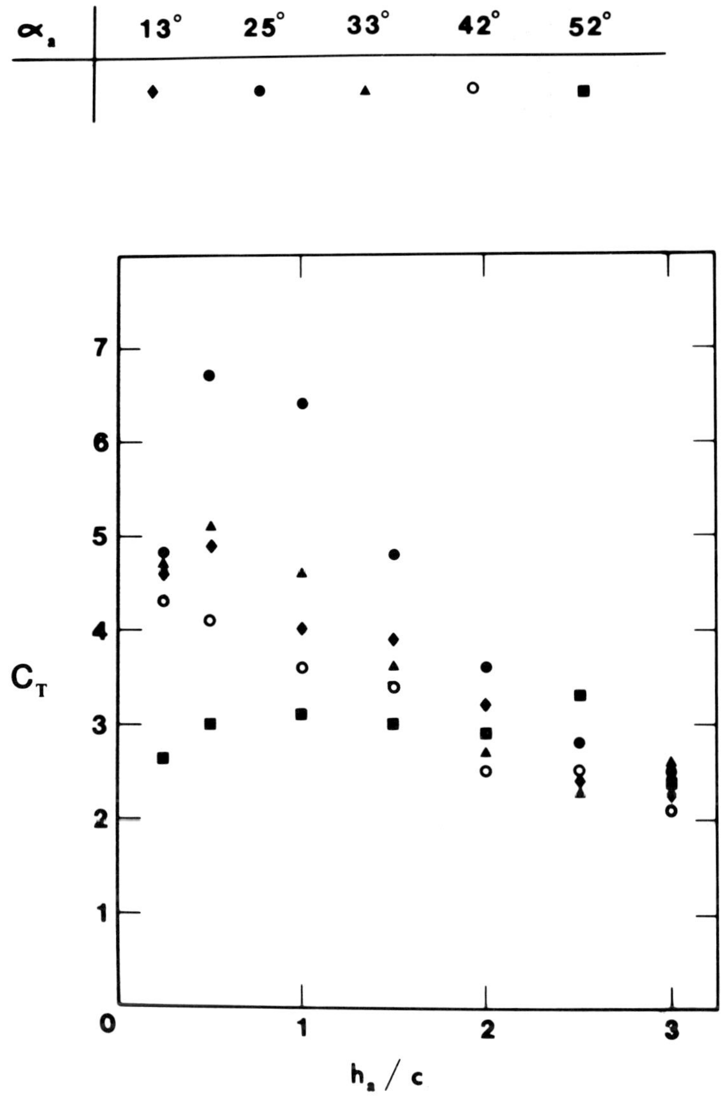

Fig. 6.8. Thrust coefficient C_T versus h_a/c and α_a for mode 2 hovering.

complex and pretty vortex tapestry [12] develops, which should however be quite inefficient in creating a highly directed jet.

Figure 6.12 shows a sequence of an entire cycle of airfoil oscillation, in mode 1 operation with $\alpha_a = 66°$ and $h_a/c = 1.5$, for which a large C_T is obtained. Frames are ordered into columns from top to bottom, columns are ordered from left to right. Time between consecutive frames is $\Delta t = \frac{1}{16}$ sec. Frame 1 shows the airfoil in its farthest right position. From there to the bottom of column 2 the airfoil moves to its farthest left position and in the process creates a clockwise rotating vortex. The last two columns show the airfoil moving to the right again creating in the process a counterclockwise rotating vortex to the left and below the previously generated clockwise vortex. This process repeats during each cycle and the result is a vortex street or hover-jet.

Fig. 6.9. Single photograph of well-tuned mode 1 hover-jet. $h_a/c = 1.5$, $\alpha_a = 66°$, $f = 1$ Hz, $R_f = 340$.

The vortex generation near the airfoil is quite complicated as a closeup sequence, taken slightly from the left, reveals in Fig. 6.13. In this case the first frame shows the airfoil passing through the $h = 0$ position from left to right, where the airfoil reaches its maximum angle of attack. This way we focus on the development of the dynamic stall vortex in the first column. In the second column this vortex partly rolls over the left edge of the airfoil and gets severed during the leftward motion of the airfoil in column 2. The severed parts of the vortex then amalgamate with the vorticity which is generated and shed from the right edge of the airfoil and form a single clockwise vortex by the end of column 2. An analogous process creates the counterclockwise vortex but would need observation slightly from the right.

The process of vortex severing is shown even more pristinely in Fig. 6.14 which represents only part of a sequence but at a time resolution $\Delta t = \frac{1}{64}$ sec. Vortex severing has been previously discovered by Ziada and Rockwell [13] in a different context.

Amalgamation of vortices of the same sense of rotation but without prior severing were also observed if the airfoil was operated in mode 2 and tuned for high-thrust coefficients. A sequence is shown in Fig. 6.15.

(Text continues on p. 156)

Fig. 6.10. Single photograph of well-tuned mode 2 hover-jet. $h_a/c = 1$, $\alpha_a = 33°$, $f = 1.3$ Hz, $R_f = 300$.

Fig. 6.11. Single photograph of mode 1 hover-jet far from optimum thrust. $h_a/c = 1.5$, $\alpha_a = 25°$, $f = 1$ Hz, $R_f = 340$.

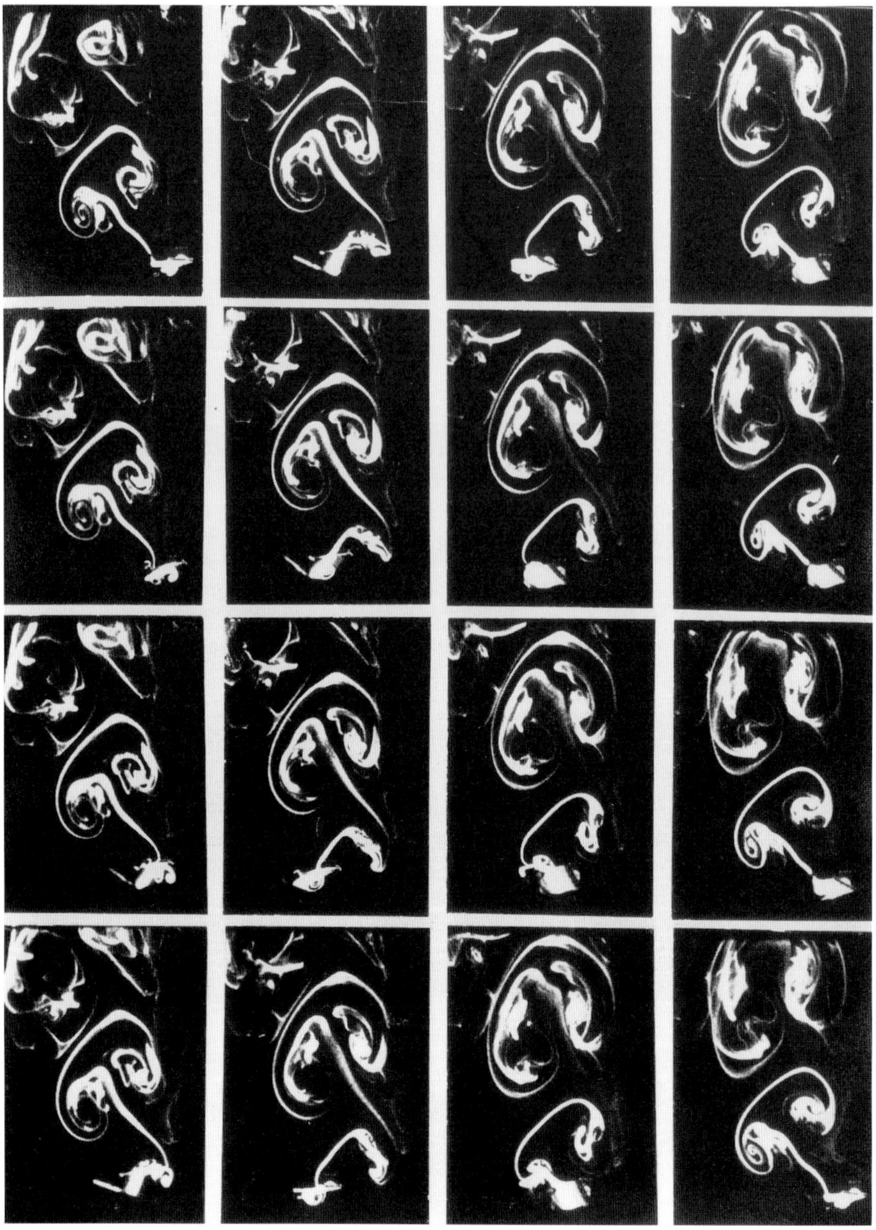

Fig. 6.12. Sequence of mode 1 hover-jet. $h_a/c = 1.5$, $\alpha_a = 66°$, $f = 1$ Hz, $R_f = 340$, $\Delta t = \frac{1}{16}$ sec.

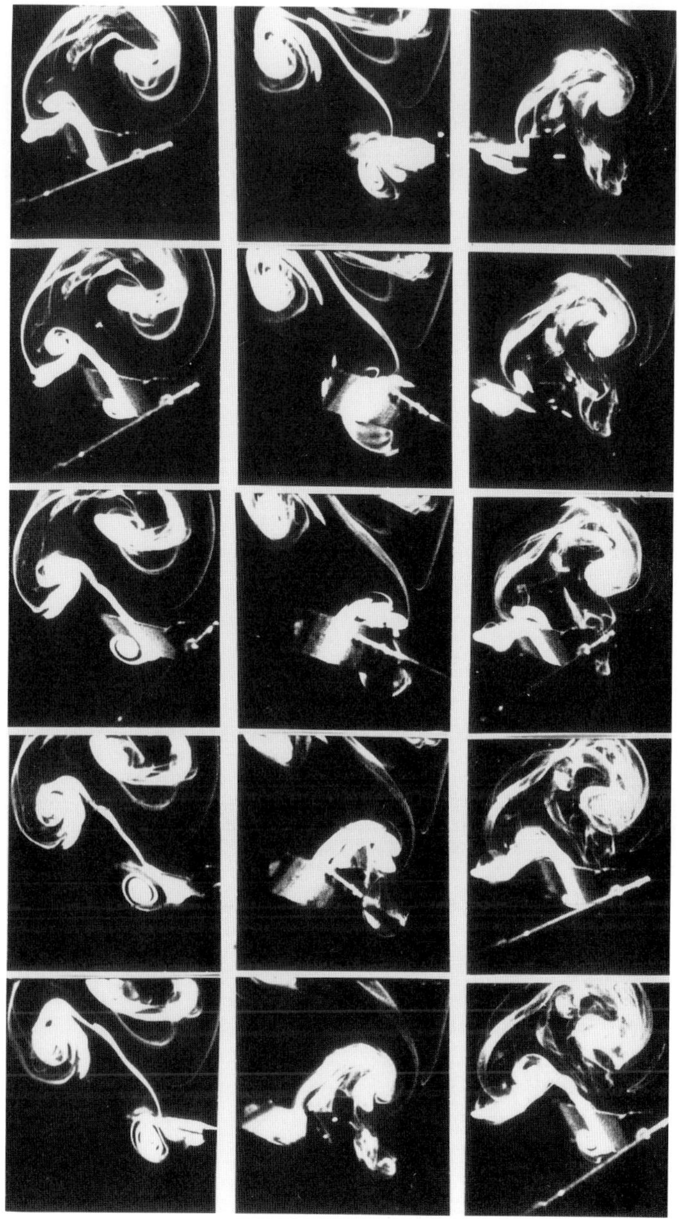

Fig. 6.13. Closeup sequence of mode 1 hover-jet. $h_a/c = 1.5$, $\alpha_a = 66°$, $f = 1.2$ Hz, $R_f = 400$, $\Delta t = \frac{1}{16}$ sec.

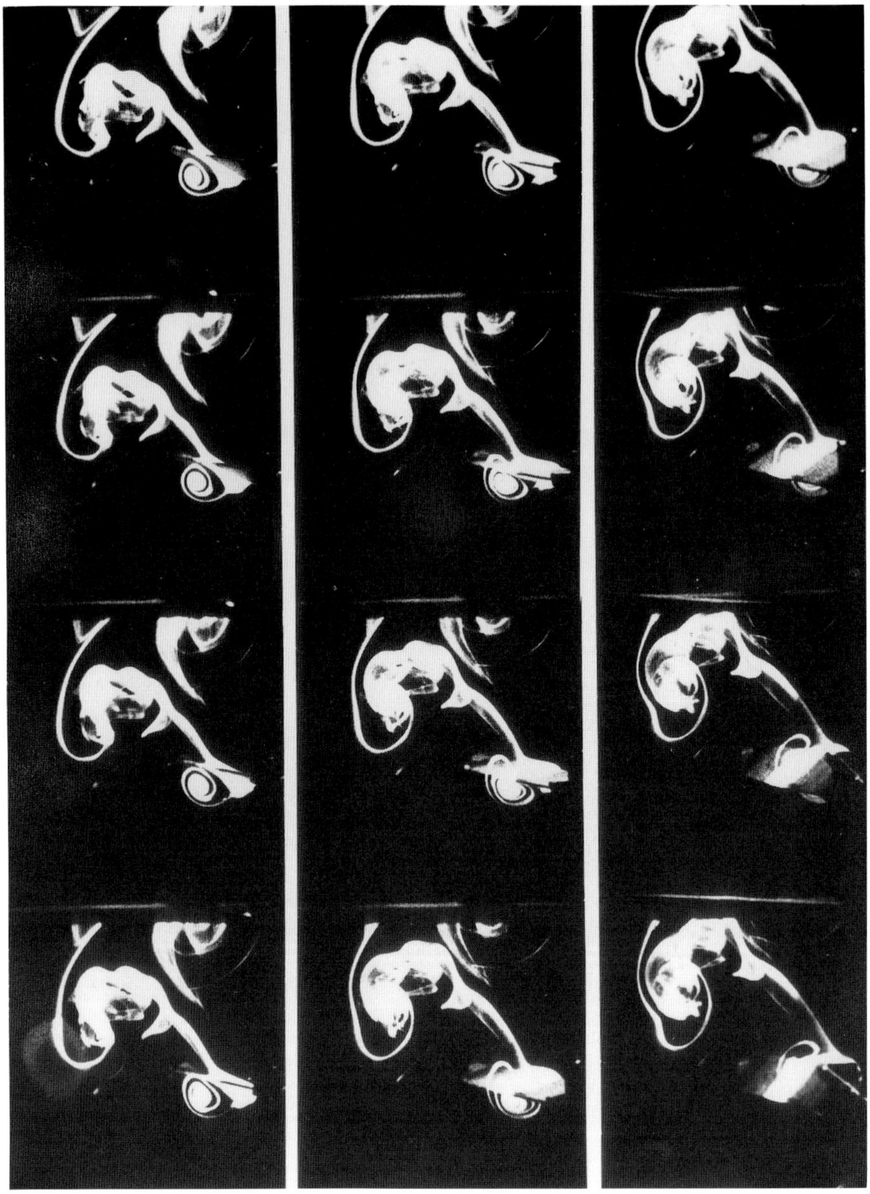

Fig. 6.14. Part of a closeup sequence of mode 1 hover-jet with increased time resolution. $h_a/c = 1.5$, $\alpha_a = 66°$, $f = 1$ Hz, $R_f = 340$, $\Delta t = \frac{1}{64}$ sec.

High-thrust generation may therefore have the following explanation. When during airfoil motion at high angle of attack a dynamic stall vortex is generated near the leading edge, it initially delivers high thrust (Kramer effect). Before the thrust force is diminished due to the development of deep stall, the airfoil turns into the opposite direction, in the process discards the dynamic stall vortex before it gets troublesome, and generates a new thrust producing stall vortex. This vortex again is dumped before deep stall occurs and so on. The result of this cyclic process should be the generation of high thrust.

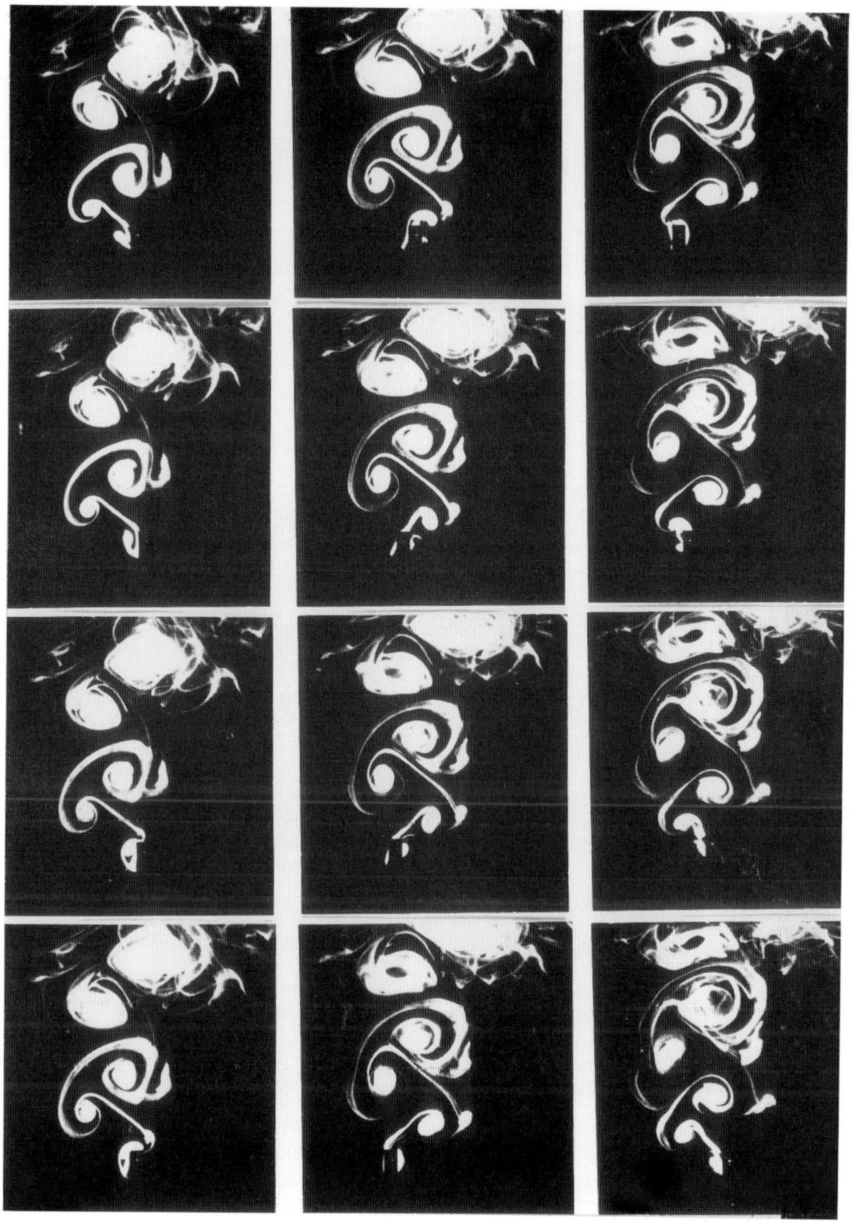

Fig. 6.15. Sequence of mode 2 hover-jet. $h_a/c = 1$, $\alpha_a = 33°$, $f = 1.3$ Hz, $R_f = 300$, $\Delta t = \frac{1}{16}$ sec.

It should be mentioned that flow visualizations shown are at fairly low Reynolds numbers where flow is mostly laminar. At the higher Reynolds numbers where thrust coefficients were measured, the hover-jet assumes a more turbulent appearance characterized by diffuseness of smoke visualization. Nevertheless, basic large-scale features seem to be the same as at lower Reynolds numbers. The sequences of tuned mode 1 and 2 jets shown in Figs. 6.16 and 6.17 may serve to support this statement.

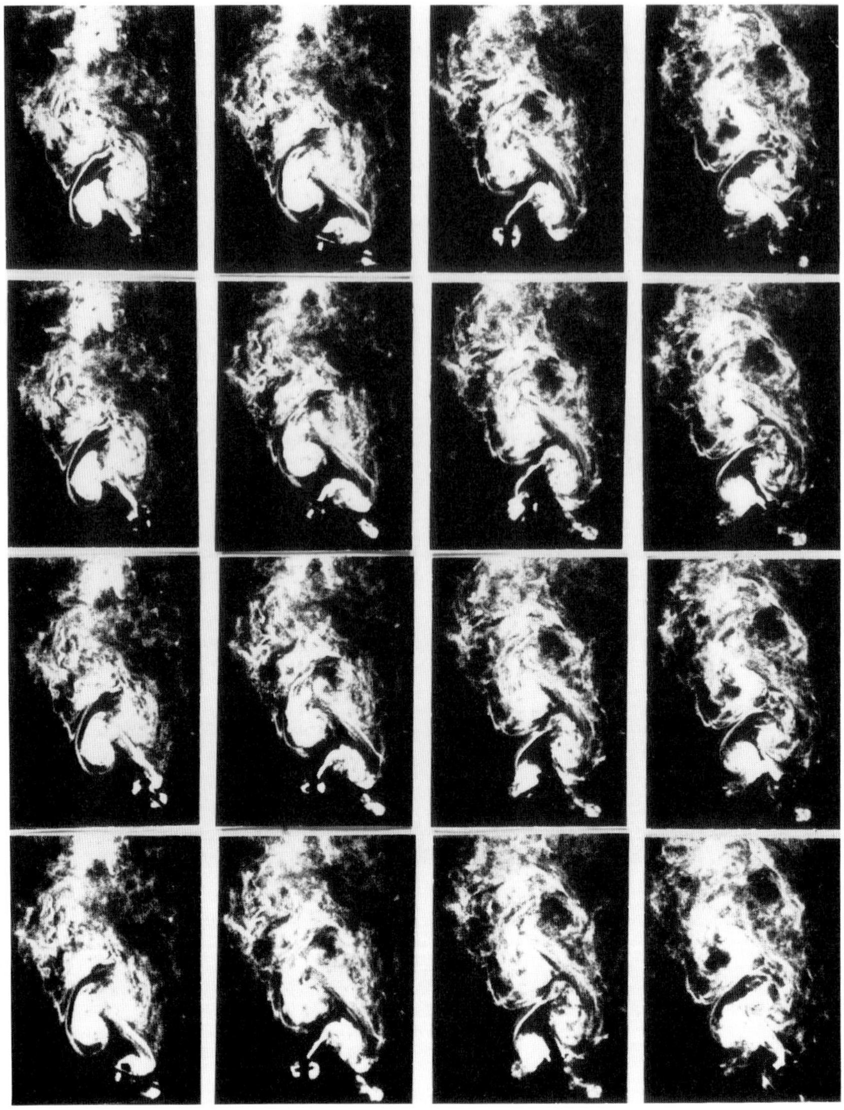

Fig. 6.16. Sequence of mode 1 hover-jet at higher Reynolds number. $h_a/c = 1.5$, $\alpha_a = 66°$, $f = 4$ Hz, $R_f = 1400$, $\Delta t = \frac{1}{64}$ sec.

Our recent finding [23] of a hover-jet oblique to the horizontal plunge plane is shown in Fig. 6.18. To obtain this mode the mean pitch angle α_0 must differ from 0° (mode 1) and from 90° (mode 2) and is a $\alpha_0 = 60°$ in the example shown. We also call this "oblique mode" of hovering the "dragonfly mode" since hovering dragonflies (and some other insects) are known [5] to have the stroke plane of their wings oblique (rather than perpendicular) to the vertical direction of the hover-jet they undoubtedly generate.

It was finally found [23] that a hover-jet in still air can also be generated by pure pitch motion of an airfoil if the pitch axis is at or near the leading edge. This hover-mode will be labeled as mode 4, or "Chinese fan mode" for obvious connotations. It will also be labeled tentatively as "butterfly mode" since butterflies with their large area wings may exploit this mode during their hovering. Butterflies with their latched

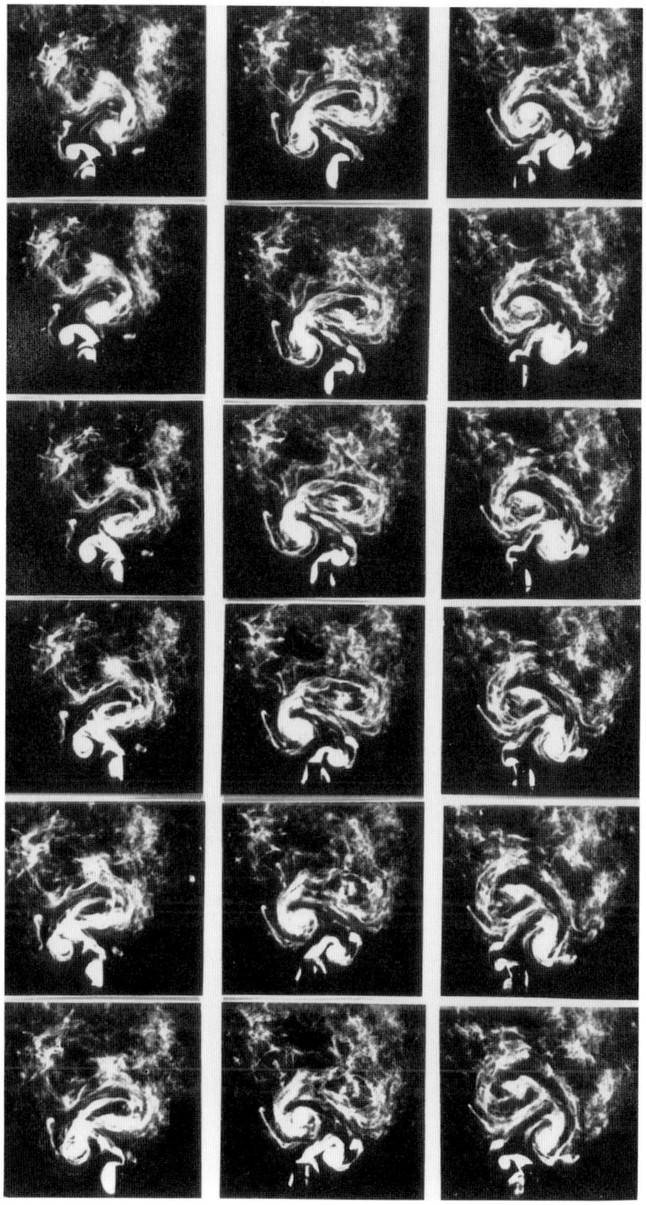

Fig. 6.17. Sequence of mode 2 hover-jet at higher Reynolds number. $h_a/c = 1$, $\alpha_a = 33°$, $f = 4$ Hz, $R_f = 900$, $\Delta t = \frac{1}{64}$ sec.

together wing pairs are perfectly capable of flapping motions around the longitudinal insect axis but pitching motions perpendicular to this axis are quite restricted. During their somewhat irregular hover flight each wing pair seems to perform flapping motions around an obliquely downward pointing mean angle. This should generate a hover-jet obliquely down and to the left on the left side of the butterfly and another one obliquely down and to the right on its right side. The composite result of these jets should be an upward lift force on the butterfly, i.e., hovering.

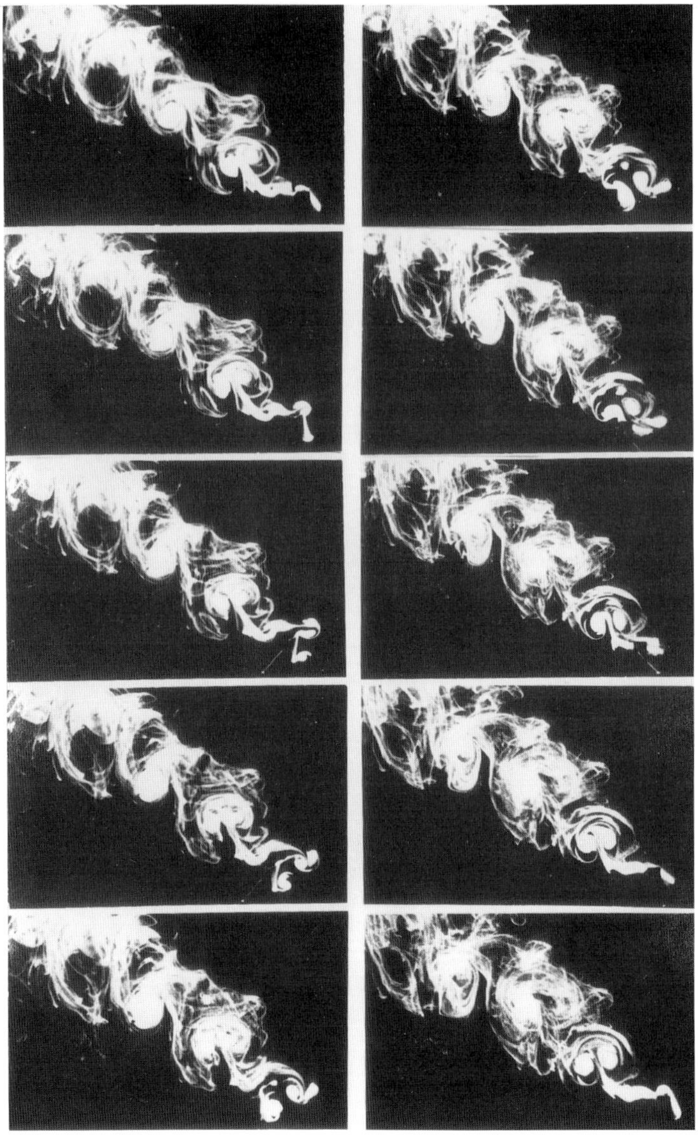

Fig. 6.18. Example of a hover-jet oblique to the horizontal plunge plane of the airfoil (the "dragonfly mode" of hovering). $\alpha_0 = 60°$, $\alpha_a = 45°$, $\phi = -90°$, $h_a/c = 1$, $R_f = 400$, $f = 1.8$ Hz, $\Delta t = \frac{1}{16}$ sec.

2. Computation

Previously [14]–[16] we have formulated and developed a technique for the study of unsteady flows past airfoils. Our goal in the work to be reported here was to extend those techniques to enable the first simulation of the vortex dynamics of the hovering mode in a still air environment. To do so requires a consideration of additional acceleration terms to accommodate a combination of rapid periodic plunge and pitch motions of an airfoil, within the geometrical setting shown in Fig. 6.19.

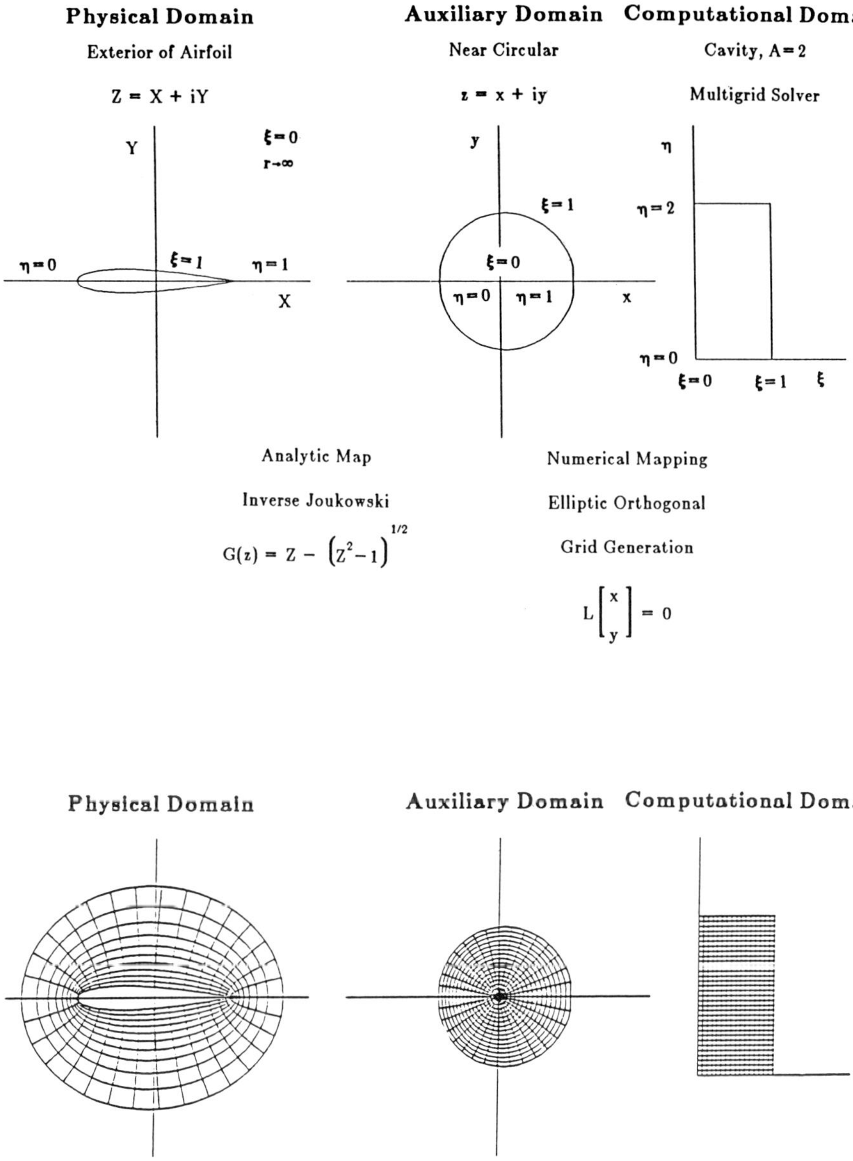

Fig. 6.19. Exterior physical domain, auxiliary near-polar interior domain, and rectangular computational domain (above). Coarse grid example of numerically generated near-orthogonal boundary fitted coordinates on those domains (below).

The preliminary computational results as shown in Color Figs. 6.1, 6.2 (see insert after p. 166) and Fig. 6.20 were found to be in quite good qualitative agreement with the physical visualizations. This good agreement of the vortex dynamics reported here bodes well for further detailed investigation at higher resolutions, and for lift computations which will be important in their own right.

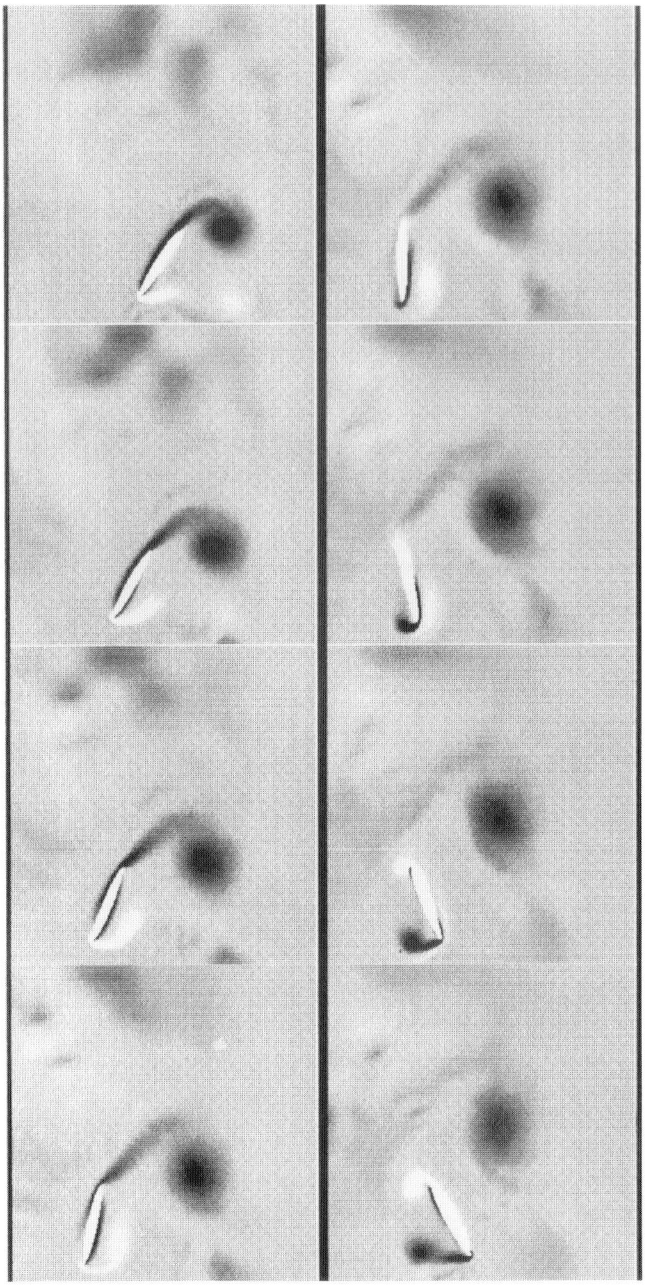

Fig. 6.20. Numerical simulation of hovering mode 2. Re = 300, f = 1.3 Hz, $\alpha_a = 33°$, $h_a/c = 1$. $\Delta t = \frac{1}{16}$ sec.

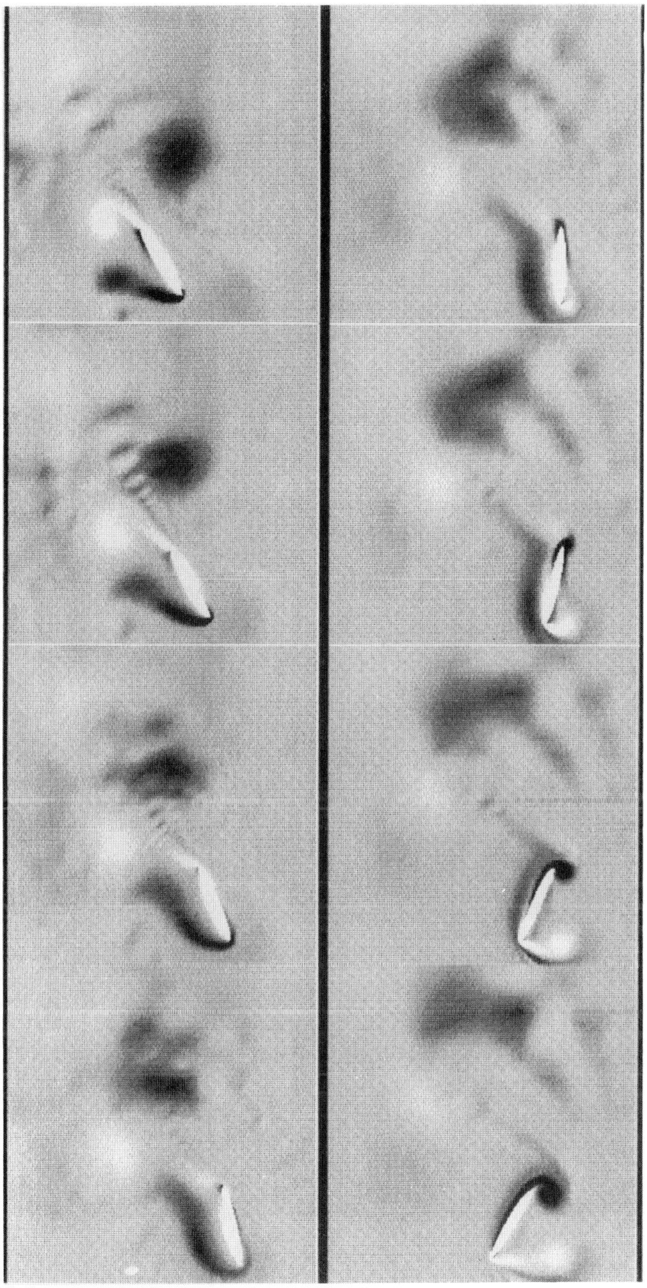

Fig. 6.20 (continued).

Unsteady Airfoil Flow Computations. An excellent early study of vortex dynamics about an impulsively started airfoil was that of Mehta and Lavan [17]. There the exterior flow field is mapped into the unit circle and the latter computational domain is split into two parts, a viscous region near the airfoil and a larger region in which the flow is assumed to be irrotational. Inertial boundary conditions are employed at the far-field cutoff to permit vortices to pass through the downstream boundary.

A later study by Lugt and Haussling [18] initiated the investigation of constantly accelerating airfoils. To handle this accelerating mode a boundary layer solution ω_B was introduced to permit more accurate force computations. A far-field uniform parallel flow was assumed.

For an indication of the state of recent work on accelerating unsteady aerodynamics see Shen [19] and McCune [20].

In our approach [15], [16] (see Fig. 6.19) we conformally map the exterior domain into an auxiliary interior, near-circular, domain. The latter is then provided with an orthogonal grid. The grid is computed in a rectangular computational domain, employing the weak constraint method of Ryskin and Leal [21]. In this way we are able to obviate the need to specify outflow boundary conditions at truncated far-field boundaries. In addition, difficulties posed by the need to specify some reasonable outflow condition at a downstream boundary are also removed.

In the first part of this paper, the physical simulation of the hovering modes was achieved by the rapid airfoil movement in still air. In our numerical simulations, the reference frame is fixed with the airfoil. To allow simulation of the hovering mode, the freestream flow relative to the airfoil is varied to model the airfoil movement.

Equations of the Motion. Let us first recall the basic viscous incompressible Navier–Stokes equations of motion in the geometry of Fig. 6.19. We assume here that the grid has already been generated, with scale factors h_1 and h_2 in the ξ and η directions, respectively. See [14]–[16] for more details about our grid generation scheme and the related considerations, which we omit here.

The equations of motions are formulated in terms of the stream function ψ and ω, the vorticity component normal to the ξ, η plane. All physical quantities have been made dimensionless by a characteristic velocity U and a characteristic length a. The characteristic length a is determined by the coordinate system in the physical plane of the grid generation mapping and is approximately equal to one half the chord length of the body. In terms of these quantities, the dimensionless governing equations on an orthogonal curvilinear coordinate system are

$$\frac{\partial \omega}{\partial t}+\frac{1}{h_1 h_1}\left(\frac{\partial h_2 u \omega}{\partial \xi}+\frac{\partial h_1 v \omega}{\partial \eta}\right)=\frac{L}{R}\nabla^2\omega, \tag{2.1}$$

$$\nabla^2\psi=-\omega, \tag{2.2}$$

where

$$\nabla^2 \equiv \frac{1}{h_1 h_2}\left(\frac{\partial}{\partial \xi}\left(\frac{h_2}{h_1}\frac{\partial}{\partial \xi}\right)+\frac{\partial}{\partial \eta}\left(\frac{h_1}{h_2}\frac{\partial}{\partial \eta}\right)\right) \tag{2.3}$$

and the Reynolds number is defined as

$$R=\frac{Uc}{\nu} \tag{2.4}$$

and c is the chord and L the dimensionless chord length. The velocities are defined by

$$u=\frac{1}{h_2}\frac{\partial\psi}{\partial\eta}, \qquad v=-\frac{1}{h_1}\frac{\partial\psi}{\partial\xi}. \tag{2.5}$$

The multigrid method previously used for the solution of the grid generation equations can also be used to solve the Poisson equation governing the stream function. However, the equation for ψ may not be solved directly since the ψ values at infinity are unbounded. Instead, the disturbance stream function ψ^* is introduced where

$$\psi^*=\psi-\psi_\infty \tag{2.6}$$

and ψ_∞ is the stream function for a uniform freestream flow. The governing equation for ψ^* remains the Poisson equation, since $\nabla^2\psi_\infty=0$,

$$\nabla^2\psi^*=-\omega \tag{2.7}$$

and is solved at each timestep using the multigrid algorithm.

The vorticity transport equation is solved using an ADI time marching scheme. Each dimensionless timestep is composed of two successive halfsteps. This formulation is first-order accurate in time and second-order accurate in space. Conservative differencing of the inertial term of the equation is used for better vortex resolution in the flow calculations. The coefficient matrices in the ξ direction are tridiagonal; in the η direction the method for periodic line relaxation must be used to calculate the solution.

The boundary conditions for ψ and ω must be set at the airfoil surface as well as at infinity.

At infinity, a uniform freestream flow at an angle of attack α is specified. This boundary condition on ψ is satisfied by ψ_∞, and is given by

$$\psi_\infty=\cos(\alpha)Y-\sin(\alpha)X, \tag{2.8}$$

where X, Y are the coordinates in the physical domain. The modified velocities due to ψ_∞ are calculated from

$$\tilde{u}_\infty=h_2u_\infty=\cos(\alpha)Y_\eta-\sin(\alpha)X_\eta, \tag{2.9}$$

$$\tilde{v}_\infty=h_1v_\infty=-\cos(\alpha)Y_\xi+\sin(\alpha)X_\xi. \tag{2.10}$$

The modified velocities used in the ADI solution of the vorticity transport equation are then given by

$$\tilde{u}=\frac{\partial\psi^*}{\partial\eta}+\tilde{u}_\infty, \qquad \tilde{v}=-\frac{\partial\psi^*}{\partial\xi}+\tilde{v}_\infty. \tag{2.11}$$

Thus, the boundary conditions on ψ at infinity and the corresponding boundary conditions on ψ^* at $\xi=0$, are

$$\psi_{\text{infinity}}=\psi_\infty, \qquad \psi^*_{\xi=0}=0. \tag{2.12}$$

Likewise, at the surface of the airfoil, the condition of an impermeable body is satisfied by

$$\psi_{\text{surface}}=0, \qquad \psi^*_{\xi=1}=-\psi_{\infty\xi=1}. \tag{2.13}$$

Boundary conditions on the vorticity ω must satisfy the no-slip condition of the viscous flow at the surface of the airfoil and reflect the irrotational flow at infinity. At infinity, the condition on the vorticity is

$$\omega_{\xi=0}=0. \tag{2.14}$$

This condition is much easier to impose than the outflow condition at a truncated domain boundary, and is one of the great advantages of the infinite domain method.

Several possible formulations of vorticity at the airfoil surface are applicable. The boundary condition which has been used for previous computations is the first-order accurate formula:

$$\omega(1,\eta)=\frac{-2}{h_1^2\,\Delta\xi^2}(\psi(1-\Delta\xi,\eta)-\psi(1,\eta)). \tag{2.15}$$

This boundary condition exhibited the best behavior at high Reynolds numbers and angles of attack when compared to second-order formulae.

Pitching and Oscillation of Airfoils in a Uniform Free Stream Flow. To undertake the study of pitching and oscillating airfoils, the Navier–Stokes equations must be recast in terms of a frame of reference which is fixed on the rotating airfoil. Nondimensionalizing the angular velocity of the airfoil as

$$\Omega=\frac{a\alpha}{U_\infty}, \tag{2.16}$$

the vorticity transport equation (2.1) can be rewritten to account for the angular motion of the inertial reference frame as

$$\frac{\partial(\omega+2\Omega)}{\partial t}+\frac{1}{h_1h_2}\left(\frac{\partial h_2u\omega}{\partial\xi}+\frac{\partial h_1v\omega}{\partial\eta}\right)=\frac{L}{R}\nabla^2\omega. \tag{2.17}$$

The stream function due to the flow at infinity must also be modified to include the effect of the angular rotation:

$$\psi_\infty=\psi_{\text{free stream}}+\psi_{\text{rotation}}=\cos(\alpha)Y-\sin(\alpha)X-\frac{\Omega}{2}(X^2+Y^2). \tag{2.18}$$

Since the vorticity field observed in the wind tunnel reference frame differs only by a constant from the rotating reference frame, we introduce the following change of variable to facilitate numerical solution of the problem:

$$\omega^*=\omega+2\Omega. \tag{2.19}$$

Likewise, we introduce the disturbance stream function

$$\psi^*=\psi-\psi_\infty. \tag{2.20}$$

The governing equation to be solved for comparison to the flow visualization experiments are

$$\frac{\partial\omega^*}{\partial t}+\frac{1}{h_1h_2}\left(\frac{\partial\tilde{u}\omega^*}{\partial\xi}+\frac{\partial\tilde{v}\omega^*}{\partial\eta}\right)=\frac{L}{R}\nabla^2\omega^*, \tag{2.21}$$

$$\nabla^2\psi^*=\omega^*, \tag{2.22}$$

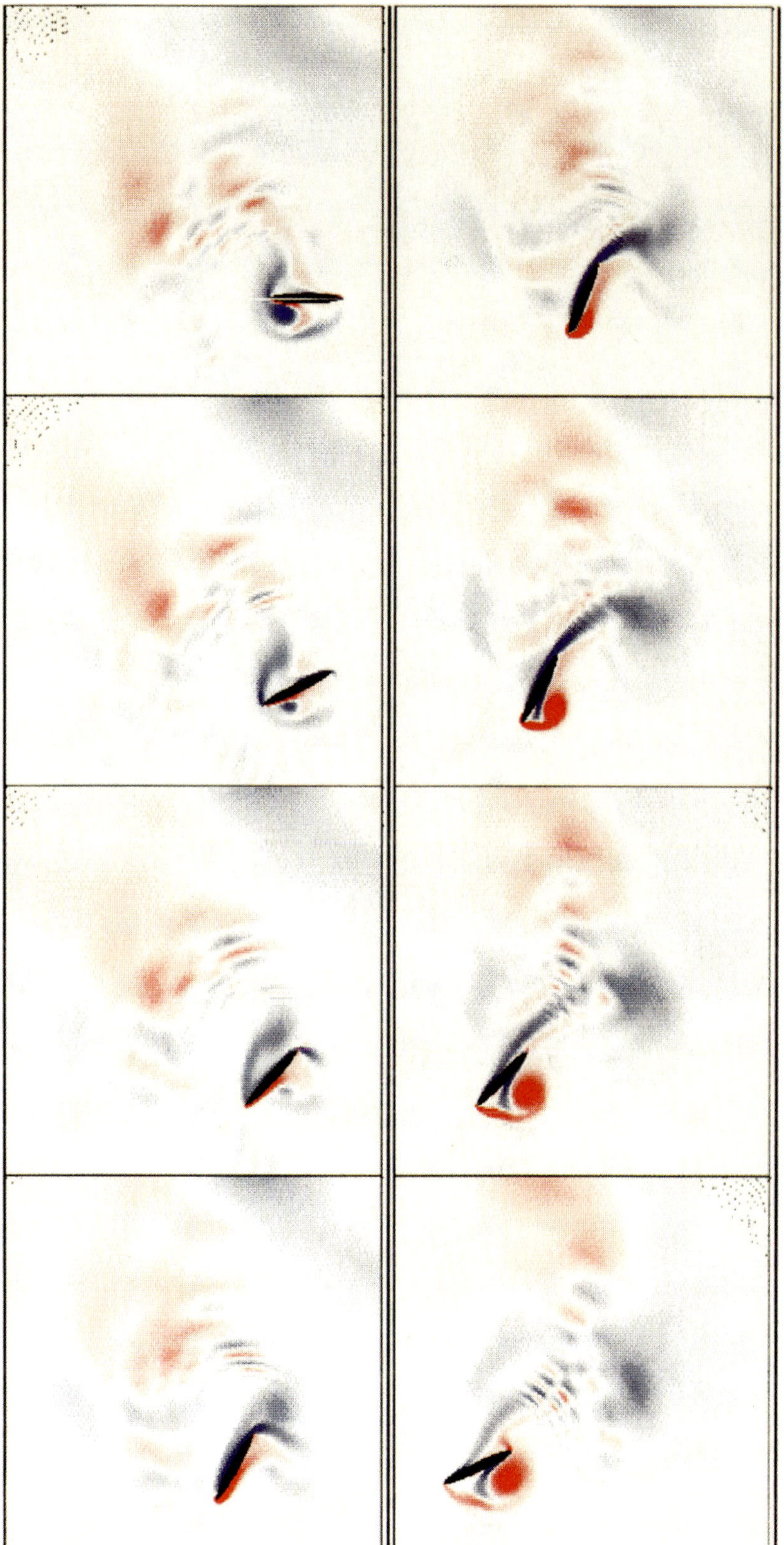

(a)

Color Fig. 6.1. Numerical simulation of hovering mode 1. Re = 350, f = 1 Hz, $\alpha_a = 66°$, $h_a/c = 1.5$, $\Delta t = \frac{1}{16}$ sec.

(b)

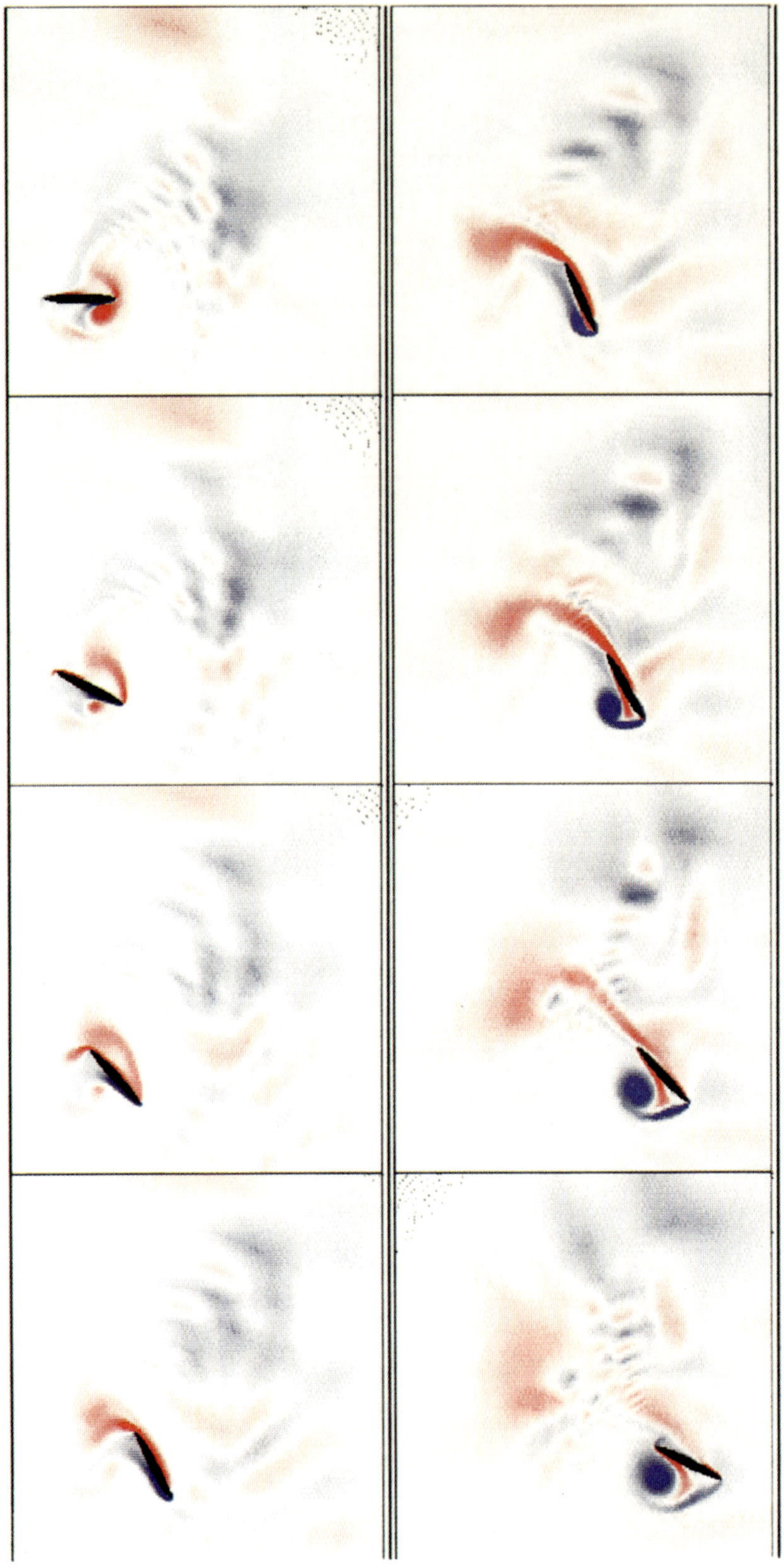

Fig. 6.1 (continued).

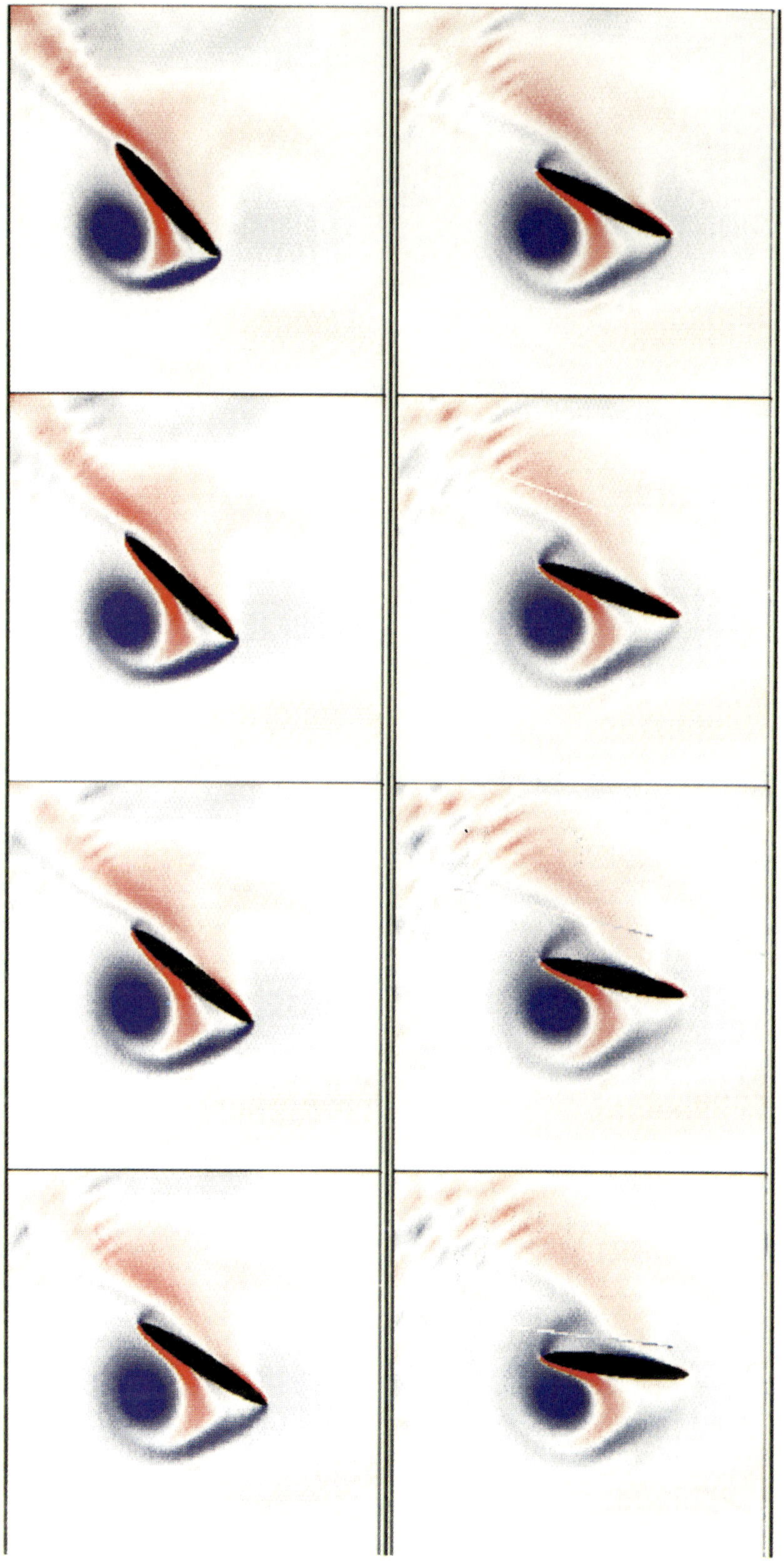

(a)

Color Fig. 6.2. Closeup of numerically simulated hovering mode 1 in the possible vortex severing interval. Re = 350, $f = 1$ Hz, $\alpha_a = 66°$, $h_a/c = 1.5$, $\Delta t = \frac{1}{64}$ sec.

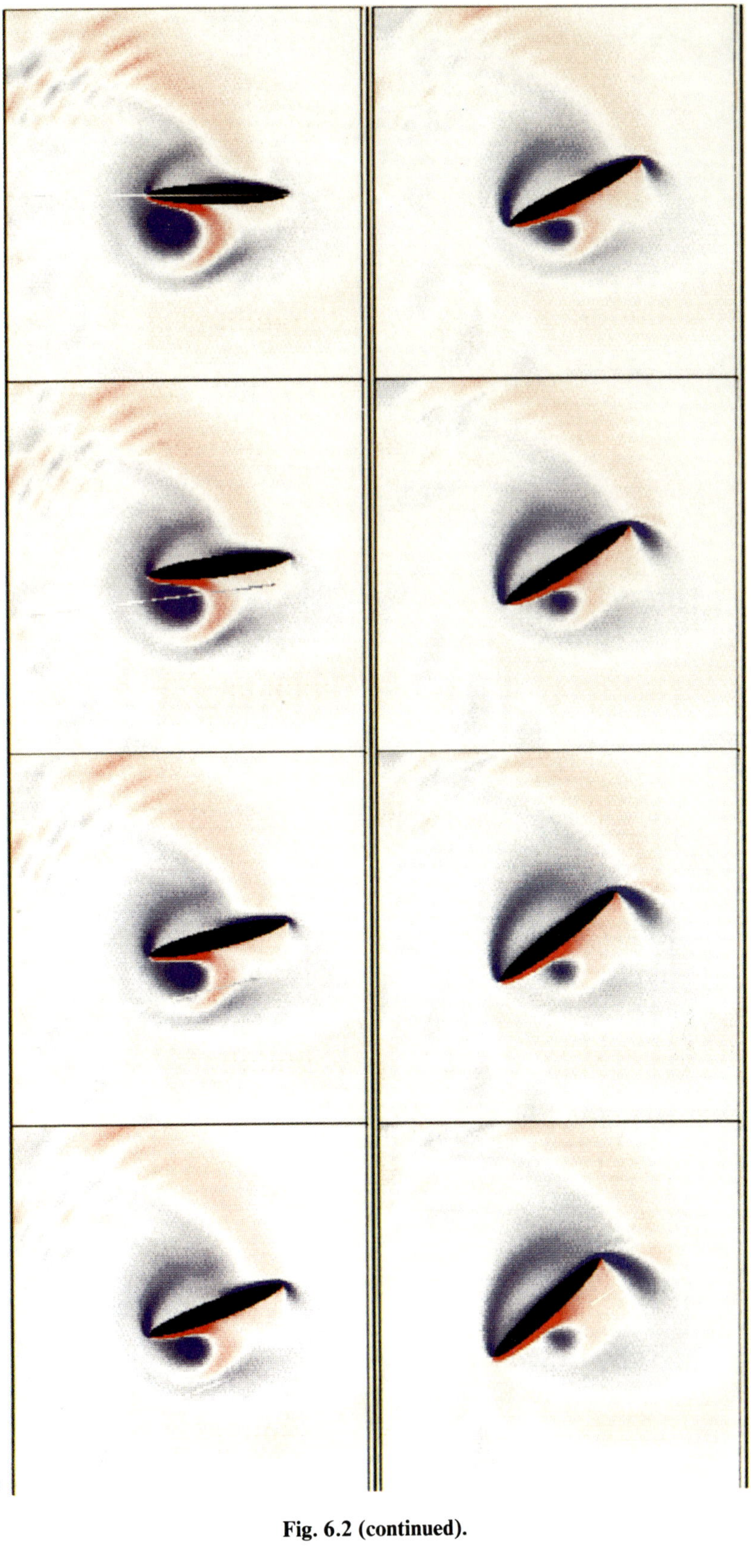

(b)

Fig. 6.2 (continued).

where

(2.23)
$$\tilde{u} = \frac{\partial \psi^*}{\partial \eta} + \tilde{u}_\infty, \qquad \tilde{v} = -\frac{\partial \psi^*}{\partial \xi} + \tilde{v}_\infty$$

and

(2.24)
$$\tilde{u}_\infty = h_2 u_\infty = \cos(\alpha) Y_\eta - \sin(\alpha) X_\eta - \Omega(X X_\eta + Y Y_\eta),$$

(2.25)
$$\tilde{v}_\infty = h_1 v_\infty = -\cos(\alpha) Y_\xi + \sin(\alpha) X_\xi + \Omega(X X_\xi + Y Y_\xi).$$

Boundary conditions for the stream function and vorticity are identical to equations (2.12)–(2.15).

Additional Considerations for the Modeling of Hovering Modes. Modeling of the hovering of the airfoil is achieved by specifying the angle of attack and velocity of the airfoil as a function of time. As described earlier, two hovering modes are of interest here: mode 1, or "water treading mode," and mode 2, or "figure eight mode." A look at Figs. 6.1 and 6.2, respectively, is very helpful at this point. The simulation (more details will be given in [22]) combines pitching and plunging (translation) as given in equations (1.1) and (1.2) to model both mode 1 and mode 2.

The preliminary runs reported here used a 65×65 mesh. This relatively coarse but boundary fitted mesh did permit a detailed resolution of fine boundary layer vortex features. However, due to its anisotropy the larger grid spacing in the far field limited the resolution there. Nonetheless there is very good qualitative agreement with what is observed in the physical laboratory.

Simulation Results. Color Fig. 6.1 (see insert following p. 166) shows the numerical simulation of hovering mode 1. We have arranged the frame and timing sequence to correspond approximately with the physical experiment shown in Fig. 6.12. The flow parameters were taken to be the same insofar as possible. The flow was in its fourth hovering motion cycle, numerically.

There is excellent qualitative agreement with the vortex dynamics recorded physically. The sheddings, amalgamations, even the regions occupied, show good correspondence.

Color Fig. 6.2 is a numerical closeup from the simulation of Color Fig. 6.1 of the interval in which the physical simulation (see Fig. 6.14) sees vortex severing. It appears from the numerical simulation that the fluid structures dynamical interaction of the shed (blue) underwing vortex and the descending underside and trailing edge of the airfoil is a more complicated and more interesting vortical event than just as a vortex severing action.

Figure 6.20 simulates hovering mode 2. Compare Fig. 6.15.

3. Conclusion and Outlook

It has been shown that an airfoil executing appropriate combined plunging and pitching motions can acquire very large thrust coefficients in a still air environment. The vortical signature of high thrust is a simple vortex street with the physically visualized character of a jet stream. Computer modeling has enabled a more detailed view of the vortex dynamics.

It is hoped that the physical and computational experiments reported here will advance the understanding of the more complex hovering flight of small birds and insects, and for application elsewhere, our understanding of the fundamentals of unsteady aerodynamics.

Acknowledgments

The first author acknowledges a visiting professorship at the F. J. Seiler Research Laboratory, sponsored by the Air Force Systems Command. An Air Force Office of Scientific Research grant F49620-84-C-0065, H. Helin program manager, provided additional assistance. The author is indebted to M. Luttges, M. Robinson, R. Reilman, and J. Walker for helpful discussions. Photocredit goes to W. Bank, the experimental apparatus was built by R. Hatfield and illustrations were provided by R. Tarasewicz.

The second and third authors acknowledge grants of computational resources on the CRAY 2 at NASA Ames Research Laboratories, on which further studies [22] will be performed.

The third author acknowledges Kelly Leutkemeyer for helpful assistance in the preparation of the images.

References

[1] P. FREYMUTH, *Propulsive vortical signature of plunging and pitching airfoils*, paper AIAA-88-0323, American Institute of Aeronautics and Astronautics, NY, 1988.

[2] ———, *Propulsive vortical signature of plunging and pitching airfoils*, AIAA J., 26 (1988), pp. 881–883.

[3] J. LIGHTHILL, *Aerodynamic aspects of animal flight*, in Swimming and Flying In Nature, Vol. II, Y. Wu, C. Brokaw, and C. Brennen, eds., Plenum Press, New York, 1975, pp. 423–491.

[4] T. WEIS-FOGH, *Flapping flight and power in birds and insects, conventional and novel mechanisms*, in Swimming and Flying in Nature, Vol. II, Y. Wu, C. Brokaw, and C. Brennen, eds., Plenum Press, New York, 1975, pp. 729–762.

[5] R. A. NORBERG, *Hovering flight of the dragonfly Aeschna Juncase L., kinematics and aerodynamics*, in Swimming and Flying in Nature, Vol. II, Y. Wu, C. Brokaw, and C. Brennen, eds., Plenum Press, New York, 1975, pp. 763–781.

[6] T. MAXWORTHY, *The fluid dynamics of insect flight*, Ann. Rev. Fluid Mech., 13 (1981), pp. 329–350.

[7] C. SOMPS AND M. LUTTGES, *Dragon flight: novel uses of unsteady separated flows*, Science, 228 (1985), pp. 1326–1329.

[8] M. A. REAVIS AND M. W. LUTTGES, *Aerodynamic forces produced by a dragonfly*, paper AIAA-88-0330, American Institute of Aeronautics and Astronautics, NY, 1988.

[9] D. SAHARON AND M. W. LUTTGES, *Dragonfly unsteady aerodynamics: the role of the wing phase relations in controlling the produced flows*, paper AIAA-89-0832, American Institute of Aeronautics and Astronautics, NY, 1989.

[10] P. FREYMUTH, W. BANK, AND M. PALMER, *Use of titanium tetrachloride for visualization of accelerating flow around airfoils*, in Flow Visualization III, Hemisphere, New York, 1985, pp. 99–105.

[11] M. C. ROBINSON AND J. B. WISSLER, *Pitch rate and Reynolds number effects on a pitching rectangular wing*, paper AIAA 88-2577-CP, American Institute of Aeronautics and Astronautics, NY, 1988.

[12] P. FREYMUTH, *Vortices*, in Handbook of Flow Visualization, W.-J. Yang, ed., Hemisphere, New York, 1989, pp. 459–479.

[13] S. ZIADA AND D. ROCKWELL, *Vortex-leading edge interaction*, J. Fluid Mech., 118 (1982), pp. 79–107.

[14] K. GUSTAFSON AND R. LEBEN, *Multigrid calculation of subvortices*, Appl. Math. Comput., 19 (1986), pp. 89–102.

[15] ———, *Vortex subdomains*, in Proc. 1st Internat. Conference on Domain Decomposition of Partial Differential Equations, R. Glowinski, G. Golub, G. Meurant, and J. Periaux, eds., Society for Industrial and Applied Mathematics, Philadelphia, PA, 1988, pp. 370–380.

[16] ———, *Robust multigrid computation and visualization of separation and vortex evolution in aerodynamic flows*, in Proc. 1st National Fluid Dynamics Congress (AIAA/ASME/SIAM/APS), Vol. 1, American Institute of Aeronautics and Astronautics, NY, 1988, pp. 174–184.

[17] U. MEHTA AND Z. LAVAN, *Starting vortex, separation bubbles, and stall: a numerical study of laminar unsteady flow around an airfoil*, J. Fluid Mech., 67 (1975), pp. 227–256.

[18] H. LUGT AND H. HAUSSLING, *The acceleration of thin cylindrical bodies in a viscous fluid*, J. Appl. Mech., 45 (1978), pp. 1–6.

[19] S. SHEN, *Considerations for analyzing separated flows over arbitrarily maneuvering bodies*, in Proc. A. F. Workshop II on Unsteady Separated Flows, J. McMichael and H. Helin, eds., FJSRL-TR-88-0004, 1988, pp. 243–247; see also Unsteady Separation Over Maneuvering Bodies, S. Shen and T. Wu, eds.; Proc. 1st National Fluid Dynamics Congress, AIAA paper 88-3542, American Institute of Aeronautics and Astronautics, NY, 1988.

[20] J. MCCUNE, *Interactive aerodynamics of wings in severe maneuver*, in Proc. A. F. Workshop II on Unsteady Separated Flows, J. McMichael and H. Helin, eds., FJSRL-TR-88-0004, 1988, pp. 161–171; see also Unsteady 3D Aerodynamics of Slender Wings in Severe Maneuvers, J. McCune and T. Tavares, eds.; Proc. 1st National Fluid Dynamics Congress, AIAA paper 88-3543, American Institute of Aeronautics and Astronautics, NY, 1988.

[21] G. RYSKIN AND L. LEAL, *Orthogonal mapping*, J. Comput. Phys., 50 (1983), pp. 71–100.

[22] K. GUSTAFSON AND R. LEBEN, *Dragonfly aerodynamics*, Comput. Phys. Comm. (1991), to appear.

[23] P. FREYMUTH, *An unsteady model of animal hovering*, Lecture Notes in Engineering, Vol. 54, Springer-Verlag, Berlin, New York, 1989, pp. 231–245.

7 TURBULENCE AND VORTICES IN SUPERFLUID HELIUM

Thomas F. Buttke

Introduction

The goal of this review is to introduce people with a moderate understanding of fluid mechanics to superfluid helium under the assumption that they have no prior knowledge of superfluid helium. We attempt to introduce the jargon of superfluids so that the reader can more easily approach the vast literature on the subject. Once we have given a minimum background on some of the general features of a superfluid we discuss vortices and turbulence in superfluid helium.

In § 1, we discuss the basic physical properties of ordinary helium as well as of superfluid helium. We briefly discuss the most important experiments which helped to elucidate the remarkable properties of superfluid helium. We introduce the two-fluid model of superfluid helium.

In § 2, we give a rather detailed derivation of the two-fluid equations. The two-fluid equations describe the flow properties of superfluid helium. We give a detailed introduction to the two-fluid equations since the equations themselves are unfamiliar to most fluid dynamicists, and we find that a thorough understanding of the basic assumptions underlying the two-fluid equations is essential to understanding the results obtained in the field of superfluid turbulence.

In § 3, we describe the different sound modes present in superfluid helium. The experimental observation of second sound in superfluid helium was one of the first validations of the two-fluid equations. We give an intuitive derivation of the sound modes so that the essential physical features of the sound modes is apparent. Finally, we give references to the validation studies investigating the nonlinear properties of superfluid helium by propagating shock waves through the liquid.

In § 4 we discuss quantized vortices. We discuss the theoretical motivation for quantized vortices and the experimental evidence for them.

In § 5 we introduce the concept of turbulence in superfluid helium. We introduce the idea of a mutual friction force and discuss in detail the Gorter–Mellink friction force. We relate the Gorter–Mellink force to the idea of vortex tangles. We conclude the paper with a discussion of the numerical simulations of vortex tangles carried out by Schwarz and the implication that the author's work has on these results.

1. Background

There are two isotopes of helium, ^{4}He and ^{3}He. The helium isotopes have the lowest boiling points of any known substance. ^{3}He boils at 3.19°K and ^{4}He boils at 4.21°K at atmospheric pressure. Both isotopes remain liquid under saturated vapor pressure down to 0°K. The liquids have a small viscosity at their normal boiling points: 19 μP (microPoise) for ^{3}He and 36 μP for ^{4}He. (It should be noted that below 1°K the viscosity of ^{3}He increases as $1/T^2$, where T is the temperature, until the temperature reaches the superfluid transition. At the superfluid transition the viscosity is on the order of 1Poise.) This is in comparison to air at room temperature (300°K) which has a viscosity of 184 μP and water at 300°K which has a viscosity of 8513 μP. Liquid ^{3}He has a density of 0.0586 g/cm^3 and ^{4}He has a density of 0.125 g/cm^3. Thus liquid ^{3}He has a kinematic viscosity of 324 μStokes, and liquid ^{4}He has a kinematic viscosity of 288 μStokes, as compared to water and air at room temperature which have kinematic viscosities of 8510 μStokes and 150,000 μStokes, respectively. The speed of sound in ^{3}He is 180 m/sec; in ^{4}He it is 240 m/sec; in water it is 1500 m/sec; and in air it is 330 m/sec. The helium liquids are more compressible than air but have kinematic viscosities which are much smaller than those of water. Liquid ^{4}He flowing through a one centimeter pipe at a velocity of 1 cm/s has a Reynolds number of 3,500, whereas water has a Reynolds number of 120 and air has a Reynolds number of 7 under the same flow conditions.

The properties of liquid ^{4}He change drastically as the liquid is cooled; a phase transition occurs at 2.17°K under saturated vapor pressure. This transition point is called the λ-point transition because the plot of the specific heat versus temperature resembles a λ at the transition point. The temperature at which the phase transition occurs is called T_λ; it is also known as the superfluid transition. The phase transition is a second-order transition, which means that there is no latent heat associated with the transition but that the specific heat becomes infinite at T_λ. Below the λ-point liquid ^{4}He has some remarkable flow properties. (Below the λ-point liquid ^{4}He is denoted as helium II and above the λ-point the liquid is denoted as helium I. This notation does not refer to isotopes and should not be confused with the ^{3}He isotope.) ^{3}He also has a superfluid transition, but the transition occurs at 1°mK, a temperature which is three orders of magnitude colder than T_λ in ^{4}He.

The superfluid transition in ^{3}He was discovered in 1972 by Osheroff et al. [1], [2], whereas the superfluid transition in ^{4}He was discovered in 1938 simultaneously by Kapitza [3] and Allen and Misener [4] after several other people had observed discontinuities of thermodynamic quantities associated with the phase transition as early as 1911 [5]. Superfluid ^{3}He has many of the analogous properties as helium II; however, due to the experimental difficulties at these ultralow temperatures there is not as much experimental data available as there is for ^{4}He. In the following we restrict ourselves to a description of the experimental properties of superfluid ^{4}He and refer the interested reader to the excellent reviews [6]–[9] for more information about superfluid ^{3}He.

We now describe a few of the experimental properties of helium II. For a detailed account we refer the reader to [9]–[13]. One of the remarkable properties of helium II is that under certain situations it flows as if it has a viscosity of less than 10^{-11} Poise; helium II flows through capillaries with no pressure drop occurring across the capillary. Another property of helium II is that the apparent thermal conductivity is 1,000 times greater than copper at room temperature.

The property of helium II which is used to determine the equations of fluid motion is called the fountain effect. The fountain effect is observed by connecting two reservoirs of helium II by a tube which is packed with a powder whose grains are the size of a few microns; this type of tube connecting two reservoirs of helium II is called a superleak. The powder size is picked small enough so that an enormous pressure gradient would be required to cause an ordinary fluid with viscosity to flow through it. For instance, if we consider flow through a pipe one centimeter long packed with a one micron powder, we find that for helium I the average velocity would be approximately 40 μm/s at a pressure difference corresponding to a height of one centimeter of liquid ^{4}He. When the temperature of one of the reservoirs is increased there is a corresponding increase in pressure in the same reservoir when the two reservoirs are in thermodynamic equilibrium. This increase in pressure is called the fountain effect or thermomechanical effect. It is called the fountain effect because one is actually able to make a fountain of liquid helium using this mechanism as the pump. The fountain pressure, the increase in pressure associated with the fountain effect, is so large that at 1.5°K a difference of only one milliKelvin between the two reservoirs produces a pressure difference equal to a column of two centimeters of liquid helium.

In contrast to the experiments in which it appears that helium II is a liquid with no viscosity, there are flow situations in which it appears that helium II does have a viscosity. For instance, in a rotating viscometer helium II appears to have a viscosity of approximately 10 μP and the viscosity varies continuously as the temperature increases through T_λ.

The fact that under different flow situations helium II behaves as two different fluids, led Tisza in 1938 [14] to introduce the two-fluid model of helium II. In this model helium II is assumed to consist of two fluids; one called the normal component which has a viscosity, and one called the superfluid component which has no viscosity. When helium II flows through capillaries the viscosity prevents the normal component from flowing; only the superfluid component flows and thus no viscosity is observed. In the rotating viscometer both components flow and thus there is an observed viscosity in helium II.

Another phenomena which can be described by the two-fluid model is the large apparent heat conductivity of helium II. The superfluid component carries no entropy and can flow with no viscosity. Thus if a temperature gradient is produced in the bulk, the fountain effect quickly causes the superfluid component to flow toward the high temperature region and the normal component, which carries all of the entropy, flows away from the high temperature region quickly equilibrating the temperature throughout the bulk of the helium II. Because of the two-fluid model this process occurs without any mass transport; this process is not the same as convective transport caused by thermal gradients in an ordinary fluid. In the next section we will develop the equations describing the two-fluid model of superfluid ^{4}He.

2. The Two-fluid Equations

We introduce the two-fluid equations in this section. Landau introduced the two-fluid equations in 1941 [15]. The basic equations are derived from the conservation laws of mass, momentum, energy, and entropy along with an equation of motion for the superfluid component derived from the experimental facts obtained from the fountain effect. We follow the development as given by Khalatnikov [16] and Putterman [17].

The basic assumption of the two-fluid model is that there are two independent velocity fields which describe the hydrodynamic state of the superfluid. These velocity fields are denoted by $\mathbf{v}_n$ and $\mathbf{v}_s$. $\mathbf{v}_n$ is associated with the normal component and is defined by the assumption that only the normal component carries entropy. $\mathbf{v}_n$ gives the velocity of entropy transport. $\mathbf{v}_s$ is associated with the superfluid component. The superfluid component has no viscosity and the velocity $\mathbf{v}_s$ is defined by requiring that the velocity be irrotational and that the low velocity limit of the equation for $\mathbf{v}_s$ be consistent with the experimental results of the fountain effect. The basic results of the fountain effect experiment are that in the steady state the chemical potentials of the two reservoirs on either side of the superleak are equal to each other. Experimentally, it is found that

$$\rho s \Delta T = \Delta p,$$

where ρ is the density of the fluid, ΔT is the temperature difference between the two reservoirs, Δp is the pressure difference between the two reservoirs, and s is the specific entropy. The thermodynamic relation between μ, T, and p for an ordinary substance is given by

$$d\mu = -s\,dT + \frac{1}{\rho}\,dp. \tag{2.1}$$

The condition for steady state between the two reservoirs can then be expressed as

$$d\mu = 0.$$

It is postulated that the velocity field $\mathbf{v}_s$ is a potential velocity field which is driven by gradients in the chemical potential. The equation of motion for $\mathbf{v}_s$ is taken to be

$$\frac{\partial \mathbf{v}_s}{\partial t} + (\mathbf{v}_s \cdot \nabla)\mathbf{v}_s = -\nabla \mu, \tag{2.2}$$

where t is time and μ is the chemical potential. We shall see shortly that μ is a function of pressure p and of temperature T as well as a function of $\mathbf{v}_n - \mathbf{v}_s$. The dependence on $\mathbf{v}_n$ and $\mathbf{v}_s$ has been ignored in (2.1) since in the two reservoirs it is assumed that $\mathbf{v}_n = \mathbf{v}_s = 0$. Equation (2.2) is the only equation in the two-fluid equations which is not derived from a conservation law; it can be written in the equivalent form

$$\frac{\partial \mathbf{v}_s}{\partial t} + \nabla\left(\frac{1}{2}\mathbf{v}_s^2\right) = -\nabla \mu \tag{2.3}$$

if we assume that $\nabla \times \mathbf{v}_s = 0$, and thus be put into a conservation form.

One point about the two-fluid equations of motion should be clarified: the two-fluid equations do not describe the motion of two fluids. The two-fluid equations describe the hydrodynamic state of a fluid in a single phase where the thermodynamic state of the fluid is a function of a vector quantity which satisfies an equation of motion. Under certain flow conditions $\mathbf{v}_n$ or $\mathbf{v}_s$ may coincide with the actual velocity of helium, but in general this is not the case. This is in contrast to the equations describing the motion of two immiscible fluids where, at least conceptually, there are two distinct velocities both of which can be measured directly. At any given point in helium II there is only a single fluid which is described by the variables T, p, $\mathbf{v}_n$, and $\mathbf{v}_s$ or appropriate functions thereof.

The remaining equations which describe the two-fluid model of helium II are derived from conservation laws. The conservation of mass implies that

$$\frac{\partial \rho}{\partial t}+\nabla\cdot\mathbf{J}=0, \tag{2.4}$$

where $\mathbf{J} = \rho\mathbf{v}$, and $\mathbf{v}$ is usual velocity of the fluid; i.e., it is the true velocity of the fluid in the sense that it is the only velocity which can be measured directly. This equation is identical to the continuity equation for an ordinary fluid. The conservation of momentum implies that

$$\frac{\partial \mathbf{J}}{\partial t}+\nabla\cdot\mathbf{\Pi}=0, \tag{2.5}$$

where $\mathbf{\Pi}$ is the momentum flux tensor. In an ordinary fluid we have that

$$\mathbf{\Pi}=\rho\mathbf{v}\mathbf{v}+p\mathbf{I},$$

where $\mathbf{I}$ is the identity tensor and p is the scalar pressure. The conservation of energy implies that

$$\frac{\partial E}{\partial t}+\nabla\cdot\mathbf{Q}=0, \tag{2.6}$$

where E is the total energy per unit volume of the fluid and $\mathbf{Q}$ is the energy flux vector of the fluid. In an ordinary fluid we have that

$$E=\tfrac{1}{2}\rho\mathbf{v}^2+\rho e,$$

$$\mathbf{Q}=\mathbf{v}(\tfrac{1}{2}\rho\mathbf{v}^2+\rho e+p),$$

where e is the internal energy per unit mass. The conservation of entropy implies that

$$\frac{\partial \rho s}{\partial t}+\nabla\cdot\mathbf{F}=0, \tag{2.7}$$

where s is the entropy per unit mass and $\mathbf{F}$ is the entropy flux. In an ordinary fluid

$$\mathbf{F}=\rho s\mathbf{v}.$$

Up to this point we have written down general conservation laws, equations (2.4)–(2.7), but we have not specified what the fluxes should be for the case of helium II. We find the fluxes by making (2.4)–(2.7) self-consistent and by determining the dependence of the thermodynamics on the quantity $\mathbf{w} \equiv \mathbf{v}_n - \mathbf{v}_s$. Equations (2.4)–(2.7) are self-consistent for an ordinary fluid with the use of the basic thermodynamic identity

$$d(\rho e)=Td(\rho s)+\mu d\rho, \tag{2.8}$$

where s is the entropy per unit mass and μ is the chemical potential. Equations (2.4)–(2.6) along with (2.8) imply (2.7) for an ordinary fluid for smooth flows; we must define the thermodynamic quantities and fluxes so that the same is true for the two-fluid equations.

We determine the dependence of the fluxes on the quantity $\mathbf{v}_s$ by considering an appropriate Galilean transformation. First we determine the ways in which the fluxes in (2.4)–(2.7) transform under a Galilean transformation. Consider the fluid equations in two reference frames which are related by a Galilean transformation such that

$$\mathbf{x}_2=\mathbf{x}_1+\mathbf{v}_r t_1 \quad\text{and}\quad t_2=t_1,$$

where the subscripts denote the quantities measured in their respective coordinate systems and $\mathbf{v}_r$ is a constant. The Galilean transformation implies that the derivatives in the two-coordinate systems satisfy the following transformations:

$$\nabla_1 = \nabla_2,$$

$$\frac{\partial}{\partial t_1} = \frac{\partial}{\partial t_2} + \mathbf{v}_r \cdot \nabla_2.$$

We now show how the continuity equation implies the way in which the mass flux $\mathbf{J}$ must transform if we assume that the density ρ is a scalar invariant. In reference frame 1 we have that

$$\frac{\partial \rho}{\partial t_1} + \nabla_1 \cdot \mathbf{J}_1 = 0.$$

If we write these derivatives in terms of the derivatives in reference frame 2 we have

$$\frac{\partial \rho}{\partial t_2} + \nabla_2 \cdot (\mathbf{J}_1 + \rho \mathbf{v}_r) = 0. \tag{2.9}$$

Galilean invariance of an equation means that the equation should have the same form in all reference frames that are related by a Galilean transformation. Thus the continuity equation in reference frame 2 should also be

$$\frac{\partial \rho}{\partial t_2} + \nabla_2 \cdot \mathbf{J}_2 = 0. \tag{2.10}$$

In order that (2.9) be identical to (2.10), we must require that

$$\mathbf{J}_2 = \mathbf{J}_1 + \rho \mathbf{v}_r.$$

We have found the transformation for the mass flux. Of course the mass flux for an ordinary fluid $\rho \mathbf{v}$ obeys this transformation. If we make the same assumption about the invariance of the entropy density we also immediately obtain that

$$\mathbf{F}_2 = \mathbf{F}_1 + \rho s \mathbf{v}_r.$$

The same procedure yields the transformation for the momentum flux tensor $\mathbf{\Pi}$:

$$\mathbf{\Pi}_2 = \mathbf{\Pi}_1 + \mathbf{J}_1 \mathbf{v}_r + \mathbf{v}_r \mathbf{J}_1.$$

Since the equation for the conservation of energy contains two quantities E and $\mathbf{Q}$, for both of which we do not know the transformation, we must assume something about the transformation of one of the quantities to obtain the transformation for the other. From the explicit form for E for an ordinary fluid assuming ρe is a scalar invariant, we find that

$$E_2 = E_1 + \mathbf{v}_r \cdot \mathbf{J}_1 + \tfrac{1}{2} \rho \mathbf{v}_r^2. \tag{2.11}$$

The transformation for $\mathbf{Q}$ can then be derived from (2.11) and the Galilean invariance of the equation of energy conservation. We find that

$$\mathbf{Q}_2 = \mathbf{Q}_1 + (E_1 + \mathbf{v}_r \cdot \mathbf{J}_1 + \tfrac{1}{2} \rho \mathbf{v}_r^2) \mathbf{v}_r + \tfrac{1}{2} \mathbf{v}_r^2 \mathbf{J}_1 + \mathbf{\Pi}_1 \cdot \mathbf{v}_r.$$

We now determine the dependence of the fluxes on $\mathbf{v}_s$. Consider a fluid in which $\mathbf{v}_n$ and $\mathbf{v}_s$ are constant throughout the fluid and consider the two reference frames related by the transformation

$$\mathbf{x}_2 = \mathbf{x}_1 + \mathbf{v}_s t.$$

The fluxes in reference frame 1 can only depend on $\mathbf{w}$ since $\mathbf{v}_s = 0$ in this frame and the fluxes in the two reference frames are related by the Galilean transformations as given above. Thus we have that

$$\mathbf{J} = \mathbf{J}_0 + \rho \mathbf{v}_s, \tag{2.12}$$

$$\mathbf{\Pi} = \mathbf{\Pi}_0 + \mathbf{J}_0 \mathbf{v}_s + \mathbf{v}_s \mathbf{J}_0, \tag{2.13}$$

$$E = E_0 + \mathbf{v}_s \cdot \mathbf{J}_0 + \tfrac{1}{2}\rho \mathbf{v}_s^2, \tag{2.14}$$

$$\mathbf{Q} = \mathbf{Q}_0 + (E_0 + \mathbf{v}_s \cdot \mathbf{J}_0 + \tfrac{1}{2}\rho \mathbf{v}_s^2)\mathbf{v}_s + \tfrac{1}{2}\mathbf{v}_s^2 \mathbf{J}_0 + \mathbf{\Pi}_0 \cdot \mathbf{v}_s, \tag{2.15}$$

$$\mathbf{F} = \mathbf{F}_0 + \rho s \mathbf{v}_s, \tag{2.16}$$

where the quantities subscripted with a zero must be Galilean invariants and thus can only depend on the quantity $\mathbf{w}$ and on other invariants.

Next we define the thermodynamic energy $E_0 \equiv \rho e$ which is a Galilean invariant. We extend the thermodynamics by writing the basic thermodynamic identity as

$$d(\rho e) = T d(\rho s) + \mu d\rho + \mathbf{w} \cdot d\mathbf{J}_0. \tag{2.17}$$

The third term expresses the fact that the velocity is the derivative of the energy with respect to momentum and can be taken as the definition of $\mathbf{v}_n$. Since the energy must be a scalar it then follows that

$$\mathbf{J}_0 = \rho_n \mathbf{w}, \tag{2.18}$$

where (2.18) is to be taken as the definition of ρ_n. The unknown scalar function ρ_n is labeled as such in anticipation of the fact that it can be associated in a certain sense with the density of the normal component. Accordingly, we define

$$\rho_s \equiv \rho - \rho_n, \tag{2.19}$$

and obtain

$$\mathbf{J} = \rho_n \mathbf{v}_n + \rho_s \mathbf{v}_s.$$

We now pick the fluxes in order to make the conservation of energy consistent with the conservation of mass, momentum, and entropy for reversible processes. We calculate the time derivative of the energy E as given in (2.14) using (2.17) and find that

$$\begin{aligned}\frac{\partial E}{\partial t} &= T\frac{\partial \rho s}{\partial t} + \left(\mu + \frac{1}{2}\mathbf{v}_s^2 - \mathbf{v}_s \cdot (\mathbf{w} + \mathbf{v}_s)\right)\frac{\partial \rho}{\partial t} \\ &\quad + (\mathbf{w} + \mathbf{v}_s) \cdot \frac{\partial \mathbf{J}}{\partial t} + (\mathbf{J} - \rho(\mathbf{w} + \mathbf{v}_s)) \cdot \frac{\partial \mathbf{v}_s}{\partial t}.\end{aligned} \tag{2.20}$$

We replace the time derivatives in (2.20) by the appropriate divergences using the conservation laws (2.4)–(2.7) and (2.2). We make the presentation slightly more axiomatic by replacing the chemical potential in (2.2) by an unknown function ϕ. If (2.2) is written with an unknown function ϕ, then the assumptions underlying (2.2) are only that $\mathbf{v}_s$ is a potential flow field and must remain a potential flow field for all times. By taking the driving term in (2.2) in the form of a gradient we guarantee that the axioms of the Kelvin circulation theorem will be satisfied and thus that $\mathbf{v}_s$ remains a potential flow field for all times.

We also break $\mathbf{\Pi}$ into the isotropic pressure and the tensor which is dependent on the momentum fluxes by writing

$$\mathbf{\Pi} = \boldsymbol{\pi} + p\mathbf{I},$$

where

$$p = -\rho e + T\rho s + \mu\rho + \mathbf{w}\cdot\mathbf{J}_0. \tag{2.21}$$

Equation (2.21) is obtained as a direct generalization of the expression for the Gibb's free energy G for an ordinary substance:

$$G \equiv \mu N = eN - TsN + pV,$$

where V is the volume, N is the number of moles of the substance, $\rho = N/V$, where $d\rho e$ is given by (2.8). The generalization to the two-fluid case is

$$G \equiv \mu N = eN - TsN - V\mathbf{w}\cdot\mathbf{J}_0 + pV,$$

where $d\rho e$ is given by (2.17).

The resultant expression for the derivative of the energy is

$$\begin{aligned}\frac{\partial E}{\partial t} &+ \nabla\cdot\left(T\mathbf{F} + \left(\mu + \frac{1}{2}\mathbf{v}_s^2 - \mathbf{v}_s\cdot\mathbf{v}_n\right)\mathbf{J} + \mathbf{v}_n\cdot\boldsymbol{\pi}\right)\\ &= (\mathbf{F} - \rho s\mathbf{v}_n)\cdot\nabla T + (\mathbf{J} - \rho\mathbf{v}_n)\cdot\nabla(\mu - \phi) + \mathbf{J}\cdot\nabla\left(\frac{1}{2}\mathbf{v}_s^2 - \mathbf{v}_s\cdot\mathbf{v}_n\right)\\ &\quad + \boldsymbol{\pi}:\nabla\mathbf{v}_n - (\mathbf{J} - \rho\mathbf{v}_n)\cdot(\mathbf{v}_s\cdot\nabla)\mathbf{v}_s - \mathbf{J}_0\cdot(\mathbf{v}_n\cdot\nabla)\mathbf{w}.\end{aligned} \tag{2.22}$$

We can simplify (2.22) by making use of the fact that

$$\mathbf{w}\mathbf{J} + (\mathbf{J} - \rho\mathbf{v}_n)\mathbf{v}_s = \mathbf{J}_0\mathbf{v}_n$$

and obtain that

$$\begin{aligned}\frac{\partial E}{\partial t} &+ \nabla\cdot\left(T\mathbf{F} + \left(\mu + \frac{1}{2}\mathbf{v}_s^2 - \mathbf{v}_s\cdot\mathbf{v}_n\right)\mathbf{J} + \mathbf{v}_n\cdot\boldsymbol{\pi}\right)\\ &= (\mathbf{F} - \rho s\mathbf{v}_n)\cdot\nabla T + (\mathbf{J} - \rho\mathbf{v}_n)\cdot\nabla(\mu - \phi) + (\boldsymbol{\pi} - \mathbf{J}_0\mathbf{v}_n - \mathbf{v}_s\mathbf{J}):\nabla\mathbf{v}_n.\end{aligned} \tag{2.23}$$

In order that (2.23) expresses the conservation of energy the right-hand side of (2.23) must vanish. Thus we find for the fluxes that

$$\mathbf{F} = \rho s\mathbf{v}_n, \tag{2.24}$$

$$\phi = \mu, \tag{2.25}$$

$$\boldsymbol{\pi} = \mathbf{J}_0\mathbf{v}_n + \mathbf{v}_s\mathbf{J} = \rho_n\mathbf{v}_n\mathbf{v}_n + \rho_s\mathbf{v}_s\mathbf{v}_s, \tag{2.26}$$

and

$$\begin{aligned}Q &= T\mathbf{F} + (\mu + \tfrac{1}{2}\mathbf{v}_s^2 - \mathbf{v}_s\cdot\mathbf{v}_n)\mathbf{J} + \mathbf{v}_n\cdot\boldsymbol{\pi}\\ &= T\mathbf{F} + (\mu + \tfrac{1}{2}\mathbf{v}_s^2)\mathbf{J} + \mathbf{v}_n(\mathbf{v}_n\cdot\mathbf{J}_0).\end{aligned} \tag{2.27}$$

For completeness we repeat (2.14):

$$E = \rho e + \mathbf{v}_s\cdot\mathbf{J}_0 + \tfrac{1}{2}\rho\mathbf{v}_s^2. \tag{2.28}$$

We note that the fluxes can be written in the forms given in (2.12)–(2.16) where

$$\boldsymbol{\Pi}_0 = \mathbf{J}_0\mathbf{w} + p\mathbf{I},$$

and

$$\mathbf{Q}_0 = T\mathbf{F}_0 + \mu\mathbf{J}_0 + \mathbf{J}_0\mathbf{w}^2;$$

these fluxes are Galilean invariants as they must be.

Equations (2.4)–(2.7) and (2.2) with the fluxes defined in (2.24)–(2.28) are the two-fluid equations. As is the case for the Euler equations, the equations are overdetermined if one uses both the conservation of energy and entropy, since as we have seen the equations were derived so that this is the case. For smooth reversible flows it does not matter whether we use the conservation of energy or the conservation of entropy equation; for flows containing shocks one should use the conservation of energy with the entropy increasing across a shock. In the next section we will discuss some applications of the two-fluid equations.

3. Validation of the Two-fluid Equations

The two-fluid equations were originally developed to describe the fact that under different experimental situations helium II behaves as if it is composed of two distinct fluids; one which carries entropy and has a viscosity, and one which carries no entropy and has no viscosity. The form of (2.2) was introduced by London [18] in 1939 to describe the fountain effect which we described in § 1; so it is not surprising that the equations describe this phenomena for small velocities. The first validation of the two-fluid equations was the fact that two modes of sound propagation are predicted by the equations; Landau pointed this out when he introduced the equations in 1941 [15]. The experimental observation of this second mode of sound propagation was observed by Peshkov [19] several years later.

In this section we will linearize the two-fluid equations and determine the two modes of sound propagation in helium II. We will also discuss another mode of sound propagation, fourth sound, which occurs when the fluid flow is restricted so that the normal component velocity is zero. The final topic in this section will be a brief discussion of shocks in helium II and the validation of the two-fluid equations through shock experiments.

We derive the equations for the propagation of sound by ignoring terms which are second order in the velocity. We will not formally linearize the equations, but give an intuitive presentation. A more detailed presentation can be found in [16] and [17]. We take the following form for the two-fluid equations in the limit of small velocities:

$$\frac{\partial \rho}{\partial t}+\nabla\cdot\mathbf{J}=0, \tag{3.1}$$

$$\frac{\partial \rho s}{\partial t}+\rho s\nabla\cdot\mathbf{v}_n=0, \tag{3.2}$$

$$\frac{\partial \mathbf{J}}{\partial t}+\nabla p=0, \tag{3.3}$$

$$\frac{\partial \mathbf{v}_s}{\partial t}+\nabla\mu=0. \tag{3.4}$$

The wave equation for ordinary sound, which we shall refer to as first sound, is obtained by considering the pressure to be a function of the density and the entropy. First sound is a constant entropy wave and thus we approximate (3.3) by

$$\frac{\partial \mathbf{J}}{\partial t}+\left(\frac{\partial p}{\partial \rho}\right)_s\nabla\rho=0, \tag{3.5}$$

where the derivative $(\partial p/\partial\rho)_s$ is evaluated at the ambient equilibrium conditions. If we take the time derivative of (3.1) and the divergence of (3.5) and combine the resultant equations we find that

$$\frac{\partial^2 \rho}{\partial t^2} = \left(\frac{\partial p}{\partial \rho}\right)_s \Delta \rho.$$

Thus we see that first sound is a pressure wave which propagates with speed c_1 where $c_1^2 = (\partial p/\partial \rho)_s$. The value of c_1 is only weakly dependent on temperature and equals 240 m/sec for helium II at saturated vapor pressure.

In helium II first sound propagation satisfies the condition that $\mathbf{v}_n = \mathbf{v}_s$. Second sound in helium II is a temperature and entropy wave and approximately satisfies the condition that $\mathbf{J} = 0$. If we assume that $\mathbf{J} = 0$, then from (3.1) and (3.3) we infer that the density and pressure are constant. Under these assumptions when we differentiate (3.2) with respect to time and keep only terms linear in the dependent quantities, we obtain

$$\left(\frac{\partial s}{\partial T}\right)_\rho \frac{\partial^2 T}{\partial t^2} + s\nabla \cdot \frac{\partial \mathbf{v}_n}{\partial t} = 0, \tag{3.6}$$

where the subscript ρ indicates the independent variable held constant in differentiating the entropy s and the coefficients of the derivatives are constants evaluated at the equilibrium values of the thermodynamic variables. Under the same assumptions and using (2.17) and (2.21) from (3.4) we obtain that

$$\frac{\rho_n}{\rho_s} \nabla \cdot \frac{\partial \mathbf{v}_n}{\partial t} + \frac{s}{\rho} \Delta T = 0, \tag{3.7}$$

where we have also used the fact that we are assuming $\mathbf{J} = 0$ to relate $\mathbf{v}_n$ and $\mathbf{v}_s$. We combine (3.6) and (3.7) to obtain that

$$\frac{\partial^2 T}{\partial t^2} = \frac{\rho_s s^2}{\rho_n (\partial s/\partial T)_\rho} \Delta T. \tag{3.8}$$

Equation (3.8) describes the propagation of the second sound mode where the speed of propagation c_2 is given by

$$c_2^2 = \frac{\rho_s s^2}{\rho_n (\partial s/\partial T)_\rho}.$$

From (2.17) we see that

$$\left(\frac{\partial s}{\partial T}\right)_\rho = \frac{C_v}{T},$$

where C_v is the specific heat of helium at constant volume. Thus we obtain the equivalent form:

$$c_2^2 = \frac{\rho_s s^2 T}{\rho_n C_v}.$$

The speed of second sound is zero at T_λ. In the temperature range from 0.9°K to 2.0°K the speed is almost independent of temperature and equal to 20 m/sec. Below 0.9°K the speed of second sound increases until it reaches the value $c_1/\sqrt{3} = 140$ m/sec at 0°K.

There is often reference to another sound mode in the literature: fourth sound. Fourth sound is not a sound mode distinct from first sound or second sound, but rather it is a combination of the two sound modes which is manifested in a specific experimental geometry. Fourth sound is observed experimentally when the normal component is *clamped*. *Clamped* means that the flow experiment is designed so that $\mathbf{v}_n = 0$. This is usually accomplished by performing the flow experiment in a tube which is packed with a micron-sized powder. The relevant linearized equations for fourth sound,

$$\frac{\partial \rho}{\partial t} + \rho_s \nabla \cdot \mathbf{v}_s = 0, \tag{3.9}$$

and

$$\frac{\partial \mathbf{v}_s}{\partial t} + \left(\frac{\partial \mu}{\partial \rho}\right)_{\rho s} \nabla \rho = 0, \tag{3.10}$$

are obtained from (3.1) and (3.4) respectively, by setting $\mathbf{v}_n = 0$. By eliminating $\mathbf{v}_s$ from (3.9) and (3.10) we obtain that

$$\frac{\partial^2 \rho}{\partial t^2} = \rho_s \left(\frac{\partial \mu}{\partial \rho}\right)_{\rho s} \Delta \rho,$$

and thus we see that

$$c_4^2 = \rho_s \left(\frac{\partial \mu}{\partial \rho}\right)_{\rho s}. \tag{3.11}$$

Under the assumption that $(\partial \rho / \partial T)_p = 0$, equation (3.11) can be expanded to obtain that

$$c_4^2 = \frac{\rho_s}{\rho} c_1^2 + \frac{\rho_n}{\rho} c_2^2. \tag{3.12}$$

Equation (3.12) is the basis for the precise experimental determination of ρ_s. Resonators can be built in which it is possible to simultaneously determine the speeds of first, second, and fourth sound. From (3.12) one can readily determine the superfluid component density when the sound speeds are known independently. The most precise thermodynamic data available for helium II has been obtained by a simultaneous measurement of the three sound speeds [20], [21]. By integrating the appropriate thermodynamic identities and using the fact that the three sound speeds c_1, c_2, and c_4 are known experimentally, one is able to obtain the chemical potential μ as a function of p, T, and $\mathbf{w}^2$, if one assumes that ρ, s, and ρ_n are independent of $\mathbf{w}^2$. In the sense described above the thermodynamics of helium II are only known in the plane $\mathbf{w}^2 = 0$. It is generally assumed that the dependence of ρ, s, and ρ_n on $\mathbf{w}^2$ is weak [22]–[24].

The prediction and observation of the three sound modes described in this section is one of the main validations of the two-fluid equations. The observation of the sound modes validates the two-fluid equations in the linear limit. There has also been some experimental effort to validate the two-fluid equations in the nonlinear regime. The work in the nonlinear region concentrates on investigating the properties of shock waves in helium II [25]–[27]. There is good quantitative and qualitative agreement in the nonlinear region for the two-fluid equations between the experimental results and theoretical predictions; however, much more experimental work needs to be done in this area.

4. Vortices in Superfluid Helium

When the two-fluid equations were originally introduced it was assumed that the velocity $\mathbf{v}_s$ was irrotational: $\nabla \times \mathbf{v}_s = 0$. In experiments in which helium II is rotated, however, the superfluid component as well as the normal component rotates. Osborne [28] observed that the free surface of rotating superfluid had the same parabolic profile as that for an ordinary fluid. Hall [29] measured the torque on a stack of rotating discs in helium II and concluded that both the normal and superfluid components were rotating. It became clear that the condition $\nabla \times \mathbf{v}_s = 0$ could not hold everywhere in the fluid. Vorticity had to be allowed in the superfield velocity $\mathbf{v}_s$.

Onsager [30] and Feynman [31] suggested that $\mathbf{v}_s$ could remain irrotational everywhere except at a finite number of singular points or vortices. Furthermore, they argued that the circulation of these vortices should be quantized. The circulation should be constrained by the quantization rule that

$$\oint \mathbf{v}_s \cdot d\boldsymbol{l} = \frac{nh}{m},$$

where the integral is the line integral taken around a closed path in the fluid, h is Planck's constant, m is the mass of a helium atom, and n is an integer. $h/m \approx 9.97 \times 10^{-5}\ \mathrm{cm}^2/\mathrm{sec}$ and so the quantization effect is difficult to observe. The fact that rotating helium II forms a parabolic free surface is explained by assuming that the quantized vortices are uniformly distributed in a plane perpendicular to the axis of rotation. This distribution of quantized vortices approximates an ordinary fluid in which the fluid rotates uniformly and the vorticity is also uniformly distributed.

In 1961, Vinen [32] provided the first experimental evidence which was consistent with the circulation being quantized in helium II. By measuring the resonant frequencies of a wire, which was stretched down the axis of a rotating cylinder of helium, Vinen observed that for small amplitude vibrations the circulation around the wire was either zero or h/m. Vinen also observed that the circulation around the wire persisted for several hours after the cylinder had been brought to rest. This result has been confirmed in similar experiments performed by Karn, Starks, and Zimmerman [33].

Another experiment which confirmed the quantization of vorticity was that performed by Rayfield and Reif [34], [35]. Rayfield and Reif studied the mobilities of ions in helium II. They found that the velocity of the ions decreased as the energy of the ions increased when the ions were passed through helium II. Furthermore, it was found that the velocity of the ions ν was proportional to $1/E$, where E is the energy of the ion. If one assumes that the ion energy is equal to the energy of a vortex ring of radius r, and core radius σ, then in the limit where $\sigma \ll r$ we have that

$$E = \frac{1}{2}\rho\kappa^2 r\left(\eta - \frac{7}{4}\right); \tag{4.1}$$

and if the velocity of the ion is the same as that of a vortex ring of radius r we have that

$$\nu = \frac{\kappa(\eta - \frac{1}{4})}{4\pi r}, \tag{4.2}$$

where

$$\eta = \ln(8r/\sigma),$$

κ is the circulation of the vortex ring, and ρ is the density of the fluid [36]. Thus we see that if we assume that the radii r of the vortex rings do not vary greatly (i.e., eta is approximately constant) and we eliminate the radius by combining (4.1) and (4.2), the relationship between the velocity and energy of a vortex ring is such that $E \propto 1/v$. Rayfield and Reif pick $\kappa \equiv h/m$ and then choose σ to give a best fit of the experimental data to (4.1) and (4.2). They find a value of σ = 1.2 Å. (It should be noted that the interatomic spacing in liquid helium is approximately 4 Å.) For the velocities obtained in this experiment it can be inferred from (4.2) that the radius r of the vortex rings varies from 500 Å to 10,000 Å and that η varies from 9 to 11.

It should be pointed out that when the experiment of Rayfield and Reif is used as evidence for the fact that quantized vortices exist, some caution must be exercised. The experiment provides evidence that ions travel through superfluid helium so that the potential energy of the ion is inversely proportional to the velocity of the ion as it moves in the fluid. One of the key assumptions in associating this motion of the ion with that of a vortex ring is that the energy of the ion is the same as that of the vortex ring. If this assumption is made, then (4.1) and (4.2) imply that

$$E = \frac{\rho\kappa^3(\eta - \frac{7}{4})(\eta - \frac{1}{4})}{8\pi v} \tag{4.3}$$

Thus we see that even when the ion motion is associated with that of a vortex ring the experiment determines the quantity $\kappa^3\eta^2$, since η is constant to first order in r and varies only by 10 percent over the experimental range of radii considered. In the analysis of Rayfield and Reif κ is chosen so that $\kappa \equiv h/m$.

One of the most remarkable experiments verifying the existence of vortices in helium II is that performed by Yarmchuk, Gordon, and Packard [37], [38]. In this experiment a cylinder 2 mm in diameter is filled to a depth of 25 mm with helium II. The cylinder is rotated and arrays of vortices are observed. If one assumes cylindrical symmetry it is possible to predict the configuration of the vortices for a varying number of vortices as Campbell and Ziff have done [39]. It is found that the observed configurations of vortices agree qualitatively quite well with the predicted configurations. The vortices were observed by trapping negative ions on the vortex cores, in a similar manner to that employed by Rayfield and Reif, and then the ions were accelerated onto a phosphor so that they could then be recorded. The experiments were performed at a temperature of 100°mK, a temperature at which $\rho_n \approx 0$. The vortices were found to be unstable in pure helium II at these temperatures and 0.8 percent of ^{3}He was added to the helium II to provide damping and stabilize the vortices. The vortices had to be stabilized since it took a while to impact enough ions onto the phosphor to produce an observable image. It should be noted that the circulation of the helium could not be determined in this experiment, however, the spacing of the vortices was in reasonable agreement to what is expected for quantized vortices for the angular velocities used in rotating the cylinder.

5. Turbulence in Superfluid Helium

In this section we will give a brief introduction into what is known as turbulence in helium II. There has been much activity in this field of research during the past 30 years. There have been several excellent review papers written in recent years which describe the modeling of turbulence in helium II in great detail. We recommend the review written by Barenghi, Donnelly, and Vinen [40] and that by Donnelly and Swanson [41] to anyone interested in more details than are presented here. We will describe only a few of the major successes and failures of the work done in the field up to the present.

The study of turbulence in helium II actually began as a study of thermal conductivity in helium II [42], [43]. The thermal conductivity which we are discussing is not the ordinary thermal conductivity of the fluid but the effective thermal conductivity produced by the fountain effect, as described in § 1. For low velocities it is found that

$$\rho s \Delta T = \Delta p. \tag{5.1}$$

Recall that (2.2) was derived so that (5.1) would be satisfied. For low velocities the normal component of the velocity of the superfluid will be laminar and proportional to the pressure difference Δp. Thus the heat flux, which is proportional to the normal component of the velocity (see (2.27)) is linearly proportional to the temperature difference as given in (5.1). As the velocity is increased, or rather as the heat flux is increased, it is observed that the temperature difference increases more rapidly than the linear relation given by (5.1). Experimentally, it is found that

$$\Delta T = D(W - W_0)^3, \tag{5.2}$$

where D is a constant which depends on T and geometry, W is the heat flux, and W_0 is a constant which is relatively small.

In most discussions of superfluid turbulence a modified form of the two-fluid equations is used. In order to discuss the turbulence models we must introduce these modified equations. The basic assumption implicit in the modified equations is that

$$\frac{\partial \rho_n}{\partial t} + \nabla \cdot (\rho_n \mathbf{v}_n) = 0. \tag{5.3}$$

Equation (5.3) together with the continuity equation (2.4) implies the same equation for the superfluid component:

$$\frac{\partial \rho_s}{\partial t} + \nabla \cdot (\rho_s \mathbf{v}_s) = 0. \tag{5.4}$$

Equations (5.3) and (5.4) change the character of the two-fluid equations. These assumptions imply that the superfluid and normal components are individually conserved and thus in this approximation the two-fluid equations are the same as if there were two distinct fluids present. If we use the identity that

$$\frac{\partial \rho \mathbf{v}}{\partial t} + \nabla \cdot (\rho \mathbf{v}\mathbf{v}) = \mathbf{v}\left(\frac{\partial \rho}{\partial t} + \nabla \cdot (\rho \mathbf{v})\right) + \rho \frac{\partial \mathbf{v}}{\partial t} + \rho(\mathbf{v} \cdot \nabla)\mathbf{v},$$

then with the substitution of (5.3) and (5.4), the momentum conservation equation, equation (2.5), becomes

$$\rho_n \frac{D_n \mathbf{v}_n}{Dt} + \rho_s \frac{D_s \mathbf{v}_s}{Dt} + \nabla p = 0, \tag{5.5}$$

where

$$\frac{D_n}{Dt} \equiv \frac{\partial}{\partial t} + \mathbf{v}_n \cdot \nabla \quad \text{and} \quad \frac{D_s}{Dt} \equiv \frac{\partial}{\partial t} + \mathbf{v}_s \cdot \nabla.$$

Substituting (2.2) into (5.5) we obtain

(5.6)
$$\rho_n \frac{D_n \mathbf{v}_n}{Dt} = -\nabla p + \rho_s \nabla \mu.$$

Thus if we assume that the individual components of the superfluid are conserved we are able to obtain an equation of motion for $\mathbf{v}_n$ which contains a dependence on $\mathbf{v}_s$ only through thermodynamic quantities.

If we substitute for μ in terms of the temperature, pressure, and relative velocity, using the thermodynamic relation

$$d\mu = \frac{1}{\rho} dp - s dT - \frac{1}{2} \frac{\rho_n}{\rho} d\mathbf{w}^2,$$

we find that

(5.7)
$$\rho_n \frac{D_n \mathbf{v}_n}{Dt} = -\frac{\rho_n}{\rho} \nabla p - \rho_s s \nabla T - \frac{1}{2} \frac{\rho_s \rho_n}{\rho} \nabla (\mathbf{v}_n - \mathbf{v}_s)^2$$

and that

(5.8)
$$\rho_s \frac{D_s \mathbf{v}_s}{Dt} = -\frac{\rho_s}{\rho} \nabla p + \rho_s s \nabla T + \frac{1}{2} \frac{\rho_s \rho_n}{\rho} \nabla (\mathbf{v}_n - \mathbf{v}_s)^2.$$

In turbulence modeling it is tacitly assumed that the thermodynamic quantities do not depend on $\mathbf{w}^2$, except for the explicit dependence shown in (5.7) and (5.8). The velocity dependent terms are replaced by general forces which depend on the velocity and which are assumed to include the explicit terms given in (5.7) and (5.8). The modified equations become

(5.9)
$$\rho_n \frac{D_n \mathbf{v}_n}{Dt} = -\frac{\rho_n}{\rho} \nabla p - \rho_s s \nabla T + \mathbf{F}_{ns}$$

and

(5.10)
$$\rho_s \frac{D_s \mathbf{v}_s}{Dt} = -\frac{\rho_s}{\rho} \nabla p + \rho_s s \nabla T - \mathbf{F}_{ns},$$

where $\mathbf{F}_{ns}$ has units of force and is called the mutual friction force. It is interpreted as the force that the superfluid component exerts on the normal component.

The modified equations introduced above greatly simplify the two-fluid equations; however, important features of the two-fluid equations are also lost. If the additional assumption that $\nabla \cdot \mathbf{v}_s = 0$ is made, then the dynamics of the superfluid component of the velocity is totally determined by the location of the vorticity and the boundaries, independently of $\mathbf{v}_n$. Thus in order to again couple $\mathbf{v}_n$ and $\mathbf{v}_s$ a heuristic force $\mathbf{F}_{ns}$ must be added to the equation of motion for the superfluid component.

We now return to our discussion of the thermal conductivity of helium II. Gorter and Mellink [42] suggested that if

(5.11)
$$\mathbf{F}_{ns} \equiv -\rho_s \rho_n A \mathbf{w}^2 \mathbf{w},$$

then the cubic relation between the temperature gradient and the heat flux could be explained. This form for $\mathbf{F}_{ns}$ is adequate for wide channels in pure counterflow; however, it does not describe all experimental situations [44]. The Gorter and Mellink force does not adequately describe helium II in bulk flow situations. Nonetheless, for one-dimensional counterflow in a tube the Gorter–Mellink force gives a temperature gradient of the form

$$\nabla T = -aW - b(T)W^3,$$

where

$$b(T) = \frac{A\rho_n}{s(\rho_s sT)^3},$$

and where a corresponds to the linear result predicted by the two-fluid equations. This result describes the dependence of the temperature gradient on the heat flux as given in (5.2).

In an attempt to give a fundamental understanding of the Gorter–Mellink force, Feynman [31] suggested that the force could be due to the creation of a tangle of quantized vortices in the tube where the counterflow occurs. In this description the resultant additional dissipation is caused by the mutual friction between the quantized vortices and the normal component of the superfluid. In principle, if one knows the forces on the quantized vortices then one could predict the dissipation.

In a series of experiments Hall and Vinen measured the mutual friction force on the quantized vortices [45], [46]. In these experiments the force on the quantized vortices was inferred by measuring the dissipation of second sound in a rotating cylinder of helium II. It was assumed that the quantized vortices were uniformly spaced across the cylinder and that the number of vortices corresponded to the total circulation of the fluid in the cylinder. At a rotation rate of 1 rad/sec the density of quantized vortices is approximately 1,000 cm^{-2}. The form for $\mathbf{F}_{ns}$ is postulated so that there is a force in the parallel and two perpendicular directions to the quantized vortex. The force parallel to the vortex is ignored and the form for the force is taken as

$$\mathbf{F}_{ns} = B\frac{\rho_s\rho_n}{\rho}\frac{\omega\times(\omega\times\mathbf{w})}{|\omega|} + B'\frac{\rho_s\rho_n}{\rho}\omega\times\mathbf{w}, \tag{5.12}$$

where ω is the angular velocity of rotation, and B and B' are constants to be determined by the experiment. If a force of the form given in (5.12) is used in the modified two-fluid equations it is found that there is a dissipation present in second sound which is proportional to the angular velocity ω. Thus by measuring the dissipation of second sound in rotating helium II Hall and Vinen were able to determine B. Hall and Vinen were not able to determine B' from their experiments but argued that it has negligible effects on the experiment.

We note that in the derivation of (5.12) the vortex is treated as if it were a solid body subject to the forces of the surrounding fluid. This treatment of a vortex is inconsistent with the motion of a vortex as governed by Euler's equation.

In another experiment Vinen [47] measured the attenuation of second sound in the presence of a heat current. He observed that at small amplitudes of the heat current there is no attenuation, but once a *critical velocity* is attained the attenuation is proportional to the heat current squared. In an attempt to explain this attenuation in terms of the vortex tangle and in order to relate this attenuation to the Gorter–Mellink force, Vinen related the attenuation present in the counterflow experiment to the attenuation present in rotating helium II which he earlier attributed to the quantized vortices.

Vinen [48] wished to characterize the attenuation in terms of the total length per unit volume L of vortices present. First Vinen argued that in the presence of the counterflow heat current the mutual force $\mathbf{F}_{ns}$ could be expressed as

$$\mathbf{F}_{ns} = -G\mathbf{w}, \tag{5.13}$$

where

$$G = \frac{B}{3}\frac{\rho_s\rho_n}{\rho}\frac{h}{m}L,$$

where B is the same coefficient that appears in (5.12). Vinen then modeled the growth of a vortex ring and by a dimensional argument he showed that a reasonable form for the rate of production of vortex line length L should be

$$\left(\frac{dL}{dt}\right)_{\text{gen.}} = \chi_1 B\frac{\rho_n}{\rho}L^{3/2}|\mathbf{w}|, \tag{5.14}$$

where χ_1 is a dimensionless parameter which depends on the geometry of the vortex lines. By an involved heuristic argument Vinen obtains that the rate of decay of the vortex line length L should take the form

$$\left(\frac{dL}{dt}\right)_{\text{decay}} = -\chi_2\frac{h}{m}L^2, \tag{5.15}$$

where χ_2 is a dimensionless parameter assumed of order unity.

It is then assumed that when a heat flux is present as in fully developed counterflow there is a dynamic balance between the creation and decay of the vortex line length L. Thus one obtains an expression for the line length L_0 present in fully developed counterflow by requiring that

$$\frac{dL}{dt} = \left(\frac{dL}{dt}\right)_{\text{gen.}} + \left(\frac{dL}{dt}\right)_{\text{decay}} = 0$$

in equilibrium. From (5.14) and (5.15) we obtain that

$$\chi_1 B\frac{\rho_n}{\rho}L_0^{3/2}|\mathbf{w}| = \chi_2\frac{h}{m}L_0^2. \tag{5.16}$$

Simplifying (5.16) we obtain that

$$L_0^{1/2} = \frac{\chi_1}{\chi_2}\frac{B\rho_n}{2\rho}\frac{m}{h}|\mathbf{w}|. \tag{5.17}$$

Equation (5.17) gives the result that $L_0 = \gamma|\mathbf{w}|^2$ to which we shall refer later.

If (5.16) is substituted into (5.12), then we see that $\mathbf{F}_{ns}$ has the same dependence on $\mathbf{w}$ as the Gorter–Mellink force as given by (5.11). Thus the Vinen model for vortex line growth and decay is consistent with the Gorter–Mellink force.

Although the Vinen model for vortex line growth gives the correct velocity dependence for the Gorter–Mellink force, it is not clear from Vinen's presentation how the vortex tangle responsible for the line length is established and maintained. In order to clarify how the vortex tangle is established Schwarz has attempted to produce vortex tangles through numerical simulations. Schwarz [49] derived a model equation for the evolution of a quantized vortex based on (5.12). We give a brief description of his derivation.

The quantized vortex is considered to be a solid cylinder with a circulation about the cylinder equal to h/m. (As mentioned previously this interpretation of a vortex is not consistent with Euler's equation.) The cylinder is assumed to be acted on by the Magnus force. The Magnus force per unit length $\mathbf{F}_m$ is given by [50]

$$\mathbf{F}_m = \rho \Gamma \mathbf{U} \times \mathbf{t}, \tag{5.18}$$

where $\mathbf{U}$ is the velocity of the fluid relative to the cylinder, Γ is the circulation about the cylinder, ρ is the density of the fluid, and $\mathbf{t}$ is a unit vector parallel to the axis of the cylinder. Equation (5.18) is only valid in two dimensions or for an infinitely long cylinder, but the result is often used in three dimensions in which case $\mathbf{U}$ is taken to be the component of velocity perpendicular to the axis of the cylinder. If it is assumed that $\mathbf{U}\cdot\mathbf{t} = 0$, (5.18) can be inverted to obtain that

$$\mathbf{U} = \frac{\mathbf{t} \times \mathbf{F}_m}{\rho \Gamma}. \tag{5.19}$$

$\mathbf{U}$ in (5.19) is interpreted as being the velocity of the vortex tube perpendicular to its axis when it is acted on by a Magnus force $\mathbf{F}_m$.

The next step is to interpret (5.12) as a force acting on an individual quantized vortex. By assuming that the angular velocity ω is uniformly distributed parallel to the axis of rotation, one can interpret the direction of ω as being parallel to the local tangent to the quantized vortex $\mathbf{t}$ and one can interpret the magnitude of ω as being an area density of circulation. With these interpretations one obtains that the force per unit length $\mathbf{F}$ on a quantized vortex is

$$\mathbf{F} = -B\frac{\kappa \rho_s \rho_n}{2\rho} \mathbf{t} \times (\mathbf{t} \times \mathbf{w}) - B' \frac{\kappa \rho_s \rho_n}{2\rho} \mathbf{t} \times \mathbf{w}. \tag{5.20}$$

where $\kappa \equiv h/m$, and $\mathbf{t}$ is the unit tangent to the vortex. The minus sign in (5.20) appears because $\mathbf{F}_{ns}$ is defined as the force exerted on the normal component. The factor of two appears because a body rotating with angular velocity ω has a vorticity equal to 2ω.

The total velocity of the vortex $\mathbf{v}_t$ is defined as

$$\mathbf{v}_t = \mathbf{v}_l + \mathbf{v}_{nc}.$$

The term $\mathbf{v}_l$ is the velocity the vortex would have if no drag force were present; it is the velocity which results by solving Euler's equation with only the quantized vortex present. The term $\mathbf{v}_{nc}$, the nonconservative velocity, is the correction to the total velocity due to the drag on the vortex; it is modeled as the velocity due to a Magnus force on the vortex tube. We use (5.19) with $\mathbf{U} = \mathbf{v}_{nc}$, $\Gamma = \kappa$, $\rho = \rho_s$ and with $\mathbf{F}$ given by (5.20) to obtain that

$$\mathbf{v}_{nc} = \frac{\rho_n B}{2\rho} \mathbf{t} \times (\mathbf{v}_n - \mathbf{v}_s - \mathbf{v}_l) - \frac{\rho_n B'}{2\rho} \mathbf{t} \times (\mathbf{t} \times (\mathbf{v}_n - \mathbf{v}_s - \mathbf{v}_l)). \tag{5.21}$$

In (5.21) $\mathbf{v}_s$ and $\mathbf{v}_n$ are interpreted to be spatial averages of the instantaneous local velocities.

$\mathbf{v}_l$ is the solution to Euler's equation given by the particular distribution of vorticity represented by the quantized vortices present in the system. Rather than solve Euler's equation exactly the self-induction approximation [51] is used and we obtain as an approximation to $\mathbf{v}_l$ that

$$\mathbf{v}_l \approx \beta \mathbf{t} \times \mathbf{t}', \tag{5.22}$$

where $\beta = \kappa/4\pi \ln(R/a_0)$, R is the radius of curvature, a_0 is the core radius ($a_0 \approx 1.2$ Å), and $\mathbf{t}'$ is the derivative with respect to arclength of the tangent to the vortex. When (5.22) is substituted into (5.21) we obtain that

$$\mathbf{v}_t = \beta \mathbf{t} \times \mathbf{t}' + \frac{\rho_n B}{2\rho} \mathbf{t} \times (\mathbf{v}_n - \mathbf{v}_s - \beta \mathbf{t} \times \mathbf{t}') - \frac{\rho_n B'}{2\rho} \mathbf{t} \times (\mathbf{t} \times (\mathbf{v}_n - \mathbf{v}_s - \beta \mathbf{t} \times \mathbf{t}')). \tag{5.23}$$

Values of B and B' can be obtained from Lucas [52]; it is found that B' is usually 10 times smaller than B.

Schwarz based some numerical simulations of vortex tangles on (5.23) [53]–[56]. In these calculations Schwarz assumed B' to be small compared to B and based his calculations on (5.23) which after simplification and setting $B' = 0$ becomes

$$\mathbf{v}_t \equiv \frac{\partial \mathbf{r}}{\partial t} = \beta \mathbf{t} \times \mathbf{t}' + \frac{\rho_n B}{2\rho} \mathbf{t} \times (\mathbf{v}_n - \mathbf{v}_s) + \frac{\beta \rho_n B}{2\rho} \mathbf{t}', \tag{5.24}$$

where $\mathbf{r}$ is the position of the vortex and thus $\mathbf{r}' = \mathbf{t}$. Schwarz also took $\mathbf{v}_n - \mathbf{v}_s$ to be a constant. Buttke [57] pointed out that there is an intrinsic length scale and timescale present in (5.24). If one sets $\mathbf{v}_n - \mathbf{v}_s = \mathbf{w}_0$, then the dimensionless time τ is defined as $\tau \equiv |\mathbf{w}_0|^2 t/\beta$ and the dimensionless length $\chi \equiv |\mathbf{w}_0| \mathbf{r}/\beta$. With these definitions of length and time (5.24) becomes

$$\frac{\partial \chi}{\partial \tau} = \chi' \times \chi'' + \alpha \chi' \times \hat{\mathbf{w}}_0 + \alpha \chi'', \tag{5.25}$$

where $\hat{\mathbf{w}}_0$ is a unit vector in the $\mathbf{w}_0$ direction and $\alpha = (B\rho_n)/(2\rho)$.

The goal of the numerical simulations of vortex tangles is to obtain the result that $L_0 = \gamma |\mathbf{w}_0|^2$, where L_0 is the line length density. Numerically, one wishes to calculate the equilibrium line length present in a given volume as one increases $|\mathbf{w}_0|$. In terms of the dimensionless (5.25) if one considers a physical cube of side a, then the dimensionless cube has the length of its side equal to $|\mathbf{w}_0| a/\beta$. Thus in order to consider a fixed physical volume equal to a^3 one must consider a volume equal to $(|\mathbf{w}_0| a/\beta)^3$ in dimensionless units when solving (5.25). If we consider a self-similar solution of (5.25) which has a constant line length density l_0 as we increase the volume $(|\mathbf{w}_0| a/\beta)^3$ of the cube, the total dimensionless length of vortex lines present is $l_0(|\mathbf{w}_0| a/\beta)^3$. In physical units this is a vortex line length equal to $l_0(|\mathbf{w}_0|/\beta)^2 a^3$; this corresponds to the equilibrium line length present in a box of volume a^3. Thus the physical line length density L_0 predicted by a self-similar solution of (5.25) satisfies

$$L_0 = l_0(|\mathbf{w}_0|/\beta)^2. \tag{5.26}$$

Equation (5.26) gives us the same relation for L_0 in terms of $|\mathbf{w}_0|$ as (5.17). This is the scaling required in order to explain the Gorter–Mellink mutual friction force in terms of a vortex tangle.

The derivation of (5.26) implicitly assumes the existence of a self-similar solution. It is not obvious a priori that such a solution exists. Because of the inherent length scale $\beta/|\mathbf{w}_0|$, the size of the box in which one calculates the vortex line density and the particular boundary conditions one imposes on the box should influence the calculated equilibrium line length density. Only in the limit that the box size goes to infinity in dimensionless units or equivalently that $\beta/|\mathbf{w}_0| \to 0$ would one expect a self-similar solution, of the type described above, to exist.

Equation (5.25) is an evolution equation for the quantized vortices which only depends on the local geometry of the vortex. In order to introduce some nonlocal interactions Schwarz introduced the ansatz into his numerical simulations that two quantized vortices should reconnect whenever they cross. The reconnection is uniquely defined since the vortex lines have a direction associated with them given by the vorticity. Schwarz numerically followed the evolution of a system of a few vortex rings, in a cube with triply periodic boundary conditions imposed, governed by (5.24) and the reconnection ansatz. Schwarz found that the system of vortices would evolve into a configuration in which the line length density fluctuated about some constant value which was independent of the particular initial conditions which were used. He also found that the calculated equilibrium line length density obeyed the scaling given by (5.26).

Schwarz's discovery had a great influence on the way quantized vortices were viewed by fluid dynamicists. The basic evolution of quantized vortices is governed by the Euler equation as given in the two-fluid equations. The only difference between the evolution of a quantized vortex and an ideal vortex, even in Schwarz's model, is the frictional force and the frictional force is a perturbation to the ideal fluid motion of the vortex. In Schwarz's model the Euler's equation is modeled by the self-induction approximation. It is known that the self-induction approximation is a poor approximation to Euler's equation for an ordinary fluid. Ideal fluid vortices stretch and fold as they evolve in an Eulerian fluid in such a complicated manner that it is difficult to calculate their evolution for even short periods of time [58]–[63]. Paradoxically, superfluid vortices seem to be modeled quite well by the self-induction approximation which does not even allow the vortices to stretch.

This paradox was resolved by Buttke [64], [65] in a paper in which he repeated the numerical experiments of Schwarz. Buttke found that the numerical results of Schwarz could be duplicated if he underresolved the numerical computations. Buttke developed a numerical algorithm independently of Schwarz's algorithm which solved the model (5.25). Buttke found that if he scaled the numerical resolution of the vortices in the same way one would physically resolve them, he could reproduce the results of Schwarz. However, Buttke showed that these results which gave the correct scaling as given in (5.26) were not solutions of (5.25). When the calculations were properly resolved the results no longer scaled as in (5.26). Buttke showed that the solutions to (5.25) with periodic boundary conditions do not produce vortex tangles which scale according to (5.26). He showed that by underresolving the vortices there is an artificial generation of vortices at the small scales which is not present when (5.25) along with the reconnection ansatz is solved correctly.

Buttke pointed out that the fact that this artificial generation of vortices at small scales is necessary in order to obtain the correct physical scaling for the vortex tangles indicates that a better approximation than the self-induction approximation is necessary in modeling the Euler equation even for superfluid vortices. Even though it is not clear what equation one is solving when one numerically underresolves (5.25), one can argue that one is solving the equation correctly for large scales or small curvatures. If this is the case, then the results of Buttke's work seem to indicate that an adequate model for the vortex tangles could be obtained by keeping the terms modeling the frictional force in their present form and by improving the approximation of the term modeling the solution of Euler's equation.

It appears likely that one may still be able to explain the Gorter–Mellink force in terms of a vortex tangle; however, it is clear that the solution will require a better approximation to the Euler's equation in the model equations. The resolution of the paradox of vortex tangles is satisfying since the indication is that the dynamics of superfluid vortices are not drastically different from the dynamics of ordinary ideal fluid vortices. Vortex tangles are not self-sustaining without the inherent stretching caused by the ideal fluid flow.

References

[1] D. D. OSHEROFF, R. C. RICHARDSON, AND D. M. LEE, *Evidence for a new phase of solid* ^3He, Phys. Rev. Lett., 28 (1972), pp. 885–888.

[2] D. D. OSHEROFF, W. J. GULLY, R. C. RICHARDSON, AND D. M. LEE, *New magnetic phenomena in liquid* ^3He *below* 3 *mK*, Phys. Rev. Lett., 29 (1972), pp. 920–923.

[3] P. L. KAPITZA, *Viscosity of liquid helium below the* λ*-point*, Nature, 141 (1938), p. 74.

[4] J. F. ALLEN AND A. D. MISENER, *Flow of liquid helium* II, Nature, 141 (1938), p. 75.

[5] H. KAMERLINGH ONNES, *Further experiments with liquid helium*, Proc. Roy. Acad. Amsterdam, 13 (1911), pp. 1093–1119.

[6] O. V. LOUNASMAA, *The superfluid phases of liquid* ^3He, Contemp. Phys., 15 (1974), pp. 353–374.

[7] E. R. DOBBS, *Superfluid helium three*, Contemp. Phys., 24 (1983), pp. 389–413.

[8] J. C. WHEATLEY, *Experimental properties of superfluid* ^3He, Rev. Modern Phys., 47 (1975), pp. 415–470.

[9] J. WILKS AND D. S. BETTS, *An Introduction to Liquid Helium*, Oxford University Press, New York, 1987.

[10] W. H. KEESOM, *Helium*, Elsevier, New York, 1942.

[11] K. R. ATKINS, *Liquid Helium*, Cambridge University Press, New York, 1959.

[12] W. E. KELLER, *Helium-*3 *and Helium-*4, Plenum Press, New York, 1969.

[13] R. J. DONNELLY, *Experimental Superfluidity*, University of Chicago Press, Chicago, IL, 1967.

[14] L. TISZA, *Transport phenomena in helium* II, Nature, 141 (1938), p. 913.

[15] L. D. LANDAU, *The theory of superfluidity of helium* II, J. Phys. (Moscow), 5 (1941), pp. 71–90.

[16] I. M. KHALATNIKOV, *An Introduction to the Theory of Superfluidity*, W. A. Benjamin, New York, 1965.

[17] S. J. PUTTERMAN, *Superfluid Hydrodynamics*, North-Holland/American Elsevier, New York, 1974.

[18] H. LONDON, *Thermodynamics of the thermomechanical effect in liquid helium* II, Proc. Roy. Soc. London Ser. A, 171 (1939), pp. 484–496.

[19] V. PESHKOV, *Determination of the velocity of propagation of the second sound in helium* II, J. Phys. (Soviet), 10 (1946), pp. 389–398.

[20] J. MAYNARD, *Determination of the thermodynamics of helium* II *from sound-velocity data*, Phys. Rev. B. (3), 14 (1976), pp. 3868–3891.

[21] J. HEISERMAN, J. P. HULIN, J. MAYNARD, AND I. RUDNICK, *Precision sound-velocity measurements in helium* II, Phys. Rev. B (3), 14 (1976), pp. 3862–3867.

[22] F. POBELL, W. SCHOEPE, AND W. VEITH, *The* λ*-point, entropy and density of rotating helium* II, Phys. Lett. A, 25 (1967), pp. 209–210.

[23] H. KOJIMA, W. VEITH, E. GUYON, AND I. RUDNICK, *Superfluid density in the presence of persistent current in superfluid* ^4He, J. Low Temp. Phys., 25 (1976), pp. 195–217.

[24] R. CAREY AND F. POBELL, *Persistent currents and critical velocities in* ^3He*-*^4He *130mixtures*, J. Low Temp. Phys., 24 (1976), pp. 449–470.

[25] H. W. LIEPMANN AND G. A. LAGUNA, *Nonlinear interactions in the fluid mechanics of helium II.* Ann. Rev. Fluid Mech., 16 (1984), p. 139.

[26] D. M. MOODY AND B. STURTEVANT, *Shock waves in superfluid helium*, Phys. Fluids, 27 (1984), pp. 1125–1137.

[27] J. R. TORCZYNSKI, D. GERTHSEN, AND T. ROESGEN, *Schlieren photography of second sound shock waves in superfluid helium*, Phys. Fluids, 27 (1984), pp. 2418–2423.

[28] D. V. OSBORNE, *The rotation of liquid helium* II, Proc. Phys. Soc. A, 63 (1950), pp. 909–912.

[29] H. E. HALL, *The angular acceleration of liquid helium* II, Philos. Trans. Roy. Soc. London Ser. A, 250 (1957), pp. 359–385.

[30] L. ONSAGER, *Discussione e osservazioni of* "*The two fluid model for helium* II," Nuovo Cimento Supplement 2, 6 (1949), p. 249.

[31] R. P. FEYNMAN, *Progress in Low Temperature Physics*, Vol. 1, C. J. Gorter, ed., North Holland, Amsterdam, 1955, p. 36.

[32] W. F. VINEN, *The detection of single quanta of circulation in liquid helium* II, Proc. Roy. Soc. London Ser. A, 260 (1961), pp. 218–236.

[33] P. W. KARN, D. R. STARKS, AND W. ZIMMERMANN, JR., *Observation of quantization of circulation in rotating superfluid* ^4He, Phys. Rev. B(3), 21 (1980), pp. 1797–1805.

[34] G. W. RAYFIELD AND F. REIF, *Evidence for the creation and motion of quantized vortex rings in superfluid helium*, Phys. Rev. Lett., 11 (1963), pp. 305–308.

[35] ———, *Quantized vortex rings in superfluid helium*, Phys. Rev. A(3), 136 (1964), pp. 1194–1208.

[36] H. LAMB, *Hydrodynamics*, Dover, New York, 1945, p. 241.

[37] E. J. YARMCHUK, M. J. GORDON, AND R. E. PACKARD, *Observation of stationary vortex arrays in rotating superfluid helium*, Phys. Rev. Lett., 43 (1979), pp. 214–217.

[38] E. J. YARMCHUK AND R. E. PACKARD, *Photographic studies of quantized vortex lines*, J. Low Temp. Phys., 46 (1982), pp. 479–515.

[39] L. J. CAMPBELL AND R. M. ZIFF, *Vortex patterns and energies in a rotating superfluid*, Phys. Rev. B(3), 20 (1979), pp. 1886–1902.

[40] C. F. BARENGHI, R. J. DONNELLY, AND W. F. VINEN, *Friction on quantized vortices in helium* II: *A review*, J. Low Temp. Phys., 52 (1983), pp. 189–247.

[41] R. J. DONNELLY AND C. E. SWANSON, *Quantum turbulence*, J. Fluid Mech., 173 (1986), pp. 387–429.

[42] C. J. GORTER AND J. H. MELLINK, *On the irreversible processes in liquid helium* II, Phys., 15 (1949), pp. 285–304.

[43] W. F. VINEN, *Mutual friction in a heat current in liquid helium* II:I. *Experiments on steady heat currents*, Proc. Roy. Soc. London Ser. A, 240 (1956), pp. 114–127.

[44] H. W. LIEPMANN AND G. A. LAGUNA, *Nonlinear interactions in the fluid mechanics of helium II*. Ann. Rev. Fluid Mech., 16 (1984), p. 139.

[45] H. E. HALL AND W. F. VINEN, *The rotation of liquid helium* II:I. *Experiments on the propagation of second sound in uniformly rotating helium* II, Proc. Roy. Soc. London Ser. A, 238 (1956), pp. 204–214.

[46] H. E. HALL AND W. F. VINEN, *The rotation of liquid helium* II:II. *The theory of mutual friction in uniformly rotating helium* II, Proc. Roy. Soc. London Ser. A, 238 (1956), pp. 215–234.

[47] W. F. VINEN, *Mutual friction in a heat current in liquid helium* II:II. *Experiments on transient effects*, Proc. Roy. Soc. London Ser. A, 240 (1957), pp. 128–143.

[48] ———, *Mutual friction in a heat current in liquid helium* II:III. *Theory of the mutual friction*, Proc. Roy. Soc. London Ser. A, 242 (1957), pp. 493–515.

[49] K. W. SCHWARZ, *Turbulence in superfluid helium*: *Steady homogeneous counterflow*, Phys. Rev. (3), 18 (1978), pp. 245–262.

[50] A. J. CHORIN AND J. MARSDEN, *A Mathematical Introduction to Fluid Mechanics*, Springer-Verlag, New York, 1979.

[51] R. J. ARMS AND F. R. HAMA, *Localized-induction concept on a curved vortex and motion of an elliptic vortex ring*, Phys. Fluids, 8 (1965), pp. 553–559.

[52] P. LUCAS, *Mutual friction measurements in uniformly rotating liquid helium*, J. Phys. C(3), 3 (1970), pp. 1180–1192.

[53] K. W. SCHWARZ, *Generation of superfluid turbulence deduced from simple dynamical rules*, Phys. Rev. Lett., 49 (1982), pp. 283–285.

[54] ———, *Critical velocity for a self-sustaining vortex tangle in superfluid helium*, Phys. Rev. Lett., 50 (1983), pp. 364–367.

[55] ———, *Three dimensional vortex dynamics in superfluid* ^4He: *Line-line and line-boundary interactions*, Phys. Rev. B(3), 31 (1985), pp. 5782–5804.

[56] ———, *Three dimensional vortex dynamics in superfluid* ^4He: *Homogeneous superfluid turbulence*, Phys. Rev. B(3), 38 (1988), pp. 2398–2417.

[57] T. F. BUTTKE, *Comment on* "*Generation of superfluid turbulence deduced from simple dynamic rules*," Phys. Rev. Lett., 59 (1987), pp. 2117.

[58] A. J. CHORIN, *Estimates of intermittency, spectra and blow-up in developed turbulence*, Comm. Pure Appl. Math., 34 (1981), pp. 853–866.

[59] ———, *The evolution of a turbulent vortex*, Comm. Math. Phys., 83 (1982), pp. 517–535.

[60] ———, *Numerical estimates of Hausdorff dimension*, J. Comput. Phys., 46 (1982), pp. 390–396.

[61] ———, *Turbulence and vortex stretching on a lattice*, Comm. Pure Appl. Math., 39 (1986), pp. s47–s65.

[62] E. SIGGIA, *Collapse and amplification of a vortex filament*, Phys. Fluids, 28 (1985), pp. 794–805.

[63] E. SIGGIA AND A. PUMIR, *Incipient singularities in the Navier-Stokes equations*, Phys. Rev. Lett., 55 (1985), pp. 1749–1752.

[64] T. F. BUTTKE, *A numerical study of superfluid turbulence in the self-induction approximation*, J. Comput. Phys., 76 (1988), p. 301.

[65] ———, *A numerical study of superfluid turbulence in self-induction approximation*, Ph.D. thesis, LBL report LBL-22086, University of California, Berkeley, CA, 1986 (unpublished).

8

VORTICES, TURBULENCE, AND STATISTICAL MECHANICS

Alexandre Joel Chorin

Introduction

Turbulence is the apparently disorderly motion of a fluid and is described by the Navier–Stokes equations. Since the solution of the Navier–Stokes equations in a subsonic regime can be approximated by the motion of a collection of vortex blobs [1], [8], [9], [11], [13], [23], [24], [36], turbulence can be thought of as a many-vortex phenomenon, similar in kind but not in detail to the many-fermion or many-boson systems whose study is a mighty branch of physics. The constraints imposed by the conservation properties of the Navier–Stokes equations evoke similarities to another well-known statistical mechanical system with constraints—a collection of polymers in a solution.

In this chapter we shall present the main properties of vortex systems, explain the similarities to other physical systems, and present an analysis of the small scales of turbulence that explains certain experimental facts and leads to a useful modeling technique for practical use.

1. The Random Vortex Equations

For the sake of completeness, we begin by writing down the equation of motion for a vortex system. We assume for simplicity that the fluid is incompressible. The Navier–Stokes equations take the form

(1.1a) $$\partial_t \xi + (\mathbf{u} \cdot \nabla)\xi - (\xi \cdot \nabla)\mathbf{u} = R^{-1}\Delta\xi,$$

(1.1b) $$\operatorname{div} \mathbf{u} = 0,$$

(1.1c) $$\xi = \operatorname{curl} \mathbf{u},$$

where $\mathbf{u}$ is the velocity, ξ is the vorticity, t is the time, ∇ is the differentiation vector,

$\Delta = \nabla \cdot \nabla$, and R is the Reynolds number. R will be large or infinite. Equations (1.1b), (1.1c) can be inverted and yield

(1.2) $$\mathbf{u} = K * \xi,$$

where $*$ denotes a convolution and K is the matrix

$$K = -\frac{1}{4\pi r^3}\begin{pmatrix} 0 & -x_3 & x_2 \\ x_3 & 0 & -x_1 \\ -x_2 & x_1 & 0 \end{pmatrix},$$

where $\mathbf{x} = (x_1, x_2, x_3)$ is the position vector and $r = |\mathbf{x}|$. Equation (1.2) is known as the Biot–Savart law. For detailed derivations, see [1], [11], [12], [44], and [55]. If $R^{-1} = 0$ equations (1.1a)–(1.1c) are known as Euler's equations.

Consider unbounded flow (the only case we shall consider in the present chapter). Suppose the initial vorticity $\xi(\mathbf{x}, t = 0)$ is broken down into a sum of vector functions of small support (= "vortex blobs"),

(1.3) $$\xi = \sum_{i=1}^{N} \xi_i.$$

Let σ be a length characteristic of the support of ξ_i. σ is the "cutoff length." Choose further a decomposition (1.3) for which $\xi_i = \bar{\xi}_i \phi(\mathbf{x} - \mathbf{x}_i)$, $\mathbf{x}_i = (x_{1i}, x_{2i}, x_{3i})$, where $\phi(\mathbf{x})$ is a function of small support, $\int \phi\, d\mathbf{x} = 1$, and $\bar{\xi}_i$ is a constant. Equations (1.1a)–(1.1c) with $R^{-1} = 0$ can then be approximated by the following set of ordinary differential equations:

(1.4a) $$d\mathbf{x}_i = (K * \xi)_{\mathbf{x} = \mathbf{x}_i} dt,$$

(1.4b) $$d\bar{\xi}_i = ((\xi \cdot \nabla)\mathbf{u})_{\mathbf{x} = \mathbf{x}_i} dt,$$

where one allows the "blobs" the move. The right-hand sides of (1.14a), (1.14b) are known functions of the locations $\mathbf{x}_i$ and "strengths" $\bar{\xi}_i$. The accuracy of the approximation depends on the choice of the decomposition (1.3) and thus on the form of ϕ. The convolution in (1.4a) can be rewritten as [38]

$$(K * \xi)_{\mathbf{x} = \mathbf{x}_i} = \sum K_\sigma(\mathbf{x} - \mathbf{x}_i)\xi_i,$$

where K_σ is a smoothed kernel, $K_\sigma = K * \phi$. The obvious constraint div ξ = div curl $\mathbf{u}$ = 0 can be imposed on (1.4b) explicitly and leads to representations by vortex tubes, or it can be left to take care of itself as a consequence of convergence; in this latter case, one can treat the blobs as independent vortex "arrows" or "vortons." For details and convergence theorems, see [1], [8], [9], [11], [15], [24], [35], and [42].

Suppose now that $R^{-1} \neq 0$. Equations (1.1a)–(1.1c) then look like Fokker–Planck equations [5] for the set of stochastic ordinary differential equations

(1.5a) $$d\mathbf{x}_i = \sum K_\sigma(\mathbf{x} - \mathbf{x}_i)\bar{\xi}_i dt + (2/R)^{1/2} d\mathbf{w}(t),$$

(1.5b) $$d\bar{\xi}_i = ((\xi \cdot \nabla)\mathbf{u})_{x = x_i} dt,$$

where $\mathbf{w}(t)$ is Brownian motion in three-dimensional space. Equations (1.5a), (1.5b) can be solved and sample the solution of equations (1.1a)–(1.1c). (For further details, see [11], [30], [35], [46], [48], [55], [57], [59], [66].)

In the special case of two-dimensional flow $\bar{\xi}_i$ = constant, $\xi_i = (0, 0, \xi_i)$, and the system (1.5a), (1.5b) reduces to

$$d\mathbf{x}_i = \sum_{j \neq i} K_\sigma(\mathbf{x} - \mathbf{x}_i)\bar{\xi}_i + (2/R)^{1/2} d\mathbf{w}(t),$$

where $\mathbf{w}(t)$ is two-dimensional Brownian motion, $K_\sigma = K * \phi$, ϕ has small support, $\mathbf{x}_i = (x_{1i}, x_{2i})$, and

$$K = \frac{1}{2\pi}\begin{pmatrix} -\partial_y \\ \partial_x \end{pmatrix} \log r_{ij}, \qquad r_{ij} = |\mathbf{x}_i - \mathbf{x}_j|.$$

For more details see, e.g., [11], [23], [24], [28], [35], and [37].

2. Fractional Step Formulation

A simplified but useful way of looking at the relation between equations (1.4a), (1.4b) and (1.5a), (1.5b) involves the notion of fractional steps. Consider an operator equation of the form

(2.1) $$u_t = (A+B)u, \qquad u(0) = u_0,$$

where A, B are operators (for example, differential operators), t is the time, and u_0 is the set of initial data sufficient to produce a unique solution $u(t)$. Let S be the solution operator for (2.1); $u(t) = Su_0$, $S = S(t)$.

Consider the "component" equations

$$v_t = Av, \qquad v(0) \text{ given},$$

$$w_t = Bw, \qquad w(0) \text{ given},$$

and let S_A, S_B be their solution operators, $v(t) = S_A v(0)$, $w(t) = S_B w(0)$. Under appropriate conditions (see e.g., [26]), one has

(2.2) $$S(t) = \lim_{n \to \infty} \left(S_A\left(\frac{t}{n}\right) S_B\left(\frac{t}{n}\right)\right)^n,$$

i.e., the solution of (2.1) can be obtained by solving the "component" equations for short times, in succession and repeatedly. Formula (2.2) is known as a product formula and its numerical implementations are usually referred to as fractional step methods. It is eminently plausible that one can solve equations (1.1a)–(1.1c) by a fractional step method in which one alternately solves the Euler equations,

$$\frac{\partial}{\partial t}\xi + \partial_j(u_j \xi) = 0, \qquad \mathbf{u} = K * \xi$$

and then the diffusion equation

$$\frac{\partial}{\partial t}\xi = \frac{1}{R}\Delta\xi;$$

we are writing the formulas in two-dimensional case for the sake of simplicity.

Euler's equations can be solved by a vortex method with no viscosity, and the diffusion equation can be solved by exploiting its relation to Brownian motion. On the other hand, the stochastic differential equations (1.5a), (1.5b) can be solved by a fractional step method in which one alternately solves the ordinary differential equations

$$\frac{d\mathbf{x}_i}{dt} = \mathbf{u}_i$$

and the simple stochastic differential equations

$$d\mathbf{x} = (2/R)^{1/2} d\mathbf{w},$$

whose solutions are

$$x(t+\Delta t) = x(t) + R^{-1/2}\mathbf{W},$$

where $\mathbf{W}$ is a two component Gaussian random variable whose components have mean zero and variance $\Delta t/2$; remember that $\mathbf{w}$ is in a process with independent Gaussian increments. It is now obvious that the two fractional step methods, the one for the partial differential equations and the one for the stochastic differential equation, approximate each other, at least formally. If a boundary is present, a vorticity creation operator must be introduced into the product formula, making the analysis much more difficult.

The construction we have just described leads to a practical numerical method that has many advantages in the analysis of flow at large R. For a general discussion of fractional step methods, see [26]; for the analysis of the Navier–Stokes case, see [48].

The set of stochastic differential equation (1.4a), (1.4b) resembles equations that appear in other branches of physics; the resemblance leads to analogies that are likely to be fruitful in the future. Here is one: The transition from inviscid to viscous flow resembles the transition from classical to quantum mechanics. One standard path to the construction of a quantum version of a classical problem proceeds as follows. Construct thc action S,

$$S = \int_{t_1}^{t_2} L\,dt,$$

where L is the Lagrangian, $L = T - V$, T is the kinetic energy, V is the potential energy, and t is time. S is a functional of the path between the time t_1 and t_2 that is stationary at the classical path. The condition that S be stationary determines the unique classical path. In quantum theory, one assigns to each conceivable path a probability proportional to $e^{(i/\hbar)S}$, where $\hbar$ is Planck's constant [54]. All paths contribute to the transmission of quantum amplitudes proportionally to their weight. The form of T creates weights with factors of the form $e^{-ix^2/t}$, i.e., "Gaussian" with an imaginary time. The effect is to smear the classical path; if $\hbar$ is small, the classical path contributes most. Clearly, there is a vague analogy between this smearing and the smearing produced when Brownian motion is added to the vortex trajectories. The analogy becomes more precise when one considers quantum statistical mechanics and replaces imaginary time by temperature; the smearing becomes Gaussian in the quantum case also, and the relation between viscosity and temperature discussed below adds further plausibility to this analogy.

Furthermore, equations (1.5a), (1.5b) have the general form of the Langevin equation that describes the fluctuations of a collection of heavy particles in contact with a system in thermal equilibrium, with temperature proportional to viscosity. The proportionality of the diffusion constant to the temperature is one of the "Einstein relations" in irreversible thermodynamics [56].

3. Vortex Tubes Stretch

Consider a vortex tube, i.e., an object of finite cross section A with boundary ∂A nowhere tangent to ξ, and sides everywhere tangent to ξ. The circulation and divergence theorem ensure that for inviscid flow the quantity

$$\Gamma = \int_A \xi \cdot d\Sigma, \qquad d\Sigma \text{ normal to } A,$$

is a constant in space and time. It is an experimental and numerical fact that vortex tubes stretch wildly in a turbulent flow. There is no good theoretical explanation of this fact, which we shall accept without much further discussion. As vortex tubes stretch they create ever smaller scales of motion with some energy in them, and this cascade is one of the salient characteristics of turbulence. In fact, this cascade is the main obstacle to the efficient numerical computation of turbulent flows, and one can consider that the goal of turbulence theory is to provide an explanation of the mechanics of the small scales in turbulence and provide an effective way to parameterize them.

An interesting argument that shows that some vortex stretching must occur can be found in [27] and [53]. As we shall see the argument is inadequate for explaining the data but it does provide an interesting connection between the increase in vortex length and an increase in entropy.

The argument shows that the distance between any two points $\mathbf{x}_1$, $\mathbf{x}_2$, $|\mathbf{x}_1 - \mathbf{x}_2|$ small, will increase in an appropriate average sense in a random incompressible flow that is statistically invariant under both translation and rotation (= "homogeneous isotropic turbulence"). Random flow will not be defined more precisely here, and we wish to appeal to one's intuitive sense that the notions of random flow and a random flow map make sense.

Consider a fluid occupying the whole three-dimensional space and at time $t = 0$ pick a point $\mathbf{x}^0$ in this field. Let ϕ_t be the flow map that we shall view as being random. For a definition of the flow map, see, e.g., [25]. Thus, as usual, $\mathbf{x}^t = \phi_t(\mathbf{x}^0)$ is the (random) location of a fluid particle that was initially at $\mathbf{x}^0$. Consider at $t = 0$ another point $\mathbf{x}^0 + \delta\mathbf{x}^0$ a small distance away from $\mathbf{x}^0$; at time t the particle at $\mathbf{x}^0 + \delta\mathbf{x}^0$ will have moved to $\mathbf{x}^t + \delta\mathbf{x}^t$ where

$$\delta\mathbf{x}^t = \phi_t(\mathbf{x}^0 + \delta\mathbf{x}^0) - \phi_t(\mathbf{x}^0) \sim A\delta\mathbf{x}^0$$

to first order in $\delta\mathbf{x}^0$; A is a random linear transformation. The length $|\delta\mathbf{x}^t|$ of $\delta\mathbf{x}^t$ satisfies

$$0 \leqq |\delta\mathbf{x}^t|^2 = (\delta\mathbf{x}^t, \delta\mathbf{x}^t) \sim (A\delta\mathbf{x}^0, A\delta\mathbf{x}^0) = (\delta\mathbf{x}^0, A^T A\delta\mathbf{x}^0),$$

where A^T is the transpose of A. Consider the operator $W = A^T A$. W is positive definite and symmetric, and thus has positive real eigenvalues λ_1, λ_2, λ_3 associated with three mutually orthogonal unit eigenvectors $\mathbf{e}_1$, $\mathbf{e}_2$, $\mathbf{e}_3$. Since the fluid is incompressible and A is the Jacobian of ϕ_t, $\det A = 1$, and thus

$$\det W = \det A \cdot \det A^T = 1 = \lambda_1\lambda_2\lambda_3.$$

Write $\delta\mathbf{x}^0 = \sum_{i=1}^3 \delta x_i^0 \mathbf{e}_i$; then

$$|\delta\mathbf{x}^t|^2 = (\delta\mathbf{x}^0, W\delta\mathbf{x}^0) = \sum \lambda_i(\delta x_i^0)^2.$$

Assume that the "turbulence" is isotropic, i.e., that ϕ_t and thus A have no preferred direction on the average. Consider a family of vectors $\delta\mathbf{x}^0$ of length δx^0 and directions uniformly distributed over the unit sphere. Denote by $\langle\ \rangle$ the average on the unit sphere at $t = 0$. Clearly, $\langle(\delta x_1^0)^2\rangle = \langle(\delta x_2^0)^2\rangle = \langle(\delta x_3^0)^2\rangle = \frac{1}{3}(\delta x^0)^2$. Thus, $\langle|\delta \mathbf{x}^t|^2\rangle = \frac{1}{3}\sum_1^3 \lambda_i(\delta x^0)^2$.

Now average over all flow maps ϕ_t:

$$\overline{\langle(\delta x^t)^2\rangle} = \left(\frac{1}{3}\overline{\sum \lambda_i}\right)(\delta x^0)^2.$$

However, since $\lambda_i > 0$, $i = 1, 2, 3$,

$$\frac{1}{3}\sum \lambda_i \geqq \lambda_1\lambda_2\lambda_3 = 1,$$

with equality holding only if all the λ_i are equal. Thus

$$\frac{1}{3}\overline{\sum \lambda_i} \geqq 1,$$

with equality holding only if $\lambda_1 = \lambda_2 = \lambda_3$ with probability one—the case of trivial turbulence. Thus, in nontrivial flow

$$\overline{\langle|\delta x^t|^2\rangle} > |\delta x^0|^2.$$

Remarks. (i) This argument gives no estimate of the rate of increase of $\overline{\langle|\delta x^t|^2\rangle}$, which may well tend to zero.

(ii) The argument does not really show that vortex lines stretch, since if lengths increase only on the average, those special vectors $\delta\mathbf{x}^0$ that are aligned with ξ may be exceptional.

(iii) The argument does not depend on the direction of time; one still gets the stretching if ϕ_t is replaced by ϕ_{-t}. The argument really shows that when information is lost as we go from known data to the outcome of a random flow, lines lengthen. Line length appears as a sort of entropy.

(iv) Most importantly, the argument does not exhibit the extremely fast stretching that is observed in solutions of Euler's equations, of which indeed no use is made. A better analytical estimate of the rate of vortex stretching remains to be discovered.

Suppose that we have a collection of vortex tubes, as in the approximation (1.5a), (1.5b). An important fact is that the rate of stretching depends on the initial smoothness and on the radii of the tubes. The fatter the tubes the more slowly they stretch. It is as if the turbulent "matter" has two possible states: "laminar," i.e., regular and well behaved, and "turbulent." Since even the "laminar" vortices stretch somewhat, at sufficiently high R the "laminar" state will eventually fall into the "turbulent" state. The difference between the two states resides in the type of care required to approximate them (see below). The transition between them is a sort of phase transition, and is very poorly understood. For some partial results on model problems, see, e.g., [2] and [29].

Note that a vortex element in equations (1.5a), (1.56) is a numerical object; a real physical vortex is represented by a collection of such objects. To the extent that we shall be making statistical assertions about the objects in (1.5b), we shall be identifying physical vortices with collective modes in a vortex "gas."

4. Vortex Tubes Absorb Energy and Fold

Consider again an inviscid flow ($R^{-1} = 0$), in an unbounded domain. The energy $T = \frac{1}{2}\int |\mathbf{u}|^2\,dx$ can be rewritten as

$$T=\int d\mathbf{x}\int d\mathbf{x}'\frac{\xi(\mathbf{x})\cdot\xi(\mathbf{x}')}{|\mathbf{x}-\mathbf{x}'|}. \tag{4.1}$$

(For a derivation, see, e.g., [43].) Formula (4.1) will be crucial in the sequel. The enstrophy Z is defined by

$$Z=\int|\xi|^2\,dx. \tag{4.2}$$

We now derive some scaling properties of the integrals T and Z, following [18], [20]. Consider a set of vertical cylinders $C(r, l)$, of height l and cross section $A(r)$, where r is a generalized radius, i.e., a length characteristic of A. Denote by A the area of the cross section as well as the cross section itself. Consider a vector $\xi=(0, 0, \xi_0)$ with support in $C(r, l)$, constant in $C(r, l)$ for fixed r and l. ξ cannot quite be identified with a vorticity since $C(r, l)$ has finite length and thus div $\xi\neq 0$ unless ξ is properly continued beyond $C(r, l)$. We wish to find the variation of the integrals

$$T(r,l)=\int_{C(r,l)}d\mathbf{x}\int_{C(r,l)}d\mathbf{x}'\frac{\xi(\mathbf{x})\cdot\xi(\mathbf{x}')}{|\mathbf{x}-\mathbf{x}'|},$$

$$Z(r,l)=\int_{C(r,l)}|\xi|^2\,dx$$

as functions of r and l for fixed "circulation" $\Gamma=A(r)\xi_0$. Assume that if r changes, then $A(r)$ changes in a self-similar fashion. Let $\alpha>0$ be a real parameter. Define the scaling functions S_1, S_2 by

$$T(r,\alpha l)=S_1(l/r_1,\alpha)T(r,l),$$

$$T(\alpha r,l)=S_2(l/r,\alpha)T(r,l).$$

Clearly, under the assumption above, S_1, S_2 are functions only of l/r and α and not of r, l, α separately. A volume preserving stretching of $C(r, l)$ transforms $C(r, l)$ into $C(r/\sqrt{\alpha}, \alpha l)$ and thus multiplies $T(r, l)$ by $S_1(l/(\sqrt{\alpha}r), \alpha l)\cdot S_2(l/r, 1/\sqrt{\alpha})$. Remember that Γ is constant, and thus $\xi_0=\xi_0(\alpha)$ with $\xi_0(\alpha)A(\alpha r)=\xi_0(1)A(r)$. An easy calculation shows that

$$T(\alpha r,\alpha l)=\alpha T(r,l).$$

On the other hand,

$$\begin{aligned}T(\alpha r,\alpha l)&=S_1(l/(\alpha r),\alpha)T(\alpha r,l)\\&=S_2(l/(\alpha r),\alpha)S_2(l/r,\alpha)T(r,l),\end{aligned}$$

and thus

$$\alpha=S_1(l/(\alpha r),\alpha)S_2(l/r,\alpha). \tag{4.3}$$

Suppose $l/r\ll 1$ (a "fat" or "pancake" vortex). Doubling l means adding another pancake on top of the pancake one already has; this is roughly equivalent to doubling ξ_0 at every point of the original pancake. Since clearly T is a quadratic function of ξ_0, one has, for a "fat" vortex,

$$S_1(l/r,\alpha)\sim\alpha^2,\qquad l/r\ll 1,$$

and by (4.3),

$$S_2(l/r,\alpha)\sim\alpha^{-1},\qquad l/r\ll 1.$$

Thus, if one doubles the length of a "fat" vortex, one quadruples its energy, while if one halves its radius (without preserving volume), one doubles the energy. At the other extreme, if $l/r \gg 1$, doubling the length l does little more than place side by side two equal nearly disjoint velocity fields, and should do little more than double the energy, i.e., we expect

$$S_1(l/r,\alpha) \sim \alpha|\log\alpha|,$$

$$S_2(l/r,\alpha) \sim |\log\alpha|^{-1}, \qquad l/r \gg 1.$$

We shall use these formulas below. In any case, stretching and compressing a vortex in a way that conserves its volume increases the energy associated with it;

$$S_1(l/(\alpha r),\alpha l)\cdot S_2(l/r,1/\sqrt{\alpha}) \sim \alpha^{5/2}$$

for $l/r \ll 1$, $\sim\alpha|\log\alpha|$ for $l/r \gg 1$. (The regime in between these extreme limits has been examined numerically in [20], where the transition is seen to be monotonic.) One consequence is that in order to stretch a vortex, work must be expended, or, in other words, stretching vortices remove energy from the flow field that strains them. A vortex tube is a little like a rubber band. Furthermore, note that when $r = 0$, $T(0, l) = \infty$, thus stretching, if unchecked, creates a singularity.

It is easy to see that

$$Z(\alpha r, l) = \frac{1}{\alpha^2} Z(r, l), \qquad Z(r, \alpha l) = \alpha Z(r, l).$$

Thus a volume preserving stretching of a vortex tube increases the associated enstrophy by α^2 if the "exchange term," i.e., the contribution of the integral

$$\int_{\text{tube}} d\mathbf{x}' \int_{\text{rest of field}} d\mathbf{x}\, \frac{\xi(\mathbf{x})\cdot\xi(\mathbf{x}')}{|\mathbf{x}-\mathbf{x}'|}$$

does not change substantially. Suppose that a long vortex tube is stretched nonuniformly. The greatest contribution to the enstrophy will be made by those portions that are stretched most.

We shall now present a heuristic model that will illustrate the proposition that if a vortex tube is stretched while the energy associated with it remains bounded, then the tube is likely to fold in a zigzag manner.

Consider a long vortex tube V of unit circulation and a small, nonzero, approximately circular cross section. We have seen that the kinetic energy T associated with the tube can be written as

$$T = \frac{1}{2}\int |\mathbf{u}|^2\, d\mathbf{x} = \frac{1}{8\pi}\int_V d\mathbf{x} \int_V d\mathbf{x}'\, \frac{\xi(\mathbf{x})\cdot\xi(\mathbf{x}')}{|\mathbf{x}-\mathbf{x}'|},$$

where $\xi = \text{curl}\, \mathbf{u}$ as usual. Suppose the support of V can be approximately covered by N circular cylinders I_i, $i = 1, \cdots, N$, of equal lengths l and some radii r_i, $i = 1, \cdots, N$. T can then be approximated by $\tilde{T}$,

$$8\pi\tilde{T} = \sum_{i=1}^{N}\sum_{\substack{j=1\\ j\neq i}}^{N} \tilde{T}_{ij} + \sum_{1}^{N} \tilde{T}_{ij}, \tag{4.4}$$

where

$$\tilde{T}_{ij} = \int_{I_i} d\mathbf{x} \int_{I_j} d\mathbf{x}' \frac{\boldsymbol{\xi}(\mathbf{x}) \cdot \boldsymbol{\xi}(\mathbf{x}')}{|\mathbf{x} - \mathbf{x}'|}.$$

Let $\mathbf{t}_i$ be a vector lying along the axis of the cylinder I_i, originating at the center of I_i, of length $|t_i| = l$, and pointing in the direction of $\boldsymbol{\xi}$ in I_i. If I_i and I_j are far from each other,

$$\tilde{T}_{ij} \sim \frac{\mathbf{t}_i \cdot \mathbf{t}_j}{|i - j|},$$

where $|i - j|$ is the distance between I_i and I_j. Pretend (4.3) holds approximately whenever $i \neq j$—clearly a gross simplification.

$\tilde{T}_{ii}$ (the case $i = j$) is a function of the radius r_i of I_i, as we have seen above, changing with r_i as r_i^{-1} when $l/r_i \ll 1$, as $|\log r_i|^{-1}$ when $l/r_i \gg 1$. Thus,

$$8\pi\tilde{T} \sim \sum_i \sum_{j \neq i} \frac{t_i \cdot t_j}{|i - j|} + \sum_i \tilde{T}_{ii}(r_i),$$

where $d\tilde{T}_{ii}/dr_i < 0$.

Suppose now that the tube V is stretched by the velocity field that includes the velocity field that it itself induces, in such a way that its volume is preserved. Suppose further that a circular cylinder I_i remains approximately a circular cylinder, i.e., the cylinders do not flatten out into sheets. This is a major assumption. One could try to reduce its significance by an argument of the following form. If the cylinders do flatten out into sheetlike objects, these objects will be unstable and will roll up into cylinderlike objects. Nevertheless, we have here a major poorly controlled assumption (for some justification, see [10]). Suppose further that the support of the new stretched tube V can still be approximated by a collection of cylinders of lengths l; their number N' will be larger than N, and most of their cross sections will be smaller than before. Thus the sum $\sum \tilde{T}_{ii}$ will increase because it will have more and larger entries. As the radii tend to zero, this sum will diverge. If $\tilde{T}$, the total energy, remains bounded, then many of the terms in the double sum over i, j must decrease, i.e., the tube must fold. Thus the vortex segments I_i arrange themselves in such a way that they shield the incipient singularity due to the singular Biot–Savart kernel. Note that this phenomenon has no analogue in two space dimensions, where there is no stretching and where "self-energies" can be safely subtracted from the total energy in defining a Hamiltonian. This shielding effect has some vague similarity to the shielding of a "bare" electron by a cloud of photons in the renormalization of field theory (see, e.g., [54]). Nearby vortices have a reduced interaction because of the folding.

A simple one-dimensional cartoon of (4.4) is

$$\tilde{T} = \sum_{i=1}^{N} \sum_{\substack{j=1 \\ j \neq i}}^{N} t_i t_j \frac{1}{|i - j|} + \sum \tilde{T}_{ii}$$

where N is finite and fixed, the t_i are Ising-like spins, i.e., vectors that can point either up ($t_i = 1$) or down ($t_i = -1$), $|i - j|$ is the distance between the position of t_i and t_j, and $\tilde{T}_{ii}$ is a function of a parameter r^{-1} that increases monotonically. The "spins" are located at the nodes of a regular lattice, most of whose nodes are empty. The "spins" can move to empty locations or flip (i.e., change signs); r^{-1} increasing is interpreted as a stretching of the "spins." In [19] a sequence of spin configurations is constructed, such that the "energy" $\tilde{T}$ remains fixed. The $\tilde{T}_{ii}$ increase and thus the double sum must decrease; this requires the "spins" to bunch up as closely as is possible on the lattice, with neighboring "spins" having opposite signs.

A more abstract form of the statement that vortex stretching requires vortex folding can be given with the help of generalized potential theory [33] (see, e.g., [15] and [18]). The conclusion of that analysis is that vorticity must arrange itself on a fractal set (a set of nonzero Hausdorff dimension) if energy is to be preserved as $t \to \infty$. The folding process is the process by which this set is generated.

The analysis of this section has several important consequences. Vortex stretching increases the energy associated with a vortex and thus requires work. Vortex stretching thus acts as an energy sink for the mean flow in a turbulent medium. (A heuristic form of this statement can be found in [63].)

Furthermore, this analysis suggests an analogy between vortex motion in a fluid and the motion of a polymer in a solution [21]. A polymer is a long molecule made up of "monomers" strung together end to end. When a polymer is immersed in a solution, its behavior is the result of entrainment by the ambient flow, of hydrodynamic interaction between its component monomers, of forces between connected neighboring monomers, and of Brownian bombardment of the monomers by the molecules of the solution [65]. All these effects have vortex counterparts. The hydrodynamical interaction is given in the vortex case by the Biot–Savart law, and the origin of the force between nearby portions of a vortex has just been demonstrated. The important point is that in the vortex case there is Brownian motion with coefficient $\sim R^{-1/2}$, while in the polymer case there is real Brownian motion whose amplitude is proportional to $\sqrt{T}$, where T is the temperature. A vortex system can thus be viewed as being in thermal equilibrium with a potential background, with temperature proportional to viscosity, as indeed is reasonable in view of the "Einstein relations" of irreversible thermodynamics (see, e.g., [56]). Turbulence is thus, like superfluidity and superconductivity, a low temperature phenomenon. Its coherent structures are collective modes or "quasi particles" in a statistical description.

5. Intermittency

Vortex lines in a turbulent fluid are not spread uniformly in the fluid and tend to congregate. This phenomenon is known as "intermittency." In two space dimensions intermittency can be displayed numerically. Its occurrence can be made plausible by noting again the qualitative analogy between a plane vortex system and an Ising model in the plane. An Ising spin system at low temperature exhibits intermittency (i.e., a clustering of spins of like sign); see, e.g., [39].

The most interesting manifestations of intermittency are, however, three-dimensional. In three space dimensions intermittency is a strong, dramatic effect. Its main source is the differential stretching of vortex lines. As we have seen, a portion of a vortex tube that is stretched a lot makes a larger contribution to the enstrophy Z than a portion that is stretched less. In addition, a shielding effect brings stretched portions close to each other. The result is a very uneven distribution of vorticity. As a consequence, Z is dominated by the vorticity in a volume much smaller than the total volume available to the fluid. (For numerical examples, see, e.g., [3], [6], [18], [19], [20].)

One would expect each component of the velocity field $\mathbf{u}(\mathbf{x})$, for fixed $\mathbf{x}$, to have a probability distribution that is close to Gaussian, simply because $\mathbf{u}$ is due to contributions from a number of separate, presumably nearly independent vortex pieces, and the sum of independent or nearly independent random variables with finite variance converges to a Gaussian variable (this is the "central limit theorem"). One would also expect $|\mathbf{u}(\mathbf{x}_1) - \mathbf{u}(\mathbf{x}_2)|$, $|\mathbf{x}_1 - \mathbf{x}_2|$ not too small, to have a distribution that is nearly Gaussian, because this difference is also generated by a number of distinct vortical contributions. These expectations are well borne out by experiment, both physical and numerical.

However, one expects the distribution of $\xi(\mathbf{x})$, for fixed $\mathbf{x}$, to be at least somewhat non-Gaussian, since ξ has a substantial probability of being small and a substantial probability of being very large. Similar statements should be true for higher derivatives of $\mathbf{u}$. The departure of a random variable η from Gaussianity can be measured by its "flatness" f,

$$f = f(\eta) = \frac{\overline{(\eta - \bar{\eta})^2}}{(\overline{\eta^2})^2}$$

and "skewness" s,

$$s = s(\eta) = \frac{\overline{(\eta - \bar{\eta})^3}}{(\bar{\eta}^2)^{3/2}}.$$

If η is Gaussian, $f = 3$ and $s = 0$. The experimental values of the flatness of $\partial^n u/\partial x^n$ increase rapidly with n [7], as one would expect from the preceding discussion.

We now present a qualitative discussion of the unevenness of the stretching of vortex tubes. The main part of the stretching is due to interactions between the vorticity in several parts of the tubes. The more vorticity there is in a neighborhood of a portion of a tube the more stretching there should be; if we label a point on the axis of a tube by a lagrangian label q, the rate of stretching at q should be proportional to the amount of stretching that had already occurred near q, and thus, by conservation of circulation, to the vorticity at q,

$$\frac{d\xi}{dt}(q) \sim b(t)\xi, \qquad \xi = |\xi(q,t)|. \tag{5.1}$$

The proportionality constant b depends on the configuration near q, which is unknown and can be viewed as random. Suppose $|\xi(q, t = 0)|$ is a constant. An integration of (5.1) yields

$$\log \xi(q,t) - \log \xi(q,0) = \int_0^t b(t)\, dt.$$

The b's at different times t can be viewed as nearly independent, and by the central limit theorem the right-hand side is approximately Gaussian; thus $\log \xi \equiv \log |\xi|$ has an approximately Gaussian (= "normal") distribution. ξ is approximately "log-normal." This conclusion is reasonably well verified by numerical experiment [15]. A log-normal variable can take on some very large as well as very small values.

Let ε be a small positive number; we define the ε-support of the vorticity to be the smallest set Λ_ε such that

$$\int_{\Lambda_\varepsilon} \xi^2\, dx = (1-\varepsilon) \int_{\text{space}} \xi^2\, dx;$$

i.e., Λ_ε is the set that carries most of the vorticity. Since ξ is stretching nonuniformly, and the vorticity that has been stretched most contributes most to $\int \xi^2\,dx$, Λ_ε shrinks in size. If ξ is smooth initially, then initially the Hausdorff (= "fractal") dimension of Λ_ε, dim Λ_ε, equals 3 for all $0 \leqq \varepsilon < 1$. As stretching occurs, and if there is no viscosity, it is conceivable that the dimension D of the limit

$$\lim_{\varepsilon \to 0} \lim_{t \to \infty} \Lambda_\varepsilon$$

is well defined, with D less than three (for definition of Hausdorff dimension; see, e.g., [49], [62]).

If this is indeed true, D is a measure of intermittency. The limiting set, if it exists, is the "essential support" of the vorticity. In the case of the model equation $u_t + (u^2)_x = 0$, $\xi = u_x$, one can readily see that lim Λ_ε exists, is the set of points where the shocks are located, and its dimension is zero.

This discussion raises the question of whether the solutions of Euler's equations are smooth. The numerical evidence is that they are not, as is the experimental evidence. A numerical scaling analysis in [14] proceeds as follows. Run an initial value problem for ξ, allowing $Z = \int |\xi|^2\,d\mathbf{x}$ to grow until it becomes larger than some threshold Z_c at a time t_1. Then take out a portion V_1 (say an eighth) of the domain available to the fluid, making sure that $\xi \not\equiv 0$ in V_1, rescale the length scale and timescale so that the energy is consistent with the velocity field in V_1. Run the smaller problem for a time t_2 until $Z > Z_c$ and repeat the process. Let $\tilde{t}_1$ be the real time interval that corresponds to t_i. If $\sum_1^\infty \tilde{t}_i < +\infty$, then in the finite time $\sum_1^\infty \tilde{t}_i$ the original volume has an infinite value of Z. This calculation was done in [14], and on a simplified lattice model, in [19], with the conclusion that Z does indeed blow up in finite time. The conclusion is of course nonrigorous, mostly because the boundary conditions at the edges of the sequence of shrinking volumes are artificial. Similar scaling analysis of spin systems and random walks can be found in [16] and [17]. Other calculations whose purpose is to analyze the blowup time for Euler's equation can be found in [52] and [61].

One can define for a homogeneous random flow an energy spectrum $E(k)$; $E(k)$ is roughly the contribution to the mean energy at a point of those Fourier components of $\mathbf{u}$ whose wavenumber $\mathbf{k}$ satisfies $k < |\mathbf{k}| < k + dk$. Thus $\langle \mathbf{u}^2 \rangle = \int_0^\infty E(k)\,dk$. $E(k)$ can be obtained from the Fourier transform of the two-point correlation function $\langle \mathbf{u}(\mathbf{x}) \cdot \mathbf{u}(\mathbf{x} + \mathbf{r}) \rangle$ by an averaging operation (see [7], [64]). Similarly, one can define a vorticity spectrum $Z(k)$, with $\langle \xi^2 \rangle = \int_0^\infty Z(k)\,dk$. An elementary calculation yields $Z(k) = k^2 E(k)$ for homogeneous flow [7].

If indeed $\gamma \sim \frac{5}{3}$, then $D < 3$ (both γ and D are assumed to exist), and furthermore the enstrophy ($\int |\xi|^2\,d\mathbf{x}$ or $\langle \xi^2 \rangle$) blows up [40]. Indeed, if $D = 3$, then ξ is locally square integrable, its spectrum is integrable, $Z(k) = o(k^{-1})$ for large k, and $E(k) = o(k^{-3})$ for large k, thus $\gamma \geqq 3$ (the equality can hold for technical reasons related to the definition of D, see [21]). A model calculation of D on a lattice, using scaling transformations similar to the ones just described for the blowup analysis, has yielded a value $D \sim 2.35$, very close to experimental values [50], but one must be aware that the lattice calculations rely on several simplifying assumptions and may well not be independent of the assumed lattice structure. Thus, if $\gamma \sim \frac{5}{3}$, then $D < 3$ and Euler's equations blow up (this argument does not necessarily show that they blow up in finite time). This argument does contradict various theories of intermittency in which it is assumed that $D = 3$ implies $\gamma = \frac{5}{3}$; references and analysis can be found in [21]. The general observation that fluid turbulence in the limit $R^{-1} \to 0$ has a fractal structure was made in [49], but the specific relation between intermittency and γ presented there is not plausible [21].

6. Structure of the Inertial Range and Turbulence Modeling

The inertial range in turbulence is the range of scales much smaller than the scales of outside stirring but large in comparison with viscous scales (the latter constraint is empty if $R^{-1} = 0$). This is the range of scales where $E(k) \sim k^{-\gamma}$, γ = Kolmogorov exponent, and that one wishes to delete after appropriate parametrization in many practical problems.

To understand what happens in that range and the origin of the Kolmogorov law, consider first a model problem: the motion of a polymer immersed in a solvent. Consider, as is usually done, that the configuration of a polymer subjected to no constraints approximates an arc of Brownian motion. Consider such an arc, containing N monomers. The distance ρ between the first and last monomers scales as $\rho \sim N^{1/2}$. The fractal dimension of the axis of the polymer is then $D = 2$ (see, e.g., [45]). Add the constraint of self-avoidance, i.e., give the chain a finite thickness and forbid it to intersect itself; then $\rho \sim N^{\mu}$; μ is the "Flory exponent," $\mu \sim \frac{3}{5}$; the fractal dimension D of the chain is $\mu^{-1} \sim \frac{5}{3}$. Generally, if $\rho \sim N^{\alpha}$, then $D = 1/\alpha$, [32], [49]. Intuitively, the dimension D measures the polymer's material content as a function of its linear extent. Self-avoidance straightens out the polymer and then reduces its material content.

The two-point correlation function for a simple polymer such that $\rho \sim N^{\alpha}$ is easily evaluated [34]; it scales as $\rho^{(1/\alpha)-3}$ for a small ρ. Its Fourier transform scales as $k^{-(1/\alpha)}$ for large k. If one views a polymer chain as being analogous to a vortex, then the "vorticity" spectrum is obtained from that Fourier transform by an averaging operation that adds a factor of k^2, and then the analogue of the energy spectrum is obtained by a division by k^2. Thus $D = 1/\alpha$, $E(k) \sim k^{-(1/\alpha)}$. If one returns to the hydrodynamical notation, $E(k) \sim k^{-\gamma}$, $\gamma = D$. (Note that $d\gamma/dD > 0$. This inequality holds with some generality for reasonable one-parameter families of turbulent flows [21], and is significant in that it excludes certain simple heuristic models of the inertial range.) If $\alpha = \mu$, the Flory exponent, then $E(k)$ in the polymer case behaves as does $k^{-5/3}$; the constraint of self-avoidance has created a "spectrum" similar to the Kolmogorov spectrum. For large k, this "spectrum" is unchanged by the presence of other polymers in a sufficiently dilute solution, since they tend to segregate [34].

Consider now a collection of vortex tubes that have been stretched, and consider further those portions that have been stretched sufficiently to contribute to the inertial range. Assume the collection is in thermal equilibrium with an ambient potential flow (this notion makes sense in view of the earlier discussion of the polymer/random vortex analogy). The exact dynamics of the vortex tubes are immaterial, and the statistical properties of the collection are determined by the constraints. The main constraints are: conservation of energy and of helicity. The helicity is $H = \int \boldsymbol{\xi} \cdot \mathbf{u}\, d\mathbf{x}$, and is conserved if vortex lines do not intersect [51]; conservation of helicity is thus analogous to self-avoidance for a polymer. The energy constraint forces the vortices to fold and thus increases the dimension of the polymer above two—as we already know should be the case.

Along the stretched vortex lines there must be folds. These folds are often referred to as "hairpins"; many of them look like hairpins but more complicated structures are also possible [20]. The hairpins resemble the "Cooper pairs" of superconductivity [58], which are pairs of electrons of opposing momentum and spin and whose bonding reduces the energy of the system. The hairpins persist because there is no energy for dissociating them; they tighten up as they absorb energy. Their existence and the role they play in absorbing energy is most easily demonstrated on a lattice model [18], [20], where the motion satisfies the hydrodynamical constraints but is otherwise random. Vortex stretching can be accommodated only as hairpins form. When the hairpins tighten sufficiently, viscosity can cancel them, with added assistance from a complicated hydrodynamical mechanism that is at present under intense scrutiny [3], [41], [42], [60], [61]. Hairpins have been well documented in physical experiments [38], [50], [59], especially in boundary layers, and in numerical studies of three-dimensional vortex motion [15], [31], [42].

If the hairpins are removed on sufficiently small scales, the remaining constraint that determines the behavior of vortex lines is the conservation of helicity, which is equivalent to self-avoidance [51]. On a rough scale, rougher than the scale of the hairpins, a vortex line in "thermal" equilibrium should look like a self-avoiding walk. If the vorticity were a scalar one could deduce that the spectrum should behave like $k^{-5/3}$; since the vorticity is a vector, one needs a more elaborate renormalization argument in order to deduce the Kolmogorov law; this argument will be presented elsewhere. It is in qualitative agreement with earlier work on the role of helicity [4].

A systematic removal of tight hairpins provides a plausible way to model the effect of very fine scales on the energy containing range. Such a removal modifies vortex interactions at small scales and thus creates a dynamical rather than a rigid smoothing of the Biot–Savart kernal. For details, see [22]; earlier suggestions along these lines were made in [6]. A different vortex model of small scale turbulence was presented in [47].

More generally, hairpin density provides a qualitative measure of the intensity of turbulence. One can draw a useful analogy between a vortex system at high Reynolds number R and a system near a critical point at zero temperature. The order parameter is the hairpin density (preserving the analogy with superconductivity). This analogy helps to explain why turbulence theories based on finite closures fail to converge [23]. This analogy is presently under investigation.

Acknowledgments

This work was supported by the Office of Energy Research and the U.S. Department of Energy, under contract DE-AC 03-76SF0098.

References

[1] C. ANDERSON AND C. GREENGARD, *On vortex methods*, SIAM J. Sci. Statist. Comput., 22 (1985), pp. 413–440.

[2] C. ANDERSON, C. GREENGARD, AND M. HENDERSON, *Instability, vortex shedding and numerical convergence*, in Vortex Methods, Proc. UCLA Workshop, Anderson and Greengard, eds., Springer-Verlag, Berlin, New York, 1988.

[3] C. ANDERSON AND C. GREENGARD, *The vortex ring merger at infinite Reynolds number*, Comm. Pure Appl. Math., 42 (1989), pp. 1123–1139.

[4] J. ANDRE AND J. LESIEUR, *Influence of helicity on the evolution of turbulence*, J. Fluid Mech., 81 (1977), pp. 187–207.

[5] L. ARNOLD, *Stochastic Differential Equations*, John Wiley, New York, 1973.

[6] S. BALLARD, *Parametrization of viscosity in three-dimensional vortex methods*, in Numerical Methods in Fluid Mechanics, K. Morton and J. Baines, eds., Clarendon Press, Oxford, 1986.

[7] G. K. BATCHELOR, *The Theory of Homogeneous Turbulence*, Cambridge University Press, 1951.

[8] T. BEALE, *A convergent 3D vortex method with grid-free stretching*, Math. Comp., 46 (1986), pp. 401–424.

[9] T. BEALE AND A. MAJDA, *Vortex methods I: convergence in three dimensions*, Math. Comp., 39 (1982), pp. 1–27.

[10] A. J. CHORIN, *Computational aspects of the turbulence problem*, in Proc. 2nd Internat. Conference on Numerical Methods in Fluid Mechanics, Springer-Verlag, Berlin, New York, 1970.

[11] ———, *Numerical study of slightly viscous flow*, J. Fluid Mech., 57 (1973), pp. 785–796.

[12] ———, *Lectures on Turbulence Theory*, Publish/Perish, Berkeley, 1975.

[13] ———, *Vortex models and boundary layer instability*, SIAM J. Sci. Statist. Comput., 1 (1980), pp. 1–21.

[14] ———, *Estimates of intermittency, spectra and blow-up in fully developed turbulence*, Comm. Pure Appl. Math., 34 (1981), pp. 853–866.

[15] ———, *The evolution of a turbulent vortex*, Comm. Math. Phys., 83 (1982), pp. 517–535.

[16] ———, *Numerical estimates of Hausdorff dimension*, J. Comput. Phys., 46 (1982), pp. 390–396.

[17] ———, *Solution of lattice models by successive linkage*, Comm. Math. Phys., 99 (1982), pp. 501–515.

[18] ———, *Turbulence and vortex stretching on a lattice*, Comm. Pure Appl. Math. (special issue), 39 (1986), pp. S47–S65.

[19] ———, *Lattice vortex models and turbulence theory*, in Wave Motion, Lax 60th birthday volume, A. Chorin and A. Majda, eds. Math Sciences Research Institute/Springer-Verlag, Berlin, New York, 1987.

[20] ———, *Scaling laws in the lattice vortex model of turbulence*, Comm. Math. Phys., 114 (1988), pp. 167–176.

[21] ———, *Spectrum, dimension and polymer analogies in fluid turbulence*, Phys. Rev. Lett., 60 (1988), pp. 1947–1949.

[22] ———, *Hairpin removal in vortex interactions*, J. Comput. Phys., 1990, in press.

[23] ———, *Notes on turbulence theory*, University of California, Berkeley, CA, 1988.

[24] A. J. CHORIN AND C. ANDERSON, *Vorticity and turbulence*, to appear.

[25] A. J. CHORIN AND J. MARSDEN, *A Mathematical Introduction to Fluid Mechanics*, Springer-Verlag, Berlin, New York, 1979.

[26] A. J. CHORIN, T. HUGHES, M. MCCRACKEN, AND J. MARSDEN, *Product formulas and numerical algorithms*, Comm. Pure Appl. Math., 31 (1978), pp. 205–256.

[27] W. J. COCKE, *Turbulent hydrodynamic line stretching: consequences of isotropy*, Phys. Fluids, 12 (1969), pp. 2488–2492.

[28] G. H. COTTET, *Methodes particulaires pour l'equation d'Euler dans le plan*, these de 3e cycle, Université Marie et Pierre Curie, Paris, France, 1982.

[29] G. DEEM AND N. ZABUSKY, *Ergodic boundaries in numerical simulations of two-dimensional turbulence*, Phys. Rev. Lett., 27 (1971), pp. 396–399.

[30] R. ESPOSITO AND M. PULVIRENTI, *Three-dimensional stochastic vortex flows*, preprint, 1987.

[31] D. FISHELOV, *Vortex methods for slightly viscous three-dimensional flow*, University of California, Berkeley, CA, 1988.

[32] K. FREED, *Renormalization Group for Macromolecules*, John Wiley, New York, 1987.

[33] O. FROSTMAN, *Potentiel d'équilibre et théorie des ensembles*, thesis, Lund, 1935.

[34] P. G. DE GENNES, *Scaling Concepts in Polymer Physics*, Cornell University Press, Ithaca, NY, 1971.

[35] J. GOODMAN, *The convergence of random vortex methods*, Comm. Pure Appl. Math., 40 (1987), pp. 189–220.

[36] C. GREENGARD, *Convergence of the vortex filament method*, Math. Comp., 47 (1986), pp. 387–398.

[37] O. HALD, *Convergence of vortex methods*, II, SIAM J. Sci. Statist. Comput. 16 (1979), pp. 726–755.

[38] M. HEAD AND P. BANDYOPADHYAY, *New aspects of boundary layer structure*, J. Fluid Mech., 107 (1981), pp. 297–338.

[39] K. HUANG, *Statistical Mechanics*, John Wiley, New York, 1963, p. 329.

[40] D. KAZHDAN, private comment, 1987.
[41] R. KERR, *Simulation of vortex reconnection*, Physica D, 37 (1989), pp. 474–484.
[42] O. KNIO AND A. GHONIEM, *Three-dimensional vortex methods*, J. Comput. Phys., 86 (1990), pp. 75–106.
[43] H. LAMB, *Hydrodynamics*, Dover, New York, 1932.
[44] A. LEONARD, *Computing three dimensional vortex flows with vortex filaments*, Ann. Rev. Fluid. Mech., 17 (1985), pp. 523–559.
[45] P. LEVY, *Le Mouvement Brownien*, Gauthiers Villars, Paris, 1935.
[46] D. W. LONG, *Convergence of random vortex methods in three dimensions*, preprint, Princeton University, Princeton, NJ, 1988.
[47] T. S. LUNDGREN, *Strained spiral vortex model for turbulent fine structure*, Phys. Fluids, 25 (1982), pp. 2193–2203.
[48] A. MAJDA, *The Mathematical Foundations of Incompressible Flow*, Lecture Notes, Princeton (1986).
[49] B. MANDELBROT, *Intermittent turbulence and fractal dimension*, in Turbulence and the Navier–Stokes Equations, R. Temam, ed., Springer-Verlag, New York, 1975.
[50] C. MENEVEAU AND K. R. SREENIVASAN, *The multifractal spectrum of the dissipation field in turbulent flows*, Nuclear Phys. B (Proc. Suppl.), 2 (1987), pp. 49–76.
[51] H. K. MOFFATT, *On the degree of knottedness of tangled vortex lines*, J. Fluid. Mech., 35 (1969), pp. 117–132.
[52] R. MORF, S. ORSZAG, AND U. FRISCH, *Spontaneous singularity in three dimensional incompressible flow*, Phys. Rev. Lett., 44 (1980), pp. 572–575.
[53] S. ORSZAG, *Comments on "turbulent hydrodynamics line stretching . . . "*, Phys. Fluids, 13 (1970), pp. 2203–2204.
[54] P. RAMOND, *Field Theory, a modern primer*, W. A. Benjamin, Boston, 1984.
[55] P. RAVIART, *An analysis of particle methods*, CIME Course, Como, Italy, 1983.
[56] F. REIF, *Fundamentals of Statistical and Thermal Physics*, McGraw-Hill, New York, 1965, p. 567.
[57] S. ROBERTS, *Convergence of random walk methods*, Ph.D. thesis, University of California, Berkeley, CA, 1986.
[58] J. SCHRIEFFER, *Theory of Superconductivity*, Addison-Wesley, Reading, MA, 1983.
[59] J. SETHIAN AND A. GHONIEM, *A validation study of vortex methods*, J. Comput. Phys., 74 (1988), pp. 283–317.
[60] E. SIGGIA AND A. PUMIR, *Incipient singularity in the Navier–Stokes equations*, Phys. Rev. Lett., 55 (1985), pp. 1749–1750.
[61] K. R. SREENIVASAN, *A unified view of the origin and morphology of turbulent boundary layer structure*, IUTAM Symposium, Bangalore, H. W. Liepmann and R. Narasimha, eds., 1987.
[62] K. R. SREENIVASAN AND C. MENEVEAU, *The fractal aspects of turbulence*, J. Fluid Mech., 173 (1986), pp. 357–386.
[63] H. TENNEKES AND J. LUMLEY, *A First Course in Turbulence*, MIT Press, Cambridge, 1972, p. 257.
[64] A. YAGLOM, *Stationary Random Functions*, Dover, New York, 1962.
[65] H. YAMAKAWA, *Modern Theory of Polymer Solutions*, Harper and Row, New York, 1971.
[66] J. J. W. VAN DER VEGT, *A variationally optimized vortex tracing algorithm for three-dimensional flows around solid bodies*, Ph.D. thesis, Delft University, the Netherlands, 1988.

Index